THE NEW MATHEMATICS   Made Simple

Patrick J. Murphy, MSc, FIMA
and
Albert F. Kempf

Made Simple Books
W. H. ALLEN   London
A Howard & Wyndham Company

© 1968 edition W. H. Allen & Co. Ltd.

Made and printed in Great Britain
by Richard Clay (The Chaucer Press), Ltd., Bungay, Suffolk
for the publishers W. H. Allen & Company Ltd.,
44 Hill Street, London W1X 8LB

First edition, November 1968
Reprinted, June 1970
Second (revised) edition, July 1971
Reprinted, January 1973
Reprinted, December 1974
Reprinted, October 1976
Reprinted, April 1977
Reprinted, March 1979
Reprinted, November 1980

ISBN 0 491 01562 3 Paperbound

Foreword

Until recently the main emphasis in the teaching of elementary mathematics has been on the mechanical aspects of computation. While this is a desirable skill, it must be admitted that the time and effort involved in its acquisition, or rejection, leaves much to be desired in the education of the average person, especially when it is then declared redundant by a desk calculator and other similar aids.

You can become very skilful in learning to add, subtract, multiply, and divide without really understanding why these mathematical processes work. Indeed, few students ever believe that there may be an occasion when such operations might fail. Yet a real understanding of mathematics is not only knowing *how* an operation works but *why*. Learning by rote has proved to be of little value to a student's mathematical development because he suffers the restriction of mere 'symbol pushing' when dealing with a problem. This means that he is seldom able to extend his experience from that problem in particular, simply because he fails to recognize the common structure of similar problems in general. For example, at school many students feel that graphical work in Mathematics, Physics, and Geography are three quite separate thinking processes or situations. One way of overcoming this is by setting problems in mathematics textbooks based on results drawn from Physics and Geography. However, 'new' thinking in mathematical teaching attempts to overcome the problem more successfully by concentrating on the basic mathematical ideas in the work. The student is encouraged to understand *why* there is only one thinking process involved by studying two fundamental ideas: sets and relations.

With the emergence of new ideas in the teaching of mathematics has come a new curriculum with which to put theory into practice. Since some parts of this curriculum are younger in time than the traditional curriculum, it has assumed the somewhat misplaced title of 'Modern Mathematics' or 'New Mathematics'. New ideas usually bring new notations for precise definition, and in this book the reader will meet many new symbols. Some of the symbols have yet to be standardized, but this is pointed out where relevant.

This book is intended for anyone interested in the 'New' Mathematics, whether teacher, parent, or student. No previous knowledge of 'new' mathematics is assumed, and the overall standard and content of the book is such that anyone studying for C.S.E. or O-level G.C.E. will find it a great help not only as a source book of information and encouragement but also as a sound basis for further study.

The authors believe that an understanding of the operations within the real number system provides the easiest method of introducing the ideas of structure in a mathematical system. This accounts for the order of the first five chapters. Any reader who chooses his own scheme of study is advised, at least, to read Chapter 1 first, then Chapters 9, 10, 11, and 12, in that order. Whatever the choice, the reader is reminded that mathematics is an activity and there is no substitute for enthusiasm, pencil, and paper. He will find that even the procedure of examining a worked example is improved by an additional sketch diagram or an attempt to anticipate the next line. To this end we have given some diagrams and tables which the reader is required to complete before reading on.

Finally, a special word of thanks to D. F. Taylor, M.A., who was kind enough to offer so much respected comment on the manuscript and to C. R. Batten, M.A., who patiently read the proofs.

P.M. and A.K.

Contents

CHAPTER ONE

SETS

In mathematics as in everyday life we are often concerned not with a single object but with a collection of objects. For example, we hear about and speak of a collection of paintings, a row of seats, a herd of cattle, a crowd of people, and a company of soldiers. Some collections are quite small, such as a pair of shoes or a pair of ear-rings, while other collections may be very large. The collection of cars on the roads during a Saturday afternoon or all the fish in the sea will both be large, for our idea of size will clearly depend on the number of objects in the collection. It would, of course, be an advantage if we used one word only to indicate a collection or gathering, instead of the many available, such as row, herd, crowd, and company. Each collection is an example of what we call a set, a word we frequently use in this sense any-way—a dinner set, a set of chessmen, and so on.

A set is simply a collection of objects.

Definition 1.1

The objects contained in a given set are called **members** or **elements** of the set. The members of a collection of paintings are the individual paintings in that set. The members of a row of chairs are the individual chairs in that row. The members of a set of crockery are the individual cups, saucers, plates, etc., in that set.

One method of naming sets is shown below:

$$A = \{\text{Bob, Bill, Tom}\}$$

This is read, 'A is the set whose members are Bob, Bill, and Tom.' Capital letters are usually used to denote sets. The braces, { }, also denote a set. The names of the members of the set are listed, separated by commas, and then enclosed within braces.

An alternative use of the brace notation is illustrated below:

$$W = \{\text{Monday, Tuesday, Wednesday, Thursday, Friday, Saturday, Sunday}\}$$
$$W = \{\text{the days of the week}\}$$

The first of these examples lists or tabulates the members of the set W. In the second example a descriptive phrase is enclosed within braces. The latter example is read, 'W is the set of days of the week.'

In some cases it is impossible to list all the members of a set. For example, it would be impossible to list all the members of *the set of numbers greater than 2*, but it is possible to use a descriptive phrase, in fact we have just done that in italics. This would be written as

$$N = \{\text{the numbers greater than 2}\}$$

The phrase which we use describes a property which all the elements of a set have in common. This is often referred to as a characteristic property of the set. In the case of set *N* the characteristic property is that all the elements are greater than 2.

Using the set *W* above, we can say:

> Monday *is a member of W*.
> Saturday *is a member of W*.

We abbreviate the phrase 'is a member of' by using the Greek letter epsilon, ∈, to stand for this phrase. Then instead of the above we write:

> Monday ∈ *W*
> Saturday ∈ *W*

The slash line or slant bar, /, is often used to negate the meaning of a mathematical symbol. For example, = means 'is equal to', ≠ means 'is not equal to'; > means 'is greater than', ≯ means 'is not greater than'. The mathematical symbol ∉ is therefore read, 'is not a member of'. For set *W* we can then say:

> John ∉ *W*, (John is not a member of *W*.)
> April ∉ *W*, (April is not a member of *W*.)

The symbols denoting the individual members of a set are generally the small letters of our alphabet, such as *a*, *b*, *c*, *d*, and so on. (Printers refer to these as 'lower case' letters.)

Exercise 1.1

Name the members of each of the following sets:

1. The set of months in the year whose names begin with J.
2. The set of 30-day months in the year.
3. The set of two-letter words in this sentence.
4. The set of letters used in the word excellent.
5. The set of elements common to the answers to questions 1 and 2.
6. The set of numbers greater than 2 which belong to the set {1, 3, 7, 8}.

Write a description of each of the following sets:

7. $A = \{a, b, c, d\}$
8. $B = \{a, e, i, o, u\}$
9. $C = \{x, y, z\}$

In questions 7, 8, and 9 we were able to give the exact number of elements in the set whose characteristic property had to be described. There are, however, some occasions on which this is not possible. In these cases a sufficient number of elements will be written so as to identify the characteristic property involved, and then the list will be ended by a run of dots or by writing, 'etc.'. Questions 10, 11, and 12 will convey this idea.

10. $D = \{$Car, Bus, Coach, Walk, . . .$\}$
11. $E = \{$Roses, Lilies, Daffodils, . . .$\}$
12. $F = \{$Sparrow, Blackbird, Robin, etc.$\}$

Substitute the symbol ∈ or ∉ for each '−' to make the resulting sentences consistent with the information given in Questions 7–9.

13. $a - A$	14. $a - B$	15. $a - C$
16. $y - A$	17. $y - B$	18. $y - C$

THE EMPTY SET

In the earlier examples of Exercise 1.1 we stated the characteristic property of the set first and then searched for those elements which belonged to the set. So far we have managed to find one or two elements at least. There are, however, some occasions on which we think of a set and then discover that it has no members at all. For example, you are sitting in a room full of people when you suddenly decide to see if there is anyone present who is completely bald. On inspection you are likely to find that no one is completely bald, that is, the set of completely bald people in the room has no members. It is reasonable to suppose that among your friends there is one who has the same weight as yourself, but when you come to check this you may well find that none of them is exactly the same weight as yourself. It is just as natural therefore, that a set has no elements as it is to have one, two, or more elements. It is certainly mathematically convenient to consider a collection containing no members, as a set, and naturally we call it an **empty** set. We shall refer to all empty sets as **the** empty set or the null set.

Definition 1.2

The empty set is the set which contains no members.

The empty set is usually denoted by ϕ (a letter from the Scandinavian alphabet). ϕ is read, 'the empty set'. We can also indicate the absence of members by denoting the empty set { }. Some authors denote the null set by the Greek letter Φ, pronounced 'fi' and spelt 'phi'. We shall write ϕ but still call it 'fi'.

Other examples of the empty set are: the set of all men who have lived for more than 300 years; or the set of weeks in our year which contain more than seven days. Typical empty sets in mathematics are: the set of odd numbers with 4 as a factor; the set of even prime numbers greater than 2; the set of all triangles having two obtuse interior angles; the set of real solutions to the equation $x^2 = -1$. Notice that the empty set is **not** written {0} because this means a set having the one element 0, and we have said that the empty set has no elements at all.

SET EQUALITY

Consider the following sets:

$$A = \{r, s, t, u\}$$
$$B = \{t, r, u, s\}$$

Since each set contains the same members, we say that set A is equal to set B and use the shorthand form $A = B$.

Definition 1.3

If A and B are names for sets, $A = B$ means that set A has the same members as set B, or that A and B are two names for the same set. Note that the order in which the members are listed does not matter. For example, $\{a, b, c\} = \{c, a, b\} = \{b, a, c\}$.

Whenever the equal sign ($=$) is used, as in $A = B$, it means that the symbols on either side of it name precisely the same thing.

Consider the following sets:

$$K = \{p, q, r, s\}, \qquad M = \{r, v, x, z\}.$$

Since K and M do not contain the same members, we say K is not equal to M, or simply $K \neq M$.

Exercise 1.2

Use the sets named below and write $=$ or $\neq$ in each '$-$' so that correct sentences result.

$A = \{1, 2, 3, 4\}$ $D = \{$the vowels in the English alphabet$\}$
$B = \{a, e, i, o, u\}$ $E = \{3, 2, 1, 4\}$
$C = \{$the first four counting numbers$\}$ $F = \{o, i, a, w\}$

1. $A - B$ 2. $A - C$ 3. $A - D$ 4. $A - E$
5. $A - F$ 6. $B - C$ 7. $B - D$ 8. $B - E$
9. $B - F$ 10. $E - F$

EQUIVALENT SETS

Suppose you had some cups and some saucers. Someone asks, 'Are there more cups than saucers?' Would you have to count the objects in each set to answer the question?

All you need do is place one cup on each saucer until all of the members of one of the sets have been used. If there are some cups left over, then there are more cups than saucers. If there are some saucers left over, then there are more saucers than cups. In each case a cup is paired with one, and only one, saucer, and each saucer is paired with one, and only one, cup, so we say the sets are matched one-to-one or that there is a one-to-one correspondence between the sets when there are an equal number of cups and saucers.

Definition 1.4

There is a **one-to-one correspondence** between sets A and B if every member of A is paired with one, and only one, member of B and every member of B is paired with one, and only one, member of A.

The following illustration shows the six ways of establishing a one-to-one correspondence between the two sets, $\{a, b, c\}$ and $\{x, y, z\}$.

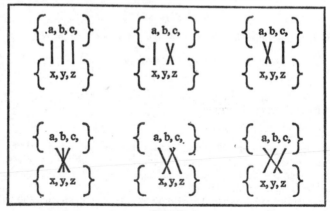

Fig. 1.0

The existence of a one-to-one correspondence between two sets has nothing to do with the way in which the pairing is done.

Definition 1.5

Two sets are said to be **equivalent** if there is a one-to-one correspondence between the two sets.

The idea of equivalent sets is not the same as that of equal sets. That is, two sets are equal if they have the same members. Two equivalent sets may have different members just so long as there exists a one-to-one correspondence between them. Clearly all equivalent sets have the same number of elements.

For example:

$$\{a, b, c, d\} \text{ is equivalent to } \{r, s, t, u\}$$
but, $\{a, b, c, d\}$ is not equal to $\{r, s, t, u\}$
$\{a, b, c, d\}$ is equal to $\{c, a, d, b\}$
and, $\{a, b, c, d\}$ is equivalent to $\{c, a, d, b\}$

Exercise 1.3

Draw matching lines as in Fig. 1.0 to show a one-to-one correspondence between the sets in each pair.

1. $\{a, b, c, d\}$
   $\{w, x, y, z\}$
2. $\{1, 2, 3, 4, 5, 6\}$
   $\{2, 4, 6, 8, 10, 12\}$

SUBSETS

It is often necessary to think of sets that are 'part of' another set or are 'sets within a set'.

The set of chairs C in a room is a set within the set of all pieces of furniture F in that room. Obviously, every chair in the room is a member of set C and also a member of set F. This and similar examples lead to the idea of a subset.

Definition 1.6

'Set A is a **subset** of set B' means that every member of set A is also a member of set B.

An alternative definition of a subset may be:

Definition 1.7

'Set A is a subset of set B' if set A contains no member that is not also in set B.

Consider the following sets:

$$R = \{a, b, c, d, e\}$$
$$S = \{a, c, e\}$$

Every member of set S is a member of set R. Hence S is a subset of R. R is not a subset of S because it contains members (b and d) which are not members of S.

Now all of the possible subsets of S are listed below, i.e. subsets which are formed by taking a selection from a, c, e. We may take 3, 2, 1 or no elements for a subset to get:

$$\{a, c, e\}; \quad \{a, c\}; \quad \{a\}; \quad \{ \ \}$$
$$\{a, e\} \quad \{c\}$$
$$\{c, e\} \quad \{e\}$$

The first and last subsets of S can lead to some general conclusions about the subset relation.

Is the empty set a subset of every set? By Definition 1.7 the empty set contains no element which is not also a member of any given set. Hence we *say* that the **empty set is a subset of every set**.

Since $S = \{a, c, e\}$, then $a, c,$ and e belong to S, so we are tempted to ask: 'Is every set a subset of itself?' Regardless of the set we choose, every member of the set is obviously a member of the set. Hence we *say* that every set is a subset of itself.

Any subset which is not also the entire set is called a **proper subset**. In the above, while $\{a, c, e\}$ is a subset of S it is not a proper subset. The other seven sets listed above (including $\{ \ \}$) are proper subsets. (Some authors also regard $\{ \ \}$ as an improper subset.)

We abbreviate the phrase 'is a subset of' by using the conventional symbol $\subseteq$. Thus $A \subseteq B$ means 'set A is a subset of set B' or simply 'A is a subset of B'. Notice that $\subseteq$ means 'contained in or possibly equal to'.

By using this symbolism we can concisely state Definition 1.6 as follows:

$$A \subseteq B \text{ if, for every } x \in A, \text{ then } x \in B$$

We abbreviate the phrase 'is a proper subset of' by using the conventional symbol $\subset$. Thus $A \subset B$ means 'set A is a proper subset of set B'. In this case set A cannot be equal to set B.

Looking back at set S and its list of possible subsets, we see that the statement

'$P \subseteq S$' has 8 possibilities for P

but, '$P \subset S$' has only 7 possibilities for P, since $P = S$ is not permissible.

It must be understood that 'membership of' is different from 'subset of'. With $S = \{a, c, e\}$ it follows that $\{a\} \subset S$ is true, but $\{a\} \in S$ is false. Of course $a \in S$, so these remarks also illustrate a difference between a and $\{a\}$.

Consider the set $Q = \{£, a, \{b\}, \text{trousers}, \text{boat}\}$. Notice that one of the elements is a set, i.e. $\{b\}$, so that we have a set within a set. Therefore:

1. $£ \in Q$, $a \in Q$, $\{b\} \in Q$, trousers $\in Q$, boat $\in Q$, are all true.
2. $\{£\} \in Q$ is false, $\{£\} \subset Q$ is true.
3. $\{\text{trousers}\} \in Q$ is false, $\{\text{trousers}\} \subset Q$ is true.
4. $\{\{b\}\} \in Q$ is false; $\{b\} \subset Q$ is false, $\{b\} \in Q$ is true.

Number 4 needs a little care, but should be clear if we realize that $\{b\}$ was listed as an **element** of Q and not b, as we have encountered up to now.

Exercise 1.4

Consider the following sets A, B and C. Then substitute the symbol $\subset$ or $\not\subset$ for each '—' so that the resulting statement is true.

$A = \{a, b, c, d, e\}$ $B = \{b, d, e, g\}$ $C = \{b, d\}$
1. $B - A$ 2. $B - C$ 3. $B - B$
4. $C - B$ 5. $C - A$ 6. $\phi - C$

List all of the possible subsets of each of the following sets:

7. $D = \{x, y\}$
8. $E = \{a, b, c, d\}$

Compare the number of subsets and the number of members of set D, E and the previously used set $S = \{a, c, e\}$.

9. ϕ has 1 subset, $\{a\}$ has 2 subsets, $\{a, b\}$ has 4 subsets, etc. Can you discover a formula for finding the number of subsets of any set? How many subsets has $\{a, b, c, d, e\}$?

10. If $P = \{t, \{t\}, u, v, \{x\}\}$, state which of the following are true (T) or false (F).

(i) $t \in P$ (ii) $\{t\} \in P$ (iii) $\{u\} \subset P$
(iv) $\{u, v\} \in P$ (v) $\{u, v\} \subset P$ (vi) $\{x\} \in P$
(vii) $\{x\} \subset P$ (viii) $\{u, v, \{x\}\} \subset P$

NUMBERS

Let us consider the collection of all sets that are equivalent to $\{a, b, c\}$. For convenience, let us denote a set by drawing a ring around the collection of objects.

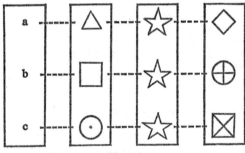

Fig. 1.1

The only thing alike about all these sets is that their members can be matched one-to-one. That is, they are equivalent sets. The thing that is common to these sets is called the number three. Of course, other sets belong to this collection too—the set of wheels on any one tricycle, the set of people in a trio, and the set of sides of a triangle. The number three has many forms—III, $2 + 1$, 3, and many more. Each of these forms is called a **numeral**. A numeral is a symbol for a number. The most common numeral for the number three is 3.

With every collection of equivalent sets is associated a number, and with each number is associated a most common numeral which we shall call the simplest numeral.

Set	Number	Simplest numeral
{ }	zero	0
{a}	one	1
{c, d}	two	2
{x, y, z}	three	3
.	.	.
.	.	.
.	.	.

The dots indicate that we can extend each of the above columns. The set of numbers so derived is called the set of cardinal numbers or the set of **whole numbers**. Thus, equivalent sets have the same cardinal number. We use the symbol $n(A)$ to mean 'the number of the set A'. Thus, if $A = \{x, y, z\}$ and $B = \{o, p, q, r\}$ we have $n(A) = 3$, $n(B) = 4$. If the elements of a further set C can be matched one-to-one with the elements of A then $n(C) = 3$, a point we have already made in defining equivalent sets.

Since we usually begin counting 'one, two, three . . .' we call 1, 2, 3, 4, 5 . . . the set of counting numbers or the set of **natural numbers**.

Set of whole numbers: $\{0, 1, 2, 3, 4, \ldots\}$

Set of natural numbers: $\{1, 2, 3, 4, \ldots\}$

Exercise 1.5

Write the simplest numeral for the number associated with each of the following sets:

1. $\{q, r, s, t, w, x, y, z\}$
2. {the days of the week}
3. $\{3, 4, 5, 6, 7, 8, 9\}$
4. {the months of the year}
5. {all three-gallon tins holding four gallons of paint}
6. {John, James, Jean, Joe}
7. $\{0\}$
8. ϕ

UNION OF SETS

We are accustomed to joining sets in our daily activities. For example, when you put some coins in your purse you are joining two sets of coins—the set of coins already in your purse and the set of coins about to be put into your purse.

Young children will often place objects into sets corresponding to colour and will unite a blue set with a red set and then separate them again without much idea of the number of objects there are. These, and many other examples, indicate that the idea of combining sets in some way is more elementary than counting. The simple combination of sets we have just mentioned is called a *union* of the two sets.

Definition 1.8

The **union** of set A and set B, denoted by $A \cup B$, is the set of all objects without repetition, which belong to either set A or set B. This does, of course, include the possibility of elements belonging to both sets.

Consider the following sets:

$$A = \{a, b, c, d\}$$
$$B = \{c, d, e, f, g\}$$
$$C = \{b, c, d\}$$
$$D = \{r, s, t\}$$

According to the definition of union, we can form the following sets:

$$A \cup B = \{a, b, c, d, e, f, g\}$$
$$A \cup C = \{a, b, c, d\}$$
$$= A$$
$$A \cup D = \{a, b, c, d, r, s, t\}$$

For $A \cup B$ there is no need to repeat the names of members c and d. For example, suppose you are referring to a set of three girls—named Jane, Mary, and Pam. Then {Jane, Mary, Pam, Mary} is correct but not preferred, since Mary is named twice and there are only 3 girls in the set. Again, we agree to write $B \cup C$ as $\{b, c, d, e, f, g\}$ and not as $\{b, c, c, d, d, e, f, g\}$.

Another way of illustrating sets and set operations is to use Venn* diagrams. A Venn diagram is merely a closed figure used to denote the set of all points within the figure. If a Venn diagram is used in this way it can only illustrate results already known, we cannot use it to *prove* results.

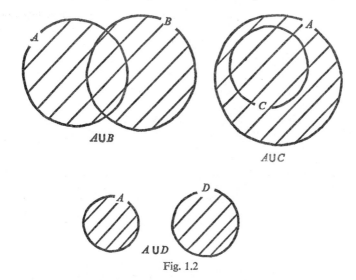

Fig. 1.2

The shaded region in each of these illustrations indicates $A \cup B$, $A \cup C$, $A \cup D$. Note that the union of two sets includes all of the members in both of the sets.

* J. Venn was a British logician (1834–1923). It is only fair to point out that Euler (1707–83) also used such diagrams, consequently they are sometimes called Venn–Euler diagrams.

Exercise 1.6

Use the following sets to form the union of each pair of sets given below:

$$K = \{3, 5, 7, 9\} \qquad M = \{2, 4, 6, 8\}$$
$$J = \{1, 2, 3\} \qquad N = \{0, 5, 9\}$$

1. $J \cup K$ 2. $K \cup M$
3. $K \cup N$ 4. $J \cup M$
5. $J \cup N$ 6. $M \cup N$
7. $N \cup M$ 8. $K \cup K$

INTERSECTION OF SETS

Suppose a teacher asked a class, 'How many of you went to the game last night?' Then several children raised their hands. Those who raised their hands are members of the set of children in the class, *and* they are also members of the set of all children who went to the game last night.

We can illustrate this situation by using a Venn diagram. Let $A = \{$all children in the class$\}$ and let $B = \{$all children who went to the game last night$\}$.

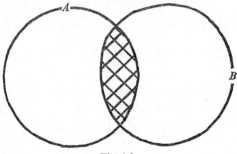

Fig. 1.3

Then C $= \{$children in A, who are also in $B\}$ is called the intersection of A and B and represented by the shaded region of Fig. 1.3.

Definition 1.9

The **intersection** of set A and set B, denoted by $A \cap B$, is the set of all objects that are members of both set A and set B.

For example, the shaded region in each of the following illustrations represents $A \cap B$, $A \cap C$, and $A \cap D$ respectively using the sets of Fig. 1.2.

What do you think the last result should be?

Consider the following sets:

$$X = \{g, h, i, j\}$$
$$Y = \{e, f, g, h\}$$
$$Z = \{a, b, c, d, e\}$$

Then: $X \cap Y = \{g, h\}$
$$X \cap Z = \phi$$
$$Y \cap Z = \{e\}$$
$$X \cap X = \{g, h, i, j\}$$

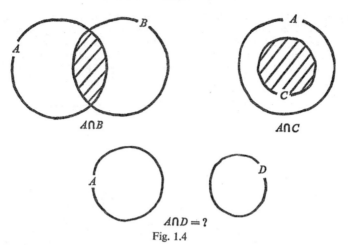

$A \cap B$ $A \cap C$

$A \cap D = ?$
Fig. 1.4

Exercise 1.7

Use the following sets to form the intersection of each pair of sets given below:

$$C = \{2, 3, 4, 5, 6\} \qquad D = \{1, 2, 3, 7, 8\}$$
$$E = \{3, 4, 5, 6\} \qquad F = \{7, 8, 9, 10\}$$

1. $C \cap D$ 2. $D \cap C$ 3. $C \cap E$ 4. $C \cap F$
5. $D \cap E$ 6. $D \cap F$ 7. $E \cap F$ 8. $E \cap E$

DISJOINT SETS

It is obvious that some sets have no members in common—such as $\{p, q, r\}$ and $\{x, y, z\}$.

Definition 1.10

Set A and set B are called disjoint sets if they have no members in common. Or set A and set B are disjoint sets if $A \cap B = \phi$. In the following diagram set A and set B do not intersect. Therefore, A and B are disjoint sets.

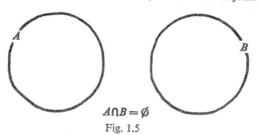

$A \cap B = \emptyset$
Fig. 1.5

Consider the following sets:

$$R = \{f, g, h, j\}$$
$$S = \{a, b, c\}$$
$$T = \{a, h, j\}$$

Sets R and S have no members in common. Hence, R and S are disjoint sets.
Sets R and T are not disjoint sets, since they both have h and j as members.
Sets S and T are not disjoint sets, since they both have a as a member.

Before going any further let us see how the information of Figs. 1.2 and 1.4 may be combined into the one diagram of Fig. 1.6, with the elements of each set inserted into the appropriate region of the diagram.

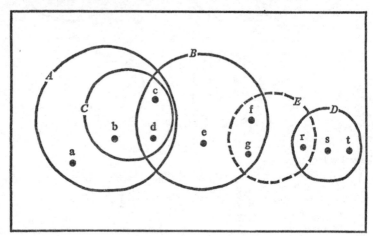

Fig. 1.6

As a test of your understanding of this diagram a new set E has been inserted.

List the elements of the sets:

(i) $B \cap E$

(ii) $E \cap D$

(iii) $E \cup D$

(iv) $A \cap E$

You will find the answers on page 312.

Exercise 1.8

Say whether each statement below is true (T) or false (F).

1. $\{q, r, s, t\}$ and $\{x, y, z\}$ are disjoint sets.
2. If $Q \cap R = \phi$, then Q and R are disjoint sets.
3. If $Q \subset R$, then Q and R are disjoint sets.
4. If $Q \subseteq R$ and $R \subseteq Q$, then $R = Q$.
5. If $R \cup Q = R$, then $Q \subseteq R$.
6. If $R \cap Q = R$, then $R \subseteq Q$.
7. If $R \cap Q = R \cup Q$, then $R = Q$.
8. If $C \cap D = \{5\}$, then $5 \in C$ and $5 \in D$.
9. If $C \cup D = \{3, 4, 5, 6, 7\}$, then $5 \in C$ and $5 \in D$.
10. If $x \in H$, then $x \in (H \cup G)$.
11. $r \in \{q, r, s, t\}$
12. $r \subset \{q, r, s, t\}$

13. $\{r\} \in \{q, r, s, t\}$
14. If $Q \cap s = \phi$, then Q and s are disjoint sets.
15. If $r \cap q = \phi$, then q and r are disjoint sets.

UNIVERSAL SETS

With the aid of Fig. 1.6 we discussed the union and intersection of the five sets A, B, C, D, and E. Obviously it was necessary to know which elements were available to make up the sets, and this must apply to any other problem involving sets. For example, in one problem we may discuss cars. It would therefore be necessary to state at the beginning whether we are going to consider all cars or only British cars and possibly to say whether they are three or four wheelers and so on. This background set which contains all the sets in the problem is called the universe of discourse or the universal set.

Definition 1.11

The set of available elements for each problem is called the **universal** set and is denoted by $\mathscr{E}$. In some books you may find the universal set denoted by U or μ. In Fig. 1.6 $\mathscr{E} = \{a, b, c, d, e, f, g, r, s, t\}$ and each of the elements belongs to one of the sets.

This may not always be the case, for consider the Venn diagram in Fig. 1.7.

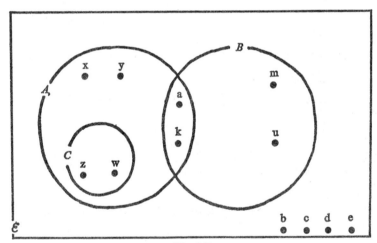

Fig. 1.7

Here we have
$$\mathscr{E} = \{a, b, c, d, e, k, m, u, w, x, y, z\}$$
$$A = \{a, k, w, x, y, z\}$$
$$B = \{a, k, m, u\}$$
$$C = \{z, w\}$$

We see that of all the available elements in $\mathscr{E}$, b, c, d, e have not been used. Notice also that when $\mathscr{E}$ is known we represent it by a rectangular region.

This illustrates clearly that each set under discussion is a subset of $\mathscr{E}$, i.e. $A \subset \mathscr{E}$, $B \subset \mathscr{E}$ and $C \subset \mathscr{E}$.

COMPLEMENT

So far we have spoken of elements which belong to sets, but as soon as we know $\mathscr{E}$ we are able to speak precisely about those elements which do not belong to a set; in other words, the **complement** of the set with respect to $\mathscr{E}$.

Definition 1.12

The **complement** of a set A is the set of elements which do not belong to A with respect to $\mathscr{E}$. We denote the complement of A by A'. (This is read 'A dash' or 'A prime'.)

In Fig. 1.7 the set $\{m, u, b, c, d, e\}$ we now denote by A'. Similarly, from Fig. 1.7 we have $B' = \{x, y, z, w, b, c, d, e\}$.

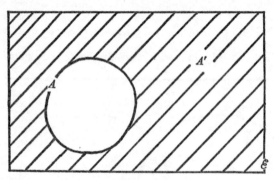

Fig. 1.8

The Venn diagram of Fig. 1.8 shows the shaded region to represent A'. It is not difficult to see that the insertion of A', B', and C' in Fig. 1.7 will need both care and a wider range of shading or colour.

Exercise 1.9

Using the information of Fig. 1.7, list the elements in the following sets:

1. $A \cap B$
2. $A \cap C$
3. $B \cap C$
4. $A \cup C$
5. The set of elements which do not belong to A.
6. The set of elements which belong to A but not to B or C.
7. $A' \cap B$
8. $A' \cup A$
9. $B' \cup B$
10. $A' \cap A$
11. Draw diagrams to illustrate De Morgan's laws that (i) $(A \cup B)' = A' \cap B'$ (ii) $(A \cap B)' = A' \cup B'$.

OPERATIONS
(THE COMMUTATIVE AND ASSOCIATIVE LAWS)

So far we have combined sets under the rules of union and intersection. The result of each combination has always been another set, sometimes only the empty set, but nevertheless a set. Whenever we combine any two objects according to a stated rule and produce a third object as the *only* result we say that we are performing an *operation*, thus union and intersection are operations, and since they involve two sets at a time, we call union and intersection *binary operations*. We have already used the symbol for the operation of union, namely ∪ and likewise ∩ for intersection.

From the Definitions 1.8 and 1.3 we see that when $A ∪ B$ has been found it does not matter what order is used in writing the elements down. This means that $A ∪ B = B ∪ A$, or the result of combining B with A is the same as combining A with B. When the result of a binary operation is independent of the order in which it is carried out we say the operation is **commutative** or obeys the **commutative law**.

Similarly, from the Definitions 1.9 and 1.3 we see that $A ∩ B = B ∩ A$, consequently the operation of intersection is also commutative.

Now quite clearly we shall not get very far by restricting our inquiries to two sets A and B. So we must ask ourselves what meaning we can give to such expressions as:

(i) $A ∪ B ∪ C$,   (ii) $A ∩ B ∩ C$,   (iii) $A ∩ B ∪ C ∪ D$, etc.

Common sense suggests that it would be to everyone's advantage if only one meaning were possible, because that alone would ensure that we all arrive at the same answer to our problems.

Taking (i) $A ∪ B ∪ C$, the difficulty is in knowing where to start! For example, shall we associate A with B first and then combine the result with C? Or shall we associate B with C first and then combine this result with A? We hope, of course, that it will not make any difference. (In fact, it does not.) Notice that we do not change the order that we are first given. In order to indicate what we are doing first we insert brackets, thus $A ∪ (B ∪ C)$ means that we will solve $B ∪ C$ first then combine the results with A. Now $B ∪ C$ is the set of elements belonging either to set B or to set C, so that $A ∪ (B ∪ C)$ is the set of elements belonging to set A or to set B or to set C. In $(A ∪ B) ∪ C$ we solve $A ∪ B$ first and then combine the results with C, this leads to the same result as before.

We have thus observed that $A ∪ (B ∪ C) = (A ∪ B) ∪ C$, and this means that since there is only the one result for $A ∪ B ∪ C$, we need not insert the brackets at all.

Consider the following sets:

$$A = \{a, b, c, d\}$$
$$B = \{b, c, d, e\}$$
$$C = \{c, d, e, f\}$$

Therefore:

$B ∪ C = \{b, c, d, e, f\}$ $B ∩ C = \{c, d, e\}$

$A ∪ B = \{a, b, c, d, e\}$ $A ∩ B = \{b, c, d\}$

Using these sets, we will illustrate the above discussion.

1. $A \cup (B \cup C) =$ ⟶ $A \cup \{b, c, d, e, f\}$ [$B \cup C$ first]
   $=$ $\{a, b, c, d\} \cup \{b, c, d, e, f\}$
   $= \{a, b, c, d, e, f\}$

2. $(A \cup B) \cup C = \{a, b, c, d, e\} \cup C$ [$A \cup B$ first]
   $= \{a, b, c, d, e\} \cup \{c, d, e, f\}$
   $= \{a, b, c, d, e, f\}$

3. Therefore $A \cup (B \cup C) = (A \cup B) \cup C$

and $A \cup B \cup C$ may be written without brackets, since it has only one meaning or result.

Since $A \cup (B \cup C) = (A \cup B) \cup C$, we say that the operation of union obeys the **associative law.**

We shall now examine $A \cap B \cap C$ in a similar manner. $B \cap C$ is the set

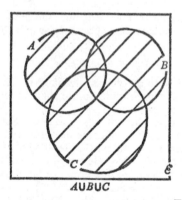

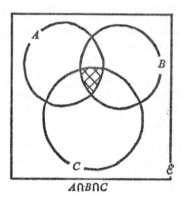

$A \cup B \cup C$ $A \cap B \cap C$

Fig. 1.9

of elements which are common to set B and set C, so that $A \cap (B \cap C)$ is the set of elements which are common to set A and set B and set C.

The same result will be obtained from $(A \cap B) \cap C$, enabling us to write

$$A \cap (B \cap C) = (A \cap B) \cap C$$

and showing that we need not insert the brackets at all because there is only the one result for $A \cap B \cap C$.

Try to illustrate this discussion similarly to that for $A \cup B \cup C$, i.e. form $A \cap (B \cap C)$ then $(A \cap B) \cap C$.

Since $A \cap (B \cap C) = (A \cap B) \cap C$, we say that the operation of intersection obeys the **associative law.**

The shaded regions of the Venn diagram in Fig. 1.9 represent the two results.

Exercise 1.10

If $A = \{a, b, c, d\}$, $B = \{c, d, e\}$, $C = \{c, e, f, g\}$, $D = \{a, b, e, f\}$
list the elements in the following:

1. $A \cup B \cup C$ 2. $A \cup B \cup C \cup D$ 3. $A \cap B \cap C$
4. $B \cap D$ 5. $A \cap B \cap D$ 6. $B \cap A \cap D \cap C$
7. $(B \cap D) \cup C$ 8. $(A \cup B) \cap C$

The last two questions in Exercise 1.10 will now be examined in order to show how important the brackets are for giving the true meaning to the expressions.

As before, we evaluate the bracketed part (or parts) first. In Question 7

this means that $(B \cap D) \cup C = \{e\} \cup C$
$$= \{e\} \cup \{c, e, f, g\}$$
$$= \{c, e, f, g\}$$

Suppose we had tried $(D \cup C)$ first?

Then, $\qquad B \cap (D \cup C) = \{c, d, e\} \cap \{a, b, c, e, f, g\}$
$$= \{c, e\}$$

Since these results are not the same, it would be foolish to leave the brackets out. That is, $B \cap D \cup C$ has two possible results, depending on where we start.

Again in Question 8 $\quad (A \cup B) \cap C = \{a, b, c, d, e\} \cap \{c, e, f, g\}$
$$= \{c, e\}$$
but, $\qquad A \cup (B \cap C) = \{a, b, c, d\} \cup \{c, e\}$
$$= \{a, b, c, d, e\}$$

This certainly shows that a mixture of $\cup$ and $\cap$ needs more attention to the brackets in order to arrive at the intended result or meaning.

Exercise 1.11

Using the sets of Exercise 1.10, list the elements in the following Questions 1–8:

1. $(A \cup B) \cap (C \cup D)$
2. $(A \cup B) \cap C \cap D$
3. $(A \cup B \cup C) \cap D$
4. $(A \cap B \cap D) \cup C$
5. Show that $A \cup (B \cap D) = (A \cup B) \cap (A \cup D)$
6. Show that $A \cup (C \cap D) = (A \cup C) \cap (A \cup D)$
7. Show that $A \cap (C \cup D) = (A \cap C) \cup (A \cap D)$
8. Show that $D \cap (B \cup C) = (D \cap B) \cup (D \cap C)$
9. State the numbers $n(A)$, $n(B)$, $n(C)$, $n(D)$
10. State the numbers $n(A \cap B)$, $n(A \cap B \cap C)$, $n(A \cup B \cup D)$

These days, with newspapers and journals frequently reporting results of opinion polls and inquiries, we are often presented with a great deal of statistical information in a numerical form which is a little difficult to analyse. A Venn diagram for the results invariably provides a clearer picture of the information and, furthermore, enables us to detect rather more quickly the faulty results of a careless inquiry.

Before discussing particular problems let us confirm the meaning to be given to the various regions of the Venn diagram in Fig. 1.10.

Each region of the diagram has been numbered. Thus, all the elements of set *A* will occur in regions (i), (ii), (iii), and (vi); all the elements of set *B* occur in regions (i), (ii), (v), and (iv), and so on.

Let us examine region (ii). This will contain elements which belong to both set *A* and set *B* but not to set *C*. Region (vii) will contain elements which belong to set *C* but do not belong to either set *A* or set *B*. Regions (vi), (iii), (vii), and (viii) contain all the elements which do not belong to set *B*, in other words elements which belong to *B'*.

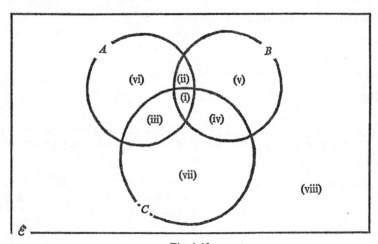

Fig. 1.10

Exercise 1.12

If $A = \{a, b, c, d, e\}$, $B = \{a, b, c, x, y, z\}$, $C = \{a, b, e, x, y, p, q\}$
$\mathscr{E} = \{a, b, c, d, e, x, y, z, p, q, r, s\}$

1. List the elements which occur in each of the eight regions of Fig. 1.10.

2. Which elements belong to set *B* or set *C* but not to set *A*? In which regions are these elements placed?

3. Which elements belong to set *A* but not to set *B* or set *C*? In which region are these elements placed?

4. Which elements do not belong to any of the sets *A*, *B*, or *C*? In which region are these elements placed?

Suppose we wish to find the number of elements in region (i); in other words, the number of elements which are common to all three sets. Clearly $n(A)$ is the number of elements belonging to regions (i), (ii), (vi), and (iii), and $n(B)$ is the number belonging to regions (i), (ii), (v), and (iv); consequently, the number of elements in regions (i) and (ii) is counted twice when we write $n(A) + n(B)$. Now $n(A) + n(B) + n(C)$ will count twice those elements in regions (ii), (iii), (iv) and count three times those elements in region (i).

Now examine the following table related to Fig. 1.10:

Number	Regions in which the elements are counted							Number
$n(A \cap B \cap C)$	(i)							2
$n(A)$	(i)	(ii)	(iii)			(vi)		5
$n(B)$	(i)	(ii)		(iv)	(v)			6
$n(C)$	(i)		(iii)	(iv)			(vii)	7
$n(A \cup B \cup C)$	(i)	(ii)	(iii)	(iv)	(v)	(vi)	(vii)	10
$n(A \cap B)$	(i)	(ii)						3
$n(B \cap C)$	(i)			(iv)				4
$n(C \cap A)$	(i)		(iii)					3

An inspection of the two halves of the table reveals the useful formula

$n(A \cup B \cup C) + n(A \cap B) + n(B \cap C) + n(C \cap A) = n(A \cap B \cap C) + n(A) + n(B) + n(C).$

We can rewrite this in a more usual form,

$n(A \cup B \cup C) = n(A) + n(B) + n(C) - n(A \cap B) - n(B \cap C) - n(C \cap A) + n(A \cap B \cap C)$ (1)

Example 1

At a Social Sports Club with 290 members it was found that 120 played Tennis, 110 played Squash, 130 played Badminton, 70 played both Tennis and Squash, 55 played both Squash and Badminton, 60 played Tennis and Badminton. It was also discovered that 75 members had joined only for the social side of the club and didn't play any of the three games. How many played all three games?

Draw a Venn diagram for the information.

We shall give two ideas for answering the problem. First we shall use the above formula (1). Let T be the set of members playing Tennis, S the set of Squash players, and B the set of Badminton players. Now $\mathscr{E} = \{$members of the club$\}$ and $n(\mathscr{E}) = 290$

$n(T \cup B \cup S) = 290 - 75 = 215; n(T) = 120, n(S) = 110, n(B) = 130$ (2)

$n(T \cap S) = 70, n(S \cap B) = 55, n(B \cap T) = 60$ (3)

Substituting in the formula (1) we have,

$215 = 120 + 110 + 130 - 70 - 55 - 60 + n(T \cap S \cap B)$

$215 = 175 + n(T \cap S \cap B)$

$n(T \cap S \cap B) = 40$

That is, there are 40 members who play all three games.

A second approach to the problem is to put $n(T \cap S \cap B) = x$ and then to complete the Venn diagram as shown on page 20.

Comparing with Fig. 1.10, the number of elements in the regions (ii), (iii), and (iv) follow from the results numbered (3) above. In order to obtain the number for region (v) we use the result $n(B) = 130$ given in (2). So far on our diagram Fig. 1.11 we have $(60 - x) + x + (55 - x) = 115 - x$ elements inserted. Therefore region (v) must contain $130 - (115 - x) = x + 15$ elements.

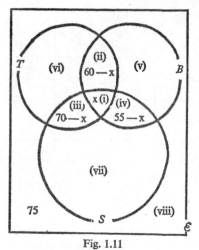

Fig. 1.11   Fig. 1.12

The completed diagram is Fig. 1.12.

We know that $n(T \cup B \cup S) = 215$. From our completed diagram we see that $n(T \cup B \cup S) = 175 + x$. Therefore $x = 40$, the same result as before.

At first sight the second method appears very long, but in fact there is not all that much difference in the work involved in the two methods. The first method does, of course, involve remembering a formula, which is not the case in the second.

The reader will no doubt choose his own preference.

Example 2

An inquiry has been carried out into the popularity of three brands of instant coffee, *A*, *B*, and *C*. Unfortunately the inquiry agent foolishly falsifies the figures in

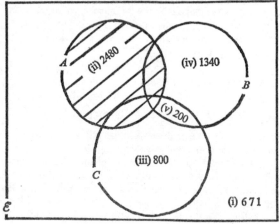

Fig. 1.13

favour of brand A without knowing anything about a Venn diagram and consequently gives himself away. Here are the results he offered:

Of 5 951 people interviewed:

(i) 671 did not drink instant coffee;
(ii) 2 480 drank brand A;
(iii) 800 drank brand C only;
(iv) 1 340 drank brand B only;
(v) only 200 drank both B and C but not A.

Clearly the favourite is brand A! (Notice that i, ii, iii, iv, v no longer refer to the regions in Fig. 1.10.)

Fig. 1.13 is the required diagram completed in the order of the statements (i)–(v).

If we examine $n(\mathscr{E})$ on the diagram we see that only 5 491 of the 5 951 interviews appear to be registered, so obviously a mistake has been made.

Example 3

The purpose of this example is to show how to use a Venn diagram for 4 overlapping sets. The following results are from a survey of 812 homes to discover the

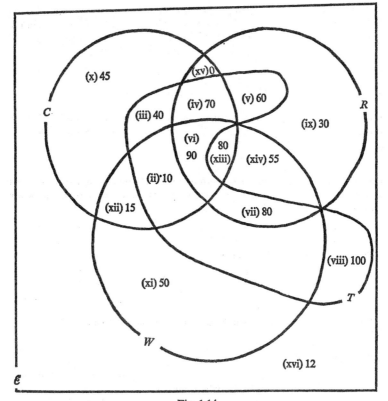

Fig. 1.14

number of people who owned cars (C), refrigerators (R), washing machines (W), and television sets (T):

(i) $n(T \cap W \cap R' \cap C') = 75$ (ii) $n(T \cap C \cap W \cap R') = 10$
(iii) $n(T \cap C \cap W' \cap R') = 40$ (iv) $n(T \cap C \cap R \cap W') = 70$
(v) $n(T \cap R \cap W' \cap C') = 60$ (vi) $n(T \cap C \cap R \cap W) = 90$
(vii) $n(T \cap R \cap W \cap C') = 80$ (viii) $n(T \cap R' \cap W' \cap C') = 100$
(ix) $n(R \cap T' \cap W' \cap C') = 30$ (x) $n(C \cap T' \cap W' \cap R') = 45$
(xi) $n(W \cap R' \cap C' \cap T') = 50$ (xii) $n(C \cap W \cap T' \cap R') = 15$
(xiii) $n(C \cap R \cap W \cap T') = 80$ (xiv) $n(R \cap W \cap T' \cap C') = 55$
(xv) $n(C \cap R \cap T' \cap W') = 0$

As in the previous example, the diagram will be completed in the order of the statements. Notice that having first drawn sets C, R, W, the perimeter for T has to be passed through all eight regions and assume a non-circular shape, as shown in Fig. 1.14. A space has been left for the reader to insert statement (i) himself.

Recall that $n(T \cap W \cap R' \cap C')$ means the number of elements common to T and W but not belonging to either R or C. Likewise $n(R \cap W \cap T' \cap C')$ is the number of elements common to R and W but not belonging to either T or C.

From the completed diagram $n(T \cup W \cup R \cup C) = 800$, so that only 12 homes were without any of these articles in region labelled (xvi).

We shall close this chapter with some examples of the use of set notation to give symbolic form to ordinary everyday statements. We recall that $C \cap D = \phi$ means that there are no elements common to the two sets C and D; in other words, the two sets C and D are disjoint. If C is the set of all car drivers and D is the set of dogs, then $C \cap D = \phi$ is a symbolic form for the statement that 'no dog drives a car'. If W is the set of women, then $C \cap W \neq \phi$ can represent 'some drivers are women'. We are, of course, not entitled to say that *all* women are drivers from this statement.

Example 4

If $\mathscr{E} = \{\text{singers}\}$, $O = \{\text{operatic singers}\}$, $J = \{\text{jazz singers}\}$ we can express the sentence 'some singers do not sing opera' symbolically as $O' \neq \phi$, i.e. the set of non-operatic singers is not empty. An alternative could be $O \neq \mathscr{E}$, i.e. 'there are other singers besides opera singers', or 'opera singers are not the only singers'.

We shall now try to fit a statement to the following symbolic forms:

(i) $O \subset J$ (ii) $O \cup J \neq \mathscr{E}$ (iii) $J \subset O'$
(iv) $O \cap J = \phi$ (v) $O \cup J = \mathscr{E}$

(i) This means O is a proper subset of J; every element of O is also an element of J. A suitable statement is therefore 'all opera singers sing jazz'.

(ii) This means that the union of O and J does not make the universal set, i.e. 'some singers do not sing either opera or jazz'.

(iii) Here J is a proper subset of the set of singers who do not sing opera. In other words 'no jazz singer sings opera'.

(iv) Here we are told that O and J are disjoint sets, so that no singer sings both opera as well as jazz.

(v) We are now told that the union of O and J is the universal set. Therefore 'all singers sing opera or jazz' or both.

Example 5

If $\mathscr{E} = \{\text{cars}\}$, $F = \{\text{foreign cars}\}$; $W = \{\text{well-designed cars}\}$

(*a*) put the following symbolic forms into statement form:

(i) $F \cap W \neq \phi$ (ii) $W \subset F$ (iii) $F \subset W$ (iv) $F' \cap W = \phi$

(*b*) write the following statements in symbolic form:

(v) Foreign cars are not well designed.

(vi) Some home-produced cars are well designed. (Home-produced means not foreign.)

(vii) There are some cars which are not well designed.

(viii) All cars are well designed.

(*a*) (i) $F \cap W \neq \phi$ means that there are some elements in $F \cap W$; in other words, 'some foreign cars are well designed'. (Possibly all of them.)

(ii) Set W is a proper subset of set F, this means 'all well-designed cars are foreign'.

(iii) Set F is a proper subset of set W; here we are saying that 'all foreign cars are well designed'.

It is worth illustrating in Fig. 1.15 that (ii) and (iii) are not the same.

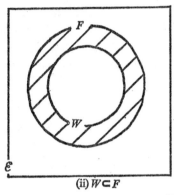
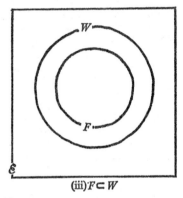

(ii) $W \subset F$ (iii) $F \subset W$

Fig. 1.15

In (ii) we observe that *some* foreign cars are not well designed; in fact, all those elements in the shaded region.

In (iii) we see clearly that *every* foreign car is well designed.

(iv) Here we see that the intersection of the set of non-foreign cars with the set of well-designed cars is empty, so our statement is 'no home-produced car is well designed'.

Before going on to part (*b*) we shall examine our answer to part (*a*) a little more closely.

(i) Notice that we were only able to say that *some* foreign cars were well designed, i.e. we were not able to say that *all* foreign cars were well designed. We can only say all when we are given $F \subseteq W$.

(iv) All the elements of $\mathscr{E}$ occur in either F or F', so that since we are told that W has no elements in common with F', the elements of W must all be elements of F. In short, $W \subseteq F$.

(*b*) (v) $W' = \{$set of cars which are not well designed$\}$, therefore our symbolic form is $F \subseteq W'$, alternatively we may write $F \cap W = \phi$.

(vi) $F' = \{$set of non-foreign cars (or home-produced cars)$\}$ and since some elements of F' belong to W, then $F' \cap W \neq \phi$.

(vii) Here we mean that there are some elements in W' or, symbolically $W' \neq \phi$.

(viii) Every car is in $\mathscr{E}$, if it is also well designed, then $W = \mathscr{E}$.

Exercise 1.13

1. If $n(A \cup B) = 51$, $n(A) = 20$, $n(B) = 44$, determine $n(A \cap B)$.

2. A sports club has facilities for Football (F), Cricket (C), and Tennis (T). An inquiry into the use of these facilities by the 274 members revealed the following results: $n(F) = 130$, $n(T) = 59$, $n(C) = 106$, $n(F \cap C) = 40$, $n(C \cap T) = 9$, $n(T \cap F) = 13$. If 38 members do not use any of these facilities at all, how many members use all three? Determine also the number $n(T \cap F \cap C')$, $n(T \cap C \cap F')$, $n(T' \cap C \cap F)$.

3. Of 174 students who have studied Mathematics, it is found that 37 studied Biology, 60 studied Chemistry, and 111 studied Physics; 29 studied Biology and Physics, 50 studied Physics and Chemistry, 13 studied Chemistry and Biology. If 45 Maths students did not study either Biology, Physics, or Chemistry, how many studied all three? How many studied both Physics and Chemistry, but not Biology?

4. Fig. 1.14 was the Venn diagram to Example 3, page 21, in which we discussed the ownership of cars (C), refrigerators (R), washing machines (W), and television (T). Since this survey was carried out the following changes have taken place:

(a) 7 people who only owned television sets have now bought refrigerators.

(b) 6 people who owned cars only have now bought television sets, 2 of them also bought a washing machine.

What changes should be made in the Venn diagram in order to bring it up to date?

5. If $\mathscr{E} = \{$people$\}$, $L = \{$lazy people$\}$, $M = \{$men$\}$, give a simple English statement for the following symbolic forms:

(i) $L \cap M \neq \phi$ (ii) $\mathscr{E} = L \cup M$ (iii) $L' = \phi$

(iv) $M \subset L$ (v) $L \subset M$ (vi) $L \cup M \neq \mathscr{E}$

6. If $T = \{$all whole numbers divisible by 3$\}$, $F = \{$all whole numbers divisible by 5$\}$, $S = \{$all whole numbers divisible by 6$\}$, give a description of the following:

(i) $T \cap F$ (ii) $S \cap F \cap T$ (iii) $S \cup T$

7. With $\mathscr{E} = \{$people$\}$, $T = \{$people who drink$\}$, $D = \{$people who drive cars$\}$, $C = \{$drivers who are careful$\}$, give a symbolic form to the following statements:

(i) Some drivers drink.
(ii) Some drivers do not drink.
(iii) No driver who drinks is careful.
(iv) All careful drivers do not drink.

Supplementary Exercise

1. If $A = \{a, b, c, d, e\}$, state whether the following are true (T), or false (F):

(i) $a \in A$ (ii) $\{a\} \in A$ (iii) $\{a, b\} \in A$

(iv) $\{a, b\} \subset A$ (v) $a \subset A$ (vi) $\{d, e\} \subset A$

(vii) $\{a, g\} \cap A = \{a\}$ (viii) $\{a, b, c\} \cup A = A$

(ix) $\{g, h\} \cap A = \phi$ (x) $\{a, b, c\} \cup \{d, e\} = A$

2. If $\mathscr{E} = \{1, 2, 3, 4, 5, 6\}$, $A = \{2, 4, 6\}$, $B = \{1, 3, 5\}$, state whether the following are true or false:

(i) $n(A) = n(B)$ (ii) $A = B$ (iii) $A' = B$

(iv) $B' = A$ (v) $A' \cup B' = \mathscr{E}$ (vi) $A \cup B = \mathscr{E}$

(vii) $A \cap B = \phi$ (viii) $B' = A'$

3. If $\mathscr{E} = \{$natural numbers$\}$, $A = \{$even numbers$\}$, $B = \{1, 3, 5, 7\}$, express the following statements in symbolic form and say whether they are true or false:

(i) All even numbers are natural numbers.
(ii) All elements of set B belong to set A.
(iii) None of the elements of set B belong to set A.
(iv) There are no even numbers in set B.
(v) A and B are disjoint sets.

4. If $\mathscr{E} = \{$Farmers$\}$, $C = \{$Farmers who keep cows$\}$, $S = \{$Farmers who keep sheep$\}$, $P = \{$Farmers who keep pigs$\}$, $G = \{$Farmers who keep goats$\}$, then $C \cap P \cap S = \{$Farmers who keep cows, pigs, and sheep$\}$. Express the following in similar form:

(i) $C \cap S \cap G$ (ii) $C \cup S$ (iii) $S \cap G \cap C'$
(iv) $S \cap G \cap C \cap P$ (v) $S \cap G \cap C' \cap P'$

5. Complaints about a works canteen fell into three main categories. Complaints about: (i) the menu (M); (ii) the food (F); (iii) the service (S).

The 173 complaints received were as follows:

$n(M) = 110$, $n(F) = 55$, $n(S) = 67$.
$n(M \cap F \cap S') = 20$, $n(M \cap S \cap F') = 11$, $n(F \cap S \cap M') = 16$.

Using the formula given for the number of elements given on page 19, determine the number of complaints about all three. How many complaints were received about two or more of menu, service, and food?

(*Hint:* Put $x = n(M \cap F \cap S)$ so that $n(M \cap F) = 20 + x$, etc.)

CHAPTER TWO

WHOLE NUMBERS

OPERATIONS WITH WHOLE NUMBERS

In Chapter One we discussed the operations of ∪ and ∩ related to sets and we encountered the important ideas of a commutative law and an associative law. Our main purpose then was to introduce the language of sets. We shall now concentrate on the mathematical operations and the laws which govern them, but, rather than restrict ourselves to ∪ and ∩, we shall widen our experience by considering the operations of +, −, ×, ÷ applied to whole numbers. The basic results are already well known to the reader, so that it should come as no surprise that $3 + 4 = 7$ or that $2 \times 5 = 10$. Since our results will be of such a simple nature it would, perhaps, be a help if we stated the aims of this chapter.

First, there are certain important properties of our number system which are often ignored in the elementary treatment we receive in our younger days. The results are taken as intuitively obvious and hardly worth mentioning. We wish to emphasize these properties using the language of sets.

Secondly, we wish to lay the foundation for a later discussion of mathematical groups.

ADDITION

A number is too difficult to define in this book, but we can still progress satisfactorily without a definition, since we are always able to deal with numbers by simply manipulating the numerals which represent them. This is rather like television. We see television programmes and we discuss them; we know that some are longer than others and so on, but most of us have no idea how television works or what it really is, yet to some engineers and physicists it is 'intuitively obvious'.

We already know that with each set is associated a number, and the symbol $n(A)$ was introduced as 'the number of set A'. We also observed that a number is not the same as a set, i.e. $n(A) \neq A$. For example

$$\text{if } A = \{a, b, c, d\}, \quad \text{then } n(A) = 4;$$
$$B = \{g, h, q, l, x\}, \quad \text{then } n(B) = 5;$$

clearly $\{a, b, c, d\} \neq 4$ and $\{g, h, q, l, x\} \neq 5$.

We obtain $n(A)$ by placing its elements in a one–one correspondence with the counting numbers taken in order starting with 1, i.e. 1, 2, 3, 4, . . . and so on. Thus for $n(A)$ above we have:

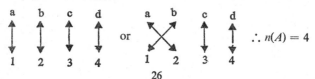

and for $n(B)$ above we have

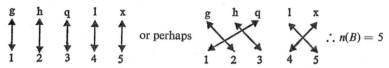

or perhaps $\therefore n(B) = 5$

The suitability of sets for defining the addition of numbers needs to be examined. When explaining how to add 5 and 4 you would probably suggest taking a set of 5 objects, say stones, and another set of 4 stones and then 'collect them together' to form a single set. A more precise explanation would be: Taking the two sets $A = \{a, b, c, d\}$ and $B = \{g, h, q, l, x\}$, we can form a new set, C which is the union of set A and set B so that $C = A \cup B$. The result of adding 4 and 5 is given by $n(A \cup B)$.

Unfortunately we soon encounter a slight difficulty. If A and B have an element or elements in common we will not obtain the required result. For example, if $A = \{a, b, c, d\}, B = \{a, l, c, y, t\}$ then our precise explanation would yield $C = A \cup B = \{a, b, c, d, l, y, t\}$

so that $n(A \cup B) = 7$

and $4 + 5 = 7.$

To get out of this difficulty we restrict our explanation to disjoint sets, i.e. sets with no common elements.

Since any number a is equal to $n(A)$ for some set A, we are now able to make the following definition.

Definition 2.1

If $a = n(A)$ and $b = n(B)$ then the sum $a + b$ of the two numbers a and b is given by $a + b = n(A \cup B)$, provided A and B are disjoint (the last phrase could have been written in the form

$$a + b = n(A \cup B); (A \cap B = \phi)$$

Caution! We add numbers, not sets. We write $4 + 5$ but we do not write $A + B$ for sets. We find the union of sets, not numbers. We write $A \cup B$, but we do not write $4 \cup 5$.

In an addition statement such as $4 + 5$ the numbers being added are called addends.

A further example. To find the sum of 6 and 4 we could think of the disjoint sets S and F such that $n(S) = 6$, $n(F) = 4$. Suitable sets might be:

$$S = \{u, v, w, x, y, z\}$$
$$F = \{p, q, r, s\}$$
$$S \cup F = \{u, v, w, x, y, z, p, q, r, s\}$$
$$n(S) + n(F) = n(S \cup F)$$
$$6 + 4 = 10$$

The numbers six and four are addends. The number ten is the sum. The numerals $6 + 4$ and 10 are two symbols for the sum. The numeral 10 is the simplest symbol for the number ten.

ADDITION IS COMMUTATIVE

If you join two sets you may wonder which set to join to which. Does the order of joining the sets change the union set?

The answer to this was no, as we saw in Chapter One, but let us look at the situation with respect to addition:

$$A = \{t, u, v\}, \, B = \{x, y, z, l\}$$

(Notice that they are disjoint sets.)

$$A \cup B = \{t, u, v, x, y, z, l\}$$
$$n(A) + n(B) = n(A \cup B)$$
$$3 + 4 = 7$$

Now let us reverse the order of joining the two sets.

$$A = \{t, u, v\}, \, B = \{x, y, z, l\}$$
$$B \cup A = \{t, u, v, x, y, z, l\}$$
$$n(B) + n(A) = n(B \cup A)$$
$$4 + 3 = 7$$

We note that $3 + 4 = 7$ and $4 + 3 = 7$
                                   or
          $3 + 4 \qquad = \quad 4 + 3$

The sum remains the same even when the order of the addends or numbers is changed. This, of course, is just the one special case only, of the sum of 3 and 4, it does not entitle us to say that this applies to all pairs of numbers. However, we do know that $A \cup B = B \cup A$ for any sets A and B, so that if these sets are disjoint, then we deduce that since

$$n(A) + n(B) = n(A \cup B)$$
and
$$n(B) + n(A) = n(B \cup A)$$
then
$$n(A) + n(B) = n(B) + n(A)$$

That is, for all whole numbers a and b

$$a + b = b + a$$

We call this idea the commutative property of addition or we say that 'addition is commutative'.

The phrase 'for all whole numbers a and b' means that a and b can be replaced by numerals for any numbers in the set of whole numbers. They may be replaced by the same numeral or by different numerals, including zero.

So even though we may not know the sum of 557 and 3 892, we know the following is true $557 + 3\,892 = 3\,892 + 557$, because addition is commutative.

Exercise 2.1

Think of doing one activity of each pair given below and then doing the other. Do the following pairs of instructions illustrate a commutative property? In other words, do they give the same final result?

1. Put on your socks; put on your shoes.
2. Take two steps forward; take two steps backward.
3. Write the letter 'O'; then write the letter 'N'.
4. Open the garage doors; back the car out.
5. Fill the swimming pool with water; dive in.

Complete each of the following sentences by using the commutative property of addition:

6. $3 + 7 = 7 +$ —   7. $9 + 1 =$ —   8. $a +$ — $= b +$ —
9. $14 + 6 =$ — $+ 14$ 10. — $+ 11 =$ — $+ 4$

IDENTITY NUMBER FOR ADDITION

Study the following union of sets.

$$\phi \cup \{x, y, z\} = \{x, y, z\}$$
$$\{x, y, z\} \cup \phi = \{x, y, z\}$$

Notice that joining the empty set to a given set, or joining a given set to the empty set, does not change the given set.

Now the addition statements which correspond to these two set operations above are:

$$0 + 3 = 3$$
$$3 + 0 = 3$$

Likewise $\begin{matrix} 0 + 22 = 22; 0 + 7 = 7 \\ 22 + 0 = 22; 7 + 0 = 7 \end{matrix}$ are two similar examples. We can generalize this result by considering

$$\phi \cup A = A$$
$$A \cup \phi = A$$

and since ϕ and A are disjoint sets, then:

$$n(\phi) + n(A) = n(A)$$
or $$0 + a = a$$
also $$a + 0 = a$$

Adding zero to any whole number a, or adding any whole number a to zero, leaves the whole number unchanged.

Since zero is the only number with this special property, the number zero is called the **identity element for addition**.

Of course, ϕ is the identity element for union of sets, since the union of ϕ with any set leaves it unchanged—a fact which we started with when we wrote $\phi \cup A = A$.

ADDITION IS ASSOCIATIVE

We have already seen that sets obey the associative law with respect to the operation of union. That is

$$(A \cup B) \cup C = A \cup (B \cup C)$$

has only the one result, where the insertion of brackets indicates which two sets are joined first. With these disjoint sets we are able to inquire into a similar property for the addition of numbers

$$n(A \cup B) + n(C) = n(A) + n(B \cup C)$$
$$(n(A) + n(B)) + n(C) = n(A) + (n(B) + n(C))$$

Using the notation $a = n(A)$, $b = n(B)$, $c = n(C)$
then we have just shown that

$$(a + b) + c = a + (b + c)$$

and hence, that addition with whole numbers is associative.

For example, if the sets

$$A = \{e, f\}; \quad B = \{g, h, k\}; \quad C = \{w, x, y, z\}$$

then 　　　　　$a = 2; \quad b = 3; \quad c = 4.$

The pattern established for joining the three sets A, B, and C enables us to write

$$2 + 3 + 4 = (2 + 3) + 4$$
$$= 5 + 4$$
$$= 9$$
$$2 + 3 + 4 = 2 + (3 + 4)$$
$$= 2 + 7$$
$$= 9$$

We have all used this rule when finding the sum of a set of numbers by collecting 9s with 1s, 8s with 2s, and so on. It is interesting to see which rules we are using on these occasions.

Suppose we find the sum of $3 + 6 + 7 + 9 + 4 + 1 + 2$? Mentally we may associate 3 with 7, 6 with 4, 9 with 1, so the rearrangement is carried out in the following manner.

$$3 + 6 + 7 + 9 + 4 + 1 + 2 = 3 + (7 + 6) + 9 + 4 + 1 + 2$$

Commutative law with $6 + 7 = 7 + 6$

$$= 3 + (7 + 6) + (4 + 9) + 1 + 2$$

Commutative law with $9 + 4 = 4 + 9$

$$= (3 + 7) + 6 + (4 + 9) + 1 + 2$$

Associative law $3 + (7 + 6) = (3 + 7) + 6$

$$= (3 + 7) + (6 + 4) + (9 + 1) + 2$$

Associative law $6 + (4 + 9) = (6 + 4) + 9$

$$= 10 + 10 + 10 + 2$$
$$= 30 + 2$$

Associative law $10 + 10 + 10 = 30$

$$= 32$$

Naturally we go through this process without a thought for the commutative and associative laws (just as well really), because our handling of whole numbers has become almost instinctive—so instinctive, in fact, that we might forget that we need to prove something about these laws when we come to consider operations with matrices and groups (Chapters Eleven and Six).

Exercise 2.2

Three things are to be combined in each exercise below. Do not change their order, only the grouping. Do the combinations show an associative property?

1. Water, lemon juice, sugar to make a drink.
2. Letter, envelope, stamp to post a letter.
3. Blue paint, red paint, green paint to make another colour of paint.
4. Boiling water, teapot, tea to make a cup of tea.

Complete each of the following sentences by using the associative property of addition:

5. $5 + (7 + 6) = (5 + \quad) +$
6. $17 + (15 + 32) = (\quad + \quad) + 32$
7. $(9 + 8) + 7 \quad = + (\quad + \quad)$
8. $\quad + (\quad + \quad) = (13 + 12) + 6$
9. $(\quad + \quad) + \quad = 72 + (31 + 46)$

Exercise 2.3

Each of the following sentences is true because of the commutative property of addition, the associative property of addition, or both. Write the letter C, A, or both C and A to tell which property or properties are used.

1. $(9 + 8) + 3 = 9 + (8 + 3)$
2. $(9 + 8) + 3 = 3 + (9 + 8)$
3. $6 + (7 + 12) = 6 + (12 + 7)$
4. $6 + (7 + 12) = (6 + 12) + 7$
5. $(13 + 5) + 14 = 14 + (5 + 13)$
6. $(32 + 9) + 8 = 9 + (32 + 8)$
7. $13 + (9 + 7) = (13 + 9) + 7$
8. $13 + (9 + 7) = (9 + 7) + 13$
9. $13 + (9 + 7) = (13 + 7) + 9$
10. $a + (b + c) = (a + b) + c$

A NUMBER LINE

It is often helpful to think of a set of numbers as corresponding to points on a line. For the set of whole numbers, $\{0, 1, 2, 3, 4, \ldots\}$ we simply draw a line and locate two points labelled 0 and 1. The arrow-heads at both ends of the picture indicate that the line extends indefinitely in both directions.

Then use the equivalent distance between the 0-point and the 1-point to locate points for 2, 3, 4, and so on.

Such a drawing is called a **number line**.

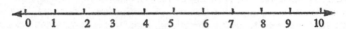

Of course, this line need not be drawn horizontally. It could just as well be in any other direction, as shown below. However, let us use the conventional horizontal arrangement.

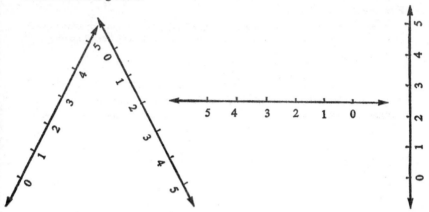

ADDITION ON A NUMBER LINE

We can picture addition by drawing arrows to represent the addends. For example, to represent 5 + 4 = ? we start by drawing an arrow from the 0-point to the 5-point to represent the addend 5. Then, starting at the head of this arrow (the 5-point), we draw another arrow extending in the same direction for 4 spaces (unit segments). The numeral for the sum is found directly below the head of the second arrow.

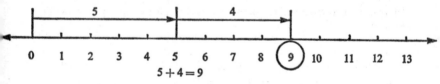

Using the conventional horizontal arrangement, addition is associated with 'moving to the right' on a number line.

We are, of course, suggesting that the movements 0 to 5, 1 to 6, 2 to 7, etc., are all equivalent. We must do this in order that the number line can be of use to us. For example, to demonstrate 4 + 5 = 5 + 4 we need to have 0 to 4 equivalent to 5 to 9 as illustrated in the diagram above.

An alternative is to regard the line as a counting device so that 5 + 4 is taken as: starting from 0 count 5 to the right and then count on, 4, to yield the equivalent to counting 9 to the right in the first place. The shorthand for this result is then 5 + 4 = 9.

Exercise 2.4

Write the addition sentence shown by each number-line drawing below.

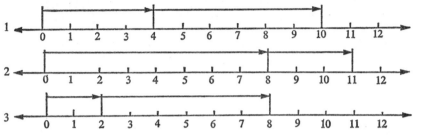

ORDER OF WHOLE NUMBERS

If two sets are not equivalent, then one set contains more members than the other. For example:

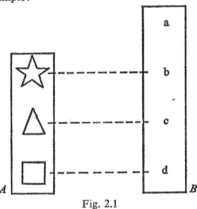

Fig. 2.1

Set *B* has some members left unmatched after all the members of set *A* have been matched.

Set *B* has more members than set *A*, or set *A* has fewer members than set *B*. We use the symbol $<$ (read: 'is less than') and the symbol $>$ (read: 'is greater than') when comparing the numbers of two sets that are not equivalent.

Thus: $\qquad n(A) < n(B) \quad \text{or} \quad n(B) > n(A)$

$$3 < 4 \quad \text{or} \quad 4 > 3$$

The order of the whole numbers is used when establishing a one-to-one correspondence between the whole numbers and selected points on a number line.

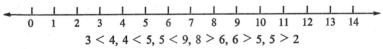

$3 < 4, 4 < 5, 5 < 9, 8 > 6, 6 > 5, 5 > 2$

On a number line, the point farther to the right corresponds to the greater of the two whole numbers.

INVERSE OPERATIONS

Many things we do can be 'undone'. If you take 2 steps backward you can return to your original position by taking 2 steps forward. If you add 6 to a number you can obtain the original number by subtracting 6 from the sum.

Any process or operation that 'undoes' another process or operation is called an **inverse operation**.

Of course, there are some activities that cannot be undone. For example, talking cannot be undone by being silent.

Exercise 2.5

For each activity given below, say how to 'undo it'.

1. Close your eyes.   2. Stand up.
3. Go to work.   4. Close this book.
5. Take 5 steps forward.   6. Untie your shoe lace.
7. Go to sleep.   8. Switch the light on.

SUBTRACTION

Three ways of thinking about the meaning of subtraction are explained below. The first two ways are helpful for interpreting a physical situation in terms of mathematics. However, their disadvantages will be pointed out. The last way defines subtraction for any mathematical situation.

1. *Removing a Subset*

John had 7 pennies and spent 4 of them. How many pennies did he have left?

We might illustrate the problem with Venn diagrams. Let H = {pennies he had} and let S = {pennies he spent}. Each circular region in the following drawing represents a distinct penny.

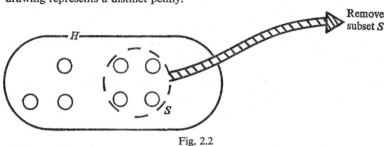

Fig. 2.2

Definition 2.2

The difference between $n(H)$ and $n(S)$, denoted by $n(H) - n(S)$, is the number of members in H but not in S.

When subset S is removed from set H only 3 pennies remain.

$$n(H) - n(S) = 3$$
$$7 - 4 = 3$$

This idea is suitable only for whole numbers and the particular type of problem illustrated.

2. Comparing Sets

Bob has 5 stamps and Jane has 9 stamps. How many more stamps does Jane have than Bob?

Let each ☐ in the following diagram represent a distinct stamp. Let $B = \{$Bob's stamps$\}$ and let $J = \{$Jane's stamps$\}$.

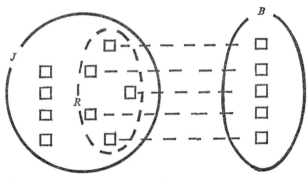

Fig. 2.3

Select a subset R of J by establishing a one-to-one correspondence between set B and subset R. Since $n(R) = n(B)$ we can treat sets J and R as in the previous method.

$$n(J) = 9, \; n(B) = n(R) = 5$$
$$n(J) - n(B) = 4$$
$$9 - 5 = 4$$

We now have a method of treating two types of subtraction problem, but as yet subtraction is not defined for *all* numbers. That is, neither of the above methods is practical for fractional numbers, only for whole numbers.

3. Inverse of Addition

By using either of the previous methods, we see that addition and subtraction are related—addition and subtraction undo each other. Addition and subtraction are inverse operations.

For the first pair of statements the relationship between the two operations can be thought of as follows:

In subtraction we 'do', i.e. we subtract 4 from 7 to obtain 3.
In addition we 'undo', i.e. we add 4 to 3 to obtain the original 7.

The relationship for the second pair of statements can be thought of as follows:

In addition we 'do', i.e. we add 8 to 1 to obtain 9.
In subtraction we 'undo', i.e. we subtract 8 from 9 to obtain the original 1.

The first example shows addition as the inverse of subtraction, while the second example shows subtraction as the inverse of addition.

Definition 2.3

For any numbers a, b, and c, if $c + b = a$, then c is the **difference** between a and b, denoted by $a - b$.

Note that $a - b$ names a number such that $(a - b) + b = a$. That is, we begin with the number a, subtract b, then add b, and the result is the number a with which we started. This shows the do–undo relationship between addition and subtraction.

Also note that $(a + b)$ names a number such that $(a + b) - b = a$. In this case subtraction undoes addition. Remember that we are still only dealing with whole numbers and, furthermore, we are for the moment subtracting the smaller number from the larger number, i.e. we are assuming that $a > b$. The reason for this last condition is to ensure that we always obtain a result which is a whole number.

Exercise 2.6

Write the simplest numeral for each of the following:

1. $(15 - 7) + 7$
2. $621 + (754 - 621)$
3. $(69 + 83) - 83$
4. $(26 - 15) + 15$
5. $57 + (756 - 57)$
6. $(39 - 17) + 17$
7. $(312 + 179) - 179$
8. $(r + s) - s$

Find the unnamed addend in each of the following:

9. $9 + ? = 15$
10. $7 + x = 11$
11. $6 + n = 14$
12. $8 + ? = 16$
13. $5 + k = 14$
14. $3 + y = 12$

SUBTRACTION ON A NUMBER LINE

Recall that addition is associated with moving to the right on a number line. Since addition and subtraction are inverse operations, we expect subtraction to be associated with moving to the left on a number line. (Alternatively, we may think of subtraction as 'counting back' as opposite to addition, which is 'counting on'.) On the following line we start at 0 and move to the right for 11 spaces and then move left for 4 spaces. The point at which we finally arrive on the line is interpreted as the result of $11 - 4$.

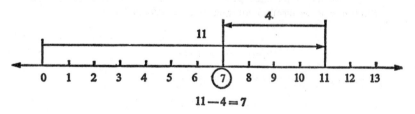

$$11 - 4 = 7$$

We are, of course, suggesting that the movements to the left, 11 to 7, 12 to 8, 13 to 9, etc., are all equivalent to any movement or counting back through 4 spaces.

Exercise 2.7

Write the subtraction sentence shown by each number-line drawing below.

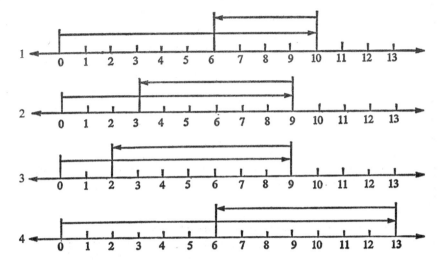

PROPERTIES OF SUBTRACTION

Is subtraction commutative? That is, can we change the order of the numbers without changing the difference or result?

$$7 - 3 = 4 \quad \text{but} \quad 3 - 7$$

does not name a whole number, let alone being equal to 4. Hence, subtraction is not commutative.

Is subtraction associative? That is, can we change the grouping of the numbers without changing the difference or result?

$$(12 - 6) - 2 = 6 - 2 = 4$$
but
$$12 - (6 - 2) = 12 - 4 = 8$$

Since $4 \neq 8$, we see that subtraction is not associative.

ZERO IN SUBTRACTION

If the empty set is removed from $A = \{a, b, c, d\}$, the result is set A.

$$n(A) - n(\phi) = n(A)$$
$$4 - 0 = 4$$

Since this is true for all sets, including disjoint sets, it is also true for all whole numbers.

If set A is removed from itself the result is the empty set.

$$n(A) - n(A) = n(\phi)$$
$$4 - 4 = 0$$

Since this is true for all sets, it is also true for all whole numbers. We can summarize these two special properties of zero as follows: For any whole number a,

$$a - 0 = a$$
and
$$a - a = 0$$

Just to summarize results so far, we see that:

1. Addition for whole numbers is commutative and associative and has zero as the identity element.
2. Subtraction for whole numbers as given in Definition 2.3 may be considered as the inverse of addition, but is neither commutative nor associative.

Many people use the fact that addition and subtraction are inverse operations when they check their results to simple problems like the following.

$$\begin{array}{rl} \text{Find the result} & 315 - 163 \\ \text{working} & 315 \text{ (i)} \\ & -163 \text{ (ii)} \\ \hline & 152 \text{ (iii)} \\ \hline \end{array}$$

the subtraction is checked by the addition of (ii) and (iii) to obtain (i).

If you are not familiar with this use of the inverse property try the following exercise.

Exercise 2.8

1.	408	2.	5382	3.	725
	−226		−3475		−537

USING SETS IN MULTIPLICATION

We have described addition of whole numbers in terms of combining disjoint sets. It is also possible to describe multiplication in this way.

A sandwich menu lists three kinds of meat: beef, ham, pork. You can have either white bread or rye bread. What are all the possible kinds of sandwiches if you choose one kind of meat and one kind of bread? The answer might be shown as follows.

(beef, white)	(ham, white)	(pork, white)
(beef, rye)	(ham, rye)	(pork, rye)

We notice that there are 3 choices of meat, 2 choices of bread, and 6 possible kinds of sandwiches. Somehow we have performed an operation on 2 and 3 to obtain 6.

Another example might involve finding the number of street intersections formed by the following situation.

{1st Ave., 2nd Ave., 3rd Ave., 4th Ave.}
{A St., B St., C St.}

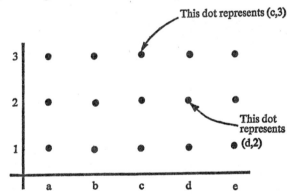

Notice that there are 4 avenues, 3 streets, and 12 intersections. Somehow we have performed an operation on 3 and 4 to obtain 12.

In a game you are to pick one letter from set A below and then pick one number from set B to make a pair.

$$A = \{a, b, c, d, e\}$$
$$B = \{1, 2, 3\}$$

To show all the possible pairs, we could construct the following array of dots. By counting we see that there are 15 possible pairs.

This dot represents (c,3)

This dot represents (d,2)

We are to choose first from set A, so let us agree to list the members of this set horizontally when making the array of dots in the above diagram. We are to choose second from set B, so let us agree to list the members of this set vertically.

In other words, the number of columns in the array is the number of the first set, and the number of rows is the number of the second set. In this case the number of columns is $n(A)$ or 5 and the number of rows is $n(B)$ or 3.

Note that the 15 dots indicate that there are 15 possible combinations for picking one letter and one number.

Exercise 2.9

Say how many dots there are in the array for picking one member from the first set below and one member from the second set.

1. {Ed, Bill, Al} and {Jo, Mary, Susan}
2. {?, *} and {a, b, c, d}
3. {cake, buns} and {coffee, tea, milk}
4. {a, b, c, d, e, f} and {7, 8, 9, 10}

DEFINITION OF MULTIPLICATION

Thinking of arrays for two sets enables us to define the operation of multiplication. Suppose we are given sets A and B such that $n(A) = a$ and $n(B) = b$. We could make an array for these sets so that it has a columns and b rows.

Definition 2.4

The product of any two whole numbers a and b, denoted by $a \times b$, is the number of dots in the array having a columns and b rows. The numbers a and b are called factors of the product. We may also describe $a \times b$ as the number of intersections of the a columns with the b rows.

MULTIPLICATION AS REPEATED ADDITION

Any array can be thought of as the union of equivalent sets.

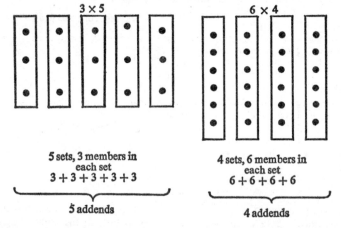

3 × 5

6 × 4

5 sets, 3 members in each set
3 + 3 + 3 + 3 + 3
5 addends

4 sets, 6 members in each set
6 + 6 + 6 + 6
4 addends

These drawings illustrate another way to think about multiplication of whole numbers.

$$3 \times 5 = 3 + 3 + 3 + 3 + 3 = 15$$
$$6 \times 4 = 6 + 6 + 6 + 6 = 24$$

This is sometimes referred to as the repeated addition description of multiplication.

Alternatively, we could have thought of these two examples in the following manner:

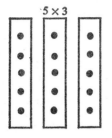

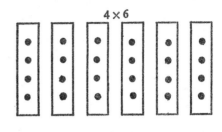

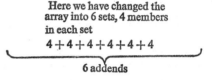

Here we have changed the array into 3 sets, 5 members in each set

$$\underbrace{5 + 5 + 5}_{\text{3 addends}}$$

Here we have changed the array into 6 sets, 4 members in each set

$$\underbrace{4 + 4 + 4 + 4 + 4 + 4}_{\text{6 addends}}$$

We notice that the order of the factors is changed, but the result remains the same.

That is for all whole numbers a and b

$$a \times b = b \times a$$

We call this idea the commutative property of multiplication. Or we say that multiplication is commutative.

Exercise 2.10

Complete each of the following sentences by using the commutative property of multiplication. Do not find any of the products.

1. $5 \times 9 = 9 \times \text{---}$
2. $\text{---} \times 8 = 8 \times 7$
3. $31 \times 7 = \text{---} \times 31$
4. $12 \times \text{---} = 6 \times 12$
5. $9 \times 17 = \text{---} \times \text{---}$
6. $p \times q = q \times \text{---}$ (p, q represent natural numbers)
7. $\square \times \triangle = \text{---} \times \square$ ($\square$, $\triangle$ represent natural numbers).

IDENTITY NUMBER OF MULTIPLICATION

Recall that zero is the identity number of addition because for any whole number a, $a + 0 = a = 0 + a$.

We would expect the identity number of multiplication to be some number such that for any whole number a, $a \times \text{---} = a = \text{---} \times a$.

Each array below has but one column, and hence the number of dots in the array is the same as the number of dots in the single column.

$$4 \times 1 = 4 \qquad 5 \times 1 = 5 \qquad 3 \times 1 = 3$$

Or we can interpret 4×1 as using 4 as an addend only once.

$$4 \times 1 = 4$$

Since multiplication is commutative, we know that $1 \times 4 = 4 \times 1$, and we conclude that $1 \times 4 = 4 = 4 \times 1$.

Multiplying any given whole number by one, or multiplying one by any given whole number, leaves the given number unchanged. Since one is the only number with this special property, we call the number one the identity number of multiplication.

For any whole number a, $a \times 1 = a = 1 \times a$.

MULTIPLICATION IS ASSOCIATIVE

Recall that the pattern of the associative property of addition is

$$(a + b) + c = a + (b + c)$$

for all whole numbers a, b, and c.

Is there such a property for multiplication? Let us look at a few examples.

$$(2 \times 3) \times 4 = 6 \times 4 = 24$$
$$2 \times (3 \times 4) = 2 \times 12 = 24$$
Therefore, $\qquad (2 \times 3) \times 4 = 2 \times (3 \times 4)$

$$(3 \times 5) \times 2 = 15 \times 2 = 30$$
$$3 \times (5 \times 2) = 3 \times 10 = 30$$
Therefore, $\qquad (3 \times 5) \times 2 = 3 \times (5 \times 2)$

When finding the product of three numbers we can group the first two factors or the last two factors and always get the same product.

This idea is called the associative property of multiplication. Or we say that multiplication is associative.

For all whole numbers a, b, and c,

$$(a \times b) \times c = a \times (b \times c)$$

Notice that when we use the associative property of multiplication the order of the factors is not changed, as it is when we use the commutative property of multiplication.

Let us consider which properties are being used at various stages of simplification in the following:

Example 1

$$(3 \times 5) \times 4 = 3 \times (5 \times 4) \qquad \text{Associative property}$$
$$= 3 \times 20$$
$$= 60$$

Example 2

$$3 \times 5 \times (9 \times 6) = 3 \times 5 \times (6 \times 9) \qquad \text{Commutative property}$$
$$= 3 \times (5 \times 6) \times 9 \qquad \text{Associative property}$$
$$= (3 \times 30) \times 9 \qquad \text{Associative property}$$
$$= 90 \times 9$$
$$= 810$$

Example 3

$$2 \times 7 \times 9 \times 5 = (7 \times 2) \times (5 \times 9) \qquad \text{Commutative}$$
$$= 7 \times (2 \times 5) \times 9 \qquad \text{Associative}$$
$$= 7 \times (10 \times 9) \qquad \text{Associative}$$
$$= 7 \times 90$$
$$= 90 \times 7 \qquad \text{Commutative}$$
$$= 630$$

Exercise 2.11

Complete each of the following sentences by using the associative property of multiplication:

1. $(3 \times 7) \times 5 = 3 \times (— \times —)$ 2. $4 \times (2 \times 3) = (— \times —) \times 3$
3. $(6 \times 3) \times 2 = — \times (— \times —)$ 4. $— \times (— \times —) = (17 \times 8) \times 5$
5. $(— \times —) \times — = 13 \times (9 \times 3)$

Each of the following is true because of the commutative property of multiplication, the associative property of multiplication, or both of these properties. Write the letter C, A, or both C and A to tell which property or properties are used.

6. $(9 \times 8) \times 3 = 9 \times (8 \times 3)$ 7. $(9 \times 8) \times 3 = 3 \times (9 \times 8)$
8. $6 \times (7 \times 12) = 6 \times (12 \times 7)$ 9. $6 \times (7 \times 12) = (6 \times 12) \times 7$
10. $(13 \times 5) \times 14 = 14 \times (5 \times 13)$ 11. $(32 \times 9) \times 8 = 9 \times (32 \times 8)$
12. $r \times (s \times t) = (r \times s) \times t$

ZERO IN MULTIPLICATION

What number is named by 0×3? By thinking of repeated addition,

$$0 \times 3 = 0 + 0 + 0 = 0$$

In order that multiplication shall be commutative for all our elements we shall define $3 \times 0 = 0 \times 3$ and that 3×0 may be taken as equal to 0.

It appears that when zero is one of the factors, then the product is zero. That is, for any whole number a,

$$a \times 0 = 0,$$
and $$0 \times a = 0.$$

What can we say about the factors if the product is zero? That is, what do we know about the factors a and b if $a \times b = 0$? The only way we can get a product of zero is to use zero as one of the factors. That is,

if $a \times b = 0$, then $a = 0$, or $b = 0$, or both factors are zero.

THE DISTRIBUTIVE PROPERTY

We now come to a law which we have not discussed before. It concerns a combination of the two operations × and +, and being a little more complicated in form, it may require a second or third reading. We shall continue to keep the examples as simple as possible in order that we may concentrate more on the operations involved.

Four boys and three girls are planning a party. Each child is to bring 2 gifts. How many gifts did they bring in all?

Two ways of thinking about solving this problem are given below:

1. There are $4 + 3$ or 7 children, each child will bring 2 gifts. Then, all together they will bring

$$(4 + 3) \times 2 \text{ or } 7 \times 2 = 14 \text{ gifts.}$$

2. Each of the 4 boys will bring 2 gifts. Then the boys will bring 4×2 gifts. Each of the 3 girls will bring 2 gifts. Then the girls will bring 3×2 gifts. All together the children will bring

$$(4 \times 2) + (3 \times 2) \text{ or } 8 + 6 = 14 \text{ gifts.}$$

From these ways of thinking about the problem we see that

$$(4 + 3) \times 2 = (4 \times 2) + (3 \times 2)$$

Since multiplication is commutative, we know that we can change the order of the factors.

Hence,
$$(4 + 3) \times 2 = 2 \times (4 + 3)$$
$$4 \times 2 = 2 \times 4,$$
and
$$3 \times 2 = 2 \times 3.$$

Then the statement
$$(4 + 3) \times 2 = (4 \times 2) + (3 \times 2)$$
can be written
$$2 \times (4 + 3) = (2 \times 4) + (2 \times 3)$$

Let us investigate such a pattern with different numbers.

$$2 \times (5 + 3) = 2 \times 8 = 16$$
$$(2 \times 5) + (2 \times 3) = 10 + 6 = 16,$$
hence
$$2 \times (5 + 3) = (2 \times 5) + (2 \times 3).$$

$$(4 + 6) \times 3 = 10 \times 3 = 30$$
$$(4 \times 3) + (6 \times 3) = 12 + 18 = 30,$$
hence
$$(4 + 6) \times 3 = (4 \times 3) + (6 \times 3).$$

This pattern is also visible in an array.

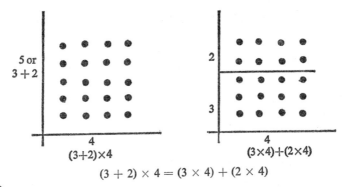

$$(3 + 2) \times 4 = (3 \times 4) + (2 \times 4)$$

By drawing a horizontal line in the array we can separate it into two arrays, one having 2 rows and 4 columns and the other having 3 rows and 4 columns. We have not discarded any of the dots, so the number of dots remains the same.

This property is called the distributive property of multiplication over addition.

For all whole numbers a, b, and c,

$$a \times (b + c) = (a \times b) + (a \times c)$$

and

$$(b + c) \times a = (b \times a) + (c \times a)$$

It is important to realize that the distributive property involves both addition and multiplication. Furthermore, it is important that we are able to 'undistribute' as follows:

$$(5 \times 3) + (5 \times 6) = 5 \times (3 + 6)$$
$$(4 \times 6) + (7 \times 6) = (4 + 7) \times 6$$

You have been using this property for some time without being able to identify it in such precise terms as we have used here.

In fact, even now the property is best illustrated by the use of unusual symbols as follows:

$$a \odot (b \boxtimes c) = (a \odot b) \boxtimes (a \odot c)$$

Try to complete the next two examples:

$$a \triangle (b \triangledown c) =$$
$$(p \odot q) \square r =$$

See end of Exercise 2.12 for solutions.

It is always useful to examine examples in which laws do not hold, as in the next case.

$$3 + (2 \times 5) = 3 + 10 = 13 \quad \text{the correct result.}$$

If we assume that $+$ is distributive over $\times$ we obtain the following result of

$$3 + (2 \times 5) = (3 + 2) \times (3 + 5)$$
$$= 5 \times 8$$
$$= 40$$

Our assumption was incorrect.

A return to Chapter One, page 17, reveals that $\cup$ is distributive over $\cap$, and vice versa.

That is $\qquad A \cup (B \cap C) = (A \cup B) \cap (A \cup C)$.

Complete this: $\qquad A \cap (B \cup C) =$

Let us see how all the properties we have discussed so far are used in the simplification of the following examples:

Example 1

$$
\begin{aligned}
34 \times 8 &= (30 + 4) \times 8 & \text{Rename 34} \\
&= (30 \times 8) + (4 \times 8) & \text{Distributive property of } \times \text{ over } + \\
&= 240 + 32 \\
&= 272
\end{aligned}
$$

Example 2

$$
\begin{aligned}
(2 + 3 + 18) \times 7 &= (3 + 2 + 18) \times 7 & \text{Commutative} \\
&= (3 + 20) \times 7 & \text{Associative} \\
&= (3 \times 7) + (20 \times 7) & \text{Distributive property} \\
&= 21 + 140 \\
&= 140 + 21 & \text{Commutative} \\
&= 161
\end{aligned}
$$

Of course, much of this simplification is carried out mentally (not necessarily in this order), and by so doing we come to take the properties of whole numbers for granted.

Let us now consider a more difficult example:

Example 3

$$
\begin{aligned}
24 \times 63 &= 24 \times (60 + 3) & \text{Rename 63} \\
&= (24 \times 60) + (24 \times 3) & \text{Distributive property of } \times \text{ over } + \\
&= (24 \times 6 \times 10) + (24 \times 3) & \text{Associative} \\
&= ((24 \times 6) \times 10) + (24 \times 3) & \text{Associative} \\
&= 1\,440 + 72 \\
&= 1\,512
\end{aligned}
$$

Exercise 2.12

Use the distributive property to complete each of the following sentences.

1. $7 \times (2 + 5) = (7 \times \text{---}) + (7 \times \text{---})$
2. $4 \times (3 + 6) = (\text{---} \times \text{---}) + (\text{---} \times \text{---})$
3. $(4 + 5) \times 3 = (\text{---} \times 3) + (\text{---} \times 3)$
4. $(8 + 9) \times 6 = (\text{---} \times \text{---}) + (\text{---} \times \text{---})$
5. $(5 \times 2) + (7 \times 2) = (\text{---} + \text{---}) \times 2$
6. $(6 \times 3) + (8 \times 3) = (\text{---} + \text{---}) \times \text{---}$

$\qquad a \triangle (b \triangledown c) = (a \triangle b) \triangledown (a \triangle c); (p \,\square\, r) \odot (q \,\square\, r)$

ESTIMATING A PRODUCT

Always we should be interested in estimates as well as the exact answers. Knowing how to multiply by multiples of powers of ten helps us to find an estimate of a product very quickly and easily. Suppose you are to find the value of n in $n = 28 \times 53$.

$$20 < 28 \quad \text{and} \quad 50 < 53,$$

so

$$20 \times 50 < n \quad \text{or} \quad 1\,000 < n$$

$$30 > 28 \quad \text{and} \quad 60 > 53,$$

so

$$30 \times 60 > n \quad \text{or} \quad 1\,800 > n.$$

Hence, we know that $1\,000 < n < 1\,800$, which is read: 1 000 is less than n, and n is less than 1 800. Another way of saying this is 'n is between 1 000 and 1 800'.

Another example might be to estimate the value of x if $x = 72 \times 587$.

$$70 < 72 \quad \text{and} \quad 500 < 587,$$

so

$$70 \times 500 = 35\,000 \text{ which is less than } x.$$

$$80 > 72 \quad \text{and} \quad 600 > 587,$$

so

$$80 \times 600 = 48\,000 \text{ which is greater than } x.$$

Hence, $35\,000 < x < 48\,000$, or the value of x is between 35 000 and 48 000.

Exercise 2.13

Find a rough estimate for n in each sentence. Then find the exact answer.

1. $n = 27 \times 65$
2. $n = 82 \times 75$
3. $n = 39 \times 58$
4. $n = 43 \times 94$
5. $n = 47 \times 367$
6. $n = 77 \times 492$
7. $n = 826 \times 52$
8. $n = 572 \times 67$

DIVISION

Division is related to multiplication in much the same way that subtraction is related to addition. When two numbers are subtracted the subtraction can be undone by addition. Similarly, when two numbers are divided the division can be undone by multiplication. Hence, multiplication and division are inverse operations.

Addition	Subtraction
$5 + 7 = 12$	$12 - 7 = 5; \quad 12 - 5 = 7$
$29 + 5 = 34$	$34 - 5 = 29; 34 - 29 = 5$

Multiplication	Division
$7 \times 6 = 42$	$42 \div 6 = 7; \quad 42 \div 7 = 6$
$24 \times 3 = 72$	$72 \div 3 = 24; 72 \div 24 = 3$

Knowing that multiplication and division are inverse operations, we can interpret $8 \div 2$ as that factor which, when multiplied by 2, yields a product of 8. That is,

$$(8 \div 2) \times 2 = 8$$

If we think of an array as we did for multiplication 8 ÷ 2 would be the number of columns in an array of 8 dots having 2 dots in each column.

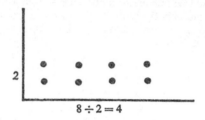

$$8 \div 2 = 4$$

Or we can think of 8 ÷ 2 as the number of disjoint subsets formed when a set of 8 objects is separated into disjoint subsets having 2 objects each.

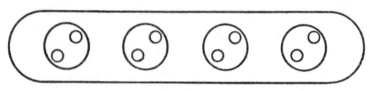

$$8 \div 2 = 4$$

Set of 8 objects separated into 4 disjoint subsets having 2 objects each.

ZERO IN DIVISION

First let us investigate a division such as 7 ÷ 0 = n. Since multiplication and division are inverse operations, the above division can be restated as a multiplication.

$$7 \div 0 = n \quad \text{so} \quad n \times 0 = 7$$

But we already know that when one of the factors is zero the product is zero. Hence there is no number n such that $n \times 0 = 7$. That is, 7 ÷ 0 does not name a number.

Now let us investigate the special case 0 ÷ 0 = n. Restate this as a multiplication.

$$0 \div 0 = n \quad \text{so} \quad n \times 0 = 0$$

In this case any number we choose for n yields a product of 0. That is, $5 \times 0 = 0$, $721 \times 0 = 0$, $9\ 075 \times 0 = 0$, and so on. If we accept 0 ÷ 0 as a name for a number, then we are forced to accept that it names every number. This is certainly not very helpful.

The fact that in the first case no number is named and in the second case every number is named is a source of difficulty in division. Let us rule out both of these cases by agreeing to this:

Division by zero is meaningless. It means that we shall not define division by zero.

Now let us investigate a case such as 0 ÷ 8 = n.

Restate this as a multiplication.

$$0 \div 8 = n \quad \text{so} \quad n \times 8 = 0$$

We already know that if the product is zero at least one of the factors must be zero. Since $8 \neq 0$, then n must be equal to zero. Hence, $0 \times 8 = 0$ and $0 \div 8 = 0$.

This is true regardless of which number we choose for a, except $a = 0$, in the following:

$$0 \div a = 0 \quad \text{if} \quad a \neq 0.$$

We can state our findings as follows. When zero is divided by any non-zero number the result is zero.

DEFINITION OF DIVISION

Now let us state a definition of division.

Definition 2.5

$$\text{If } a \times b = c \text{ and } b \neq 0, \text{ then } c \div b = a.$$

We read $c \div b$ as 'c divided by b'.

The number named by $c \div b$ is called the **quotient**, the number named by b is called the **divisor**, the number named by c is called the **dividend**.

We can show the relationship between these numbers and the numbers in a multiplication as follows:

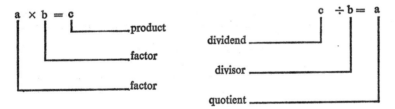

Hence, we see that in a division we are given the product and one of the factors, and we are to find the other factor.

From this definition, and knowing the basic multiplication facts, we can determine the basic division facts.

Exercise 2.14

Find each quotient:

1. $42 \div 7$	2. $35 \div 5$	3. $48 \div 8$	4. $81 \div 9$
5. $36 \div 4$	6. $21 \div 3$	7. $45 \div 9$	8. $32 \div 8$
9. $25 \div 5$	10. $49 \div 7$	11. $27 \div 3$	12. $28 \div 7$

PROPERTIES OF DIVISION

Is division commutative? That is, do $12 \div 4$ and $4 \div 12$ name the same number? $12 \div 4 = 3$, but $4 \div 12$ does not name a whole number, let alone three. Hence, $12 \div 4 \neq 4 \div 12$. Division is not commutative.

Is division associative? That is, do $(12 \div 6) \div 2$ and $12 \div (6 \div 2)$ name the same number?

$$(12 \div 6) \div 2 = 2 \div 2 = 1,$$
$$12 \div (6 \div 2) = 12 \div 3 = 4,$$

hence
$$(12 \div 6) \div 2 \neq 12 \div (6 \div 2).$$

Division is not associative.

Since division is not commutative, we know that $8 \div 1 \neq 1 \div 8$. But let us see what happens when the divisor is 1.

$$8 \div 1 = n \quad \text{so} \quad n \times 1 = 8$$

Since 1 is the identity number of multiplication, we see that $n = 8$ and $8 \div 1 = 8$. That is, when the divisor is 1 the dividend and the quotient are the same. Hence, for all whole numbers a,

$$a \div 1 = a.$$

If we are to find the value of n in $12 \div 3 = n$ we might rename 12 as $(9 + 3)$. Could it be that division distributes over addition? Let us try it.

$$(9 + 3) \div 3 = 12 \div 3 = 4$$
$$(9 \div 3) + (3 \div 3) = 3 + 1 = 4$$

Hence,
$$(9 + 3) \div 3 = (9 \div 3) + (3 \div 3)$$

We might be tempted to try the other pattern of the distributive property. That is, rename 3 as $2 + 1$ and write $12 \div 3$ as $12 \div (2 + 1)$.

$$12 \div (2 + 1) = 12 \div 3 = 4$$
$$(12 \div 2) + (12 \div 1) = 6 + 12 = 18$$

Hence,
$$12 \div (2 + 1) \neq (12 \div 2) + (12 \div 1)$$

However, it is important to remember that for all whole numbers a, b, and c, where $c \neq 0$,

$$(a + b) \div c = (a \div c) + (b \div c)$$

We say that division distributes over addition from the right only.
Study the following examples, which show the use of this property.

Example 1

$$
\begin{aligned}
32 \div 4 &= (20 + 12) \div 4 \\
&= (20 \div 4) + (12 \div 4) \\
&= 5 + 3 \\
&= 8
\end{aligned}
$$

Example 2

$$75 \div 5 = (40 + 35) \div 5$$
$$= (40 \div 5) + (35 \div 5)$$
$$= 8 + 7$$
$$= 15$$

Example 3

$$75 \div 5 = (50 + 25) \div 5$$
$$= (50 \div 5) + (25 \div 5)$$
$$= 10 + 5$$
$$= 15$$

We remark once again that in many cases division of one whole number by another will not result in a whole number. For example, $4 \div 12$ does not result in a whole number, $7 \div 5$ likewise does not result in a whole number. There is nothing wrong with our definition of division, because we were discussing only whole numbers when we said that

$$\text{if } a \times b = c, b \neq 0, \text{ then } c \div b = a.$$

If we try to fit $7 \div 5$ into this statement we see that we must really find the whole number a such that

$$a \times 5 = 7$$

Since $$1 \times 5 = 5 \quad \text{and} \quad 2 \times 5 = 10$$

we are unable to obtain a whole number for a because the value of a clearly lies between 1 and 2.

If we wish to retain division as an operation which 'works' for all whole numbers, then we must supplement the whole numbers with new ones such as $\frac{7}{5}$ or $\frac{4}{12}$. These numbers are called **rational numbers.**

We have already encountered similar restrictions in the case of subtraction. In this chapter for whole numbers you will have noticed we ensured that we always subtracted the smaller number from the larger. If, on the other hand, we are given $2 - 10$ the answer will not be a whole number and, as in the case of division, we must supplement the set of whole numbers. In this case we supplement the whole numbers with negative numbers if we wish that subtraction should always 'work'.

By seeking to satisfy the laws of operation that we have discussed so far, mathematicians have gradually enlarged the idea of number. Already in our own little way we have seen that if we are only allowed to use whole numbers for our problems and their solutions our mathematics will only apply to a small part of the world around us.

With respect to division, we may still express an answer to a problem such as $17 \div 5$ in whole numbers by agreeing to accept the answer in two parts, as you will see in the next section.

Exercise 2.15

Rename the dividend in each of the following and use the distributive property of division over addition to find each quotient as illustrated in the three examples above.

1. $16 \div 2$	2. $65 \div 5$	3. $39 \div 3$	4. $44 \div 4$
5. $84 \div 7$	6. $24 \div 4$	7. $52 \div 4$	8. $72 \div 6$
9. $95 \div 5$			

REMAINDERS IN DIVISION

If we think of separating a set into disjoint equivalent subsets we find that some divisions do not yield a whole number as a quotient.

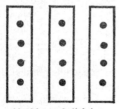

12 objects, 3 disjoint
sets, 4 objects in
each subset
12 ÷ 4 = 3

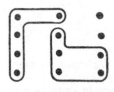

12 objects, 2 disjoint
sets of 5 objects each,
and 2 objects left over
$12 = (5 \times 2) + 2$
quotient⟶ ↗ remainder

As long as we are operating only with whole numbers, we shall give the remainder as such. Later in this book we will extend the number system so that we can carry out division without having to use remainders.

A final comment regarding the remainder is the following:

$$20 = (4 \times 5) + 0$$

so
$$20 \div 5 = 4 \text{ with a remainder } 0$$

Let us agree that every division of whole numbers has a remainder. That is, for whole numbers a and b, $b \neq 0$, the division $a \div b$ can be stated as

$$a = (q \times b) + r$$

where q is the quotient and r is the remainder. Furthermore, $r = 0$ or $r > 0$ and $r < b$. This means that the remainder is either zero or some whole number between 0 and b.

Exercise 2.16

Find each quotient and remainder:

1. $17 \div 5$ 2. $21 \div 6$ 3. $15 \div 7$ 4. $31 \div 5$
5. $40 \div 9$ 6. $33 \div 6$ 7. $55 \div 4$ 8. $73 \div 8$
9. $67 \div 6$

Summary for operations within the set of whole numbers.

1. Addition:

    (i) is commutative $a + b = b + a$
    (ii) is associative $a + (b + c) = (a + b) + c$
    (iii) the result is always a whole number
    (iv) zero is the identity element for addition $a + 0 = 0 = 0 + a$

2. Subtraction:

    (i) is not commutative $a - b \neq b - a$
    (ii) is not associative $(a - b) - c \neq a - (b - c)$
    (iii) the result is not always a whole number, e.g. $3 - 8$

3. Multiplication:

    (i) is commutative $a \times b = b \times a$
    (ii) is associative $a \times (b \times c) = (a \times b) \times c$
    (iii) the result is always a whole number
    (iv) 1   is the identity element for multiplication

$$1 \times a = a = a \times 1$$

4. Division:

    (i) is not commutative $a \div b \neq b \div a$
    (ii) is not associative $(a \div b) \div c \neq a \div (b \div c)$
    (iii) the result is not always a whole number
    (iv) by zero is meaningless

The reader should now be aware of the sort of inquiry it is necessary to carry out in order to determine whether operations within a set obey the laws we have been discussing. If we were now to examine another set of elements, for example, the set of Martian numbers, and in so doing we discovered that Martian numbers obeyed the same laws in the same way as our whole numbers, then we deduce that both sets have the same mathematical structure, and so for mathematical purposes there is no distinction between them. In other words, our inquiry reveals that we do not need to learn a new set of rules for operating with Martian numbers.

CHAPTER THREE

THE SET OF INTEGERS

CLOSURE UNDER ADDITION

We have observed that the sum of two whole numbers is always a whole number. For example, $6 + 2 = 8$, $29 + 15 = 44$. Never is the sum different from a whole number, such as $3\frac{1}{2}$ or $7\frac{3}{4}$. We express this idea by saying that the set of whole numbers is **closed under addition**, or closed with respect to addition.

Notice that this idea involves both a set of numbers and an operation. Hence, we cannot say that closure is a property of a set or of an operation, since it depends on both of them.

Definition 3.1

If a and b are *any* two numbers of set A and $+$ denotes the operation of addition, then set A is closed under addition if $a + b$ is a number of set A.

In the above case set A is the set of whole numbers, and by definition we can say that the set of whole numbers is closed under addition. It is natural to ask, is the set of whole numbers closed under subtraction? If it is, then for all whole numbers a and b, $a - b$ should be a whole number. But, we already know that $3 - 5$, $8 - 36$, etc., do not name whole numbers which you recall were the numbers of the set $\{0, 1, 2, 3, \ldots\}$. Thus, the set of whole numbers is not closed under subtraction.

In other words, we can state certain subtraction problems, or write certain equations, such as $6 + n = 0$ or $4 - x = 7$, which cannot be solved by using only the set of whole numbers.

To obtain such solutions we construct a set of numbers called the integers.

THE INTEGERS

Let us consider the equation $6 + n = 0$. Since there is no such whole number n for which $6 + n = 0$ is true, let us invent one. Let us call this new number the *negative of six* and denote it by the symbol (-6). Then we can write $6 + (-6) = 0$, since (-6) was invented especially for this purpose. Obviously for every whole number n (except 0) we can invent a new number $(-n)$ so that $n + (-n) = 0$, and therefore call $(-n)$ the negative of the whole number n.

Zero is excluded because $0 + (-0) = 0$ is equivalent to $0 + 0 = 0$, so why burden ourselves with two symbols for the number zero?

Unfortunately, having defined the negative of a whole number in this notation, we now see that sentences such as $6 + (-6) = 0$ and $6 - 6 = 0$ contain a confusing mixture of ideas. The symbol $-$ is used in two ways. One way is to indicate the negative of a number, such as (-6). The other way is to indicate subtraction, such as $6 - 6$.

Furthermore, because we want addition to remain commutative we need to examine $(-5) + 5 = 0$. Does this mean that 5 is the negative of (-5)?

If so, we ought to write this as (+5). In other words, the symbol + will be used in two ways as well. One way is to indicate addition and the other is to indicate the 'positive' of a number. Later we will see a definite relationship between these two uses, and no confusion should arise. In the meantime we make the following definitions:

Definition 3.2

The set of numbers . . ., (−4), (−3), (−2), (−1) is called **the set of negative integers.**

The set of numbers (+1), (+2), (+3), (+4), . . . is called **the set of positive integers.**

The set of numbers 0, (+1), (+2), (+3), . . . is called **the set of non-negative integers.**

The set of numbers . . ., (−4), (−3), (−2), (−1), 0 is called **the set of non-positive integers.**

Notice how we accommodate zero in these definitions; it is neither positive nor negative, but it is an integer. Also, we have written the positive integers in a similar manner to the negative integers by using the brackets. This is to ensure for the moment that we do not confuse whole numbers with integers, it also means that in the set of integers we write (+6) + (−6) = 0, and in this sense we refer to (−6) and (+6) as being opposites. Consider the following examples:

Example 1

Subtract five from nine.
In the set of whole numbers we write this as 9 − 5 = 4.
In the set of integers we write this as (+9) − (+5) = (+4).

Example 2

Subtract nine from five.
In the set of whole numbers 5 − 9 has no result or meaning.

In the set of integers this becomes (+5) − (+9) = (−4), a type of problem we shall examine in the next section.

THE NUMBER LINE

When using the number line to show addition of whole numbers you may have wondered about extending the number line to the left. Certainly we can locate points to the left of the 0-point just as we do to the right of the 0-point. Let us label them with the numerals for the integers, as shown below.

Since a number such as (+2) and its negative (−2) are on the opposite sides of the 0-point, you may think of them as being opposites of each other. Note, too, that the distance from the 0-point for any given number is the same as the distance from the 0-point to the opposite of the given number.

The integers are sometimes referred to as **directed numbers** because they not only indicate 'how many' but also in 'what direction'.

Consequently, we identify the integers with counting along the number line. Thus (+3) means that we count to the right through a distance of 3 units, while (−7) means that we count to the left through a distance of 7 units.

Many situations serve to make us aware of the need and use of such numbers. We refer to temperature readings above zero and below zero, distances above or below sea-level. Hence, a change of direction is implied by the words 'above' and 'below'.

Or someone gains £100 on a business transaction and loses £50 on another business transaction. In this case the change of direction is indicated by the words 'gain' and 'loss', and the £ amounts indicate the extent of the gain or loss.

ADDITION ON A NUMBER LINE

Now let us investigate the use of moves on a number line to help us understand the addition of integers. The sum of two positive integers is shown in the same way as the sum of two whole numbers. The answer as before is identified by the label attached to the final point of arrival on the number line, e.g. in the case of (+2) + (+3) = (+5) the (+5) is the label attached to the final point of arrival.

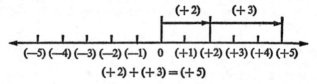

$$(+2) + (+3) = (+5)$$

This reads, 2 units right combined with 3 units right.

The following number line shows how to find the sum of (−2) and (+3).

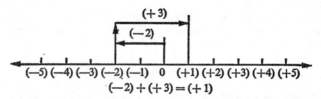

$$(-2) + (+3) = (+1)$$

This reads, 2 units left combined with 3 units right.

Since addition of integers is commutative, we know that (+3) + (−2) = (+1), as shown on the following number line:

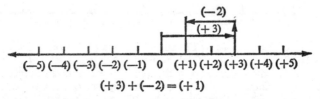

$$(+3) + (-2) = (+1)$$

This reads, 3 units right combined with 2 units left.

Another example of this type is $(-5) + (+3)$, as shown below.

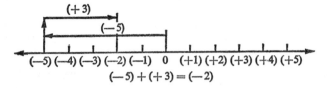

$$(-5) + (+3) = (-2)$$

This reads, 5 units left combined with 3 units right.

Finally, let us find the sum when both of the addends are negative integers, such as $(-3) + (-2) = (-5)$.

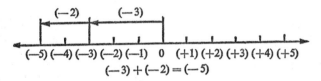

$$(-3) + (-2) = (-5)$$

This last result reads, 3 units left combined with 2 more units left.

As similarly in the case of whole numbers we are considering the following moves as equivalent $(+2)$ to $(+3)$, $(+3)$ to $(+4)$, etc., and also (-2) to $(+2)$, (-1) to $(+3)$, 0 to $(+4)$ as equivalent. The final result is obtained by reading off the labels on the line as being the simplest equivalent move to the combination of moves expressed in the addition.

These number lines give us an idea about adding integers. However, we certainly do not want to draw a number line every time we have to add integers; the number lines merely demonstrate how certain sums are found and do not necessarily prove how to add in all cases. We should like to develop the addition of integers according to the properties we have found for addition of whole numbers and without having to use a number line.

ORDER OF THE INTEGERS

In Chapter Two an order relation, $<$, was defined in the set of whole numbers. Since the set of integers is an extension of the set of whole numbers, it would be advisable to define (assuming this is possible) a similar and consistent order relation for the integers.

The set of positive integers corresponds to the set of whole numbers so that the statements $(+8) < (+13)$ in positive integers corresponds to $8 < 13$ in whole numbers.

But how can we order $(+2)$, and (-5)? Which is the 'smaller' of the two integers?

Consider the number line of the previous section. On the number line the point farther to the right corresponds to the greater of two integers. Therefore

$(+5)$ is said to be greater than $(+3)$ because $(+5)$ lies to the right of $(+3)$

or $(+3)$ is said to be less than $(+5)$ because $(+3)$ lies to the left of $(+5)$

which we abbreviate to $(+3) < (+5)$.

Also since $(+5)$ is farther to the right than (-4), then we say

$(+5)$ is greater than (-4)

i.e.　　　$(+5) > (-4)$

or　　　　(-4) is less than $(+5)$

i.e.　　　$(-4) < (+5)$

Once again we stress that the signs $>$ or $<$ are related to the relative positions of the integers on the number line.

ADDITIVE INVERSES

To invent the negative integers, we assumed that for every non-zero whole number a there exists the negative of a, denoted by $(-a)$ such that $a + (-a) = 0$.

Later on corresponding to the whole number n we wrote the positive integer $(+n)$ so that $(+n) + (-n) = 0$.

Definition 3.2

In the statement $(+n) + (-n) = 0$, $(+n)$ and $(-n)$ are called additive inverses of each other.

This means that $(+n)$ is the additive inverse of $(-n)$ and also that $(-n)$ is the additive inverse of $(+n)$. The term 'additive inverse' certainly stems from the use of addition and the fact that $(+n) + (-n)$ names the identity number of addition.

Exercise 3.1

Which numeral should replace x in each of the following so that each sentence becomes true?

1. $(+7) + x = 0$ 　　　　　　2. $(+15) + x = 0$
3. $(+9) + (-9) = x$ 　　　　　4. $x + (+12) = 0$
5. $(+129) + (-129) = x$ 　　　6. $(+306) + x = 0$

ADDITION OF INTEGERS

Definition 3.3

This merely invents the integers. In order for the integers to be useful, we must know how to calculate with them. Notice that while we defined addition and multiplication with whole numbers, we have yet to do the same for integers. We want the fundamental properties of the whole numbers and the properties of the operations on them to be true also for the set of integers. In order to discover definitions for the various operations on the integers, let us assume that addition and multiplication of integers obey the same properties that addition and multiplication of whole numbers obey. In other words, let us assume the following for all integers a, b, and c.

1. The set of integers is closed under addition, multiplication (and subtraction when discussed).

2. Associative properties:

$$(a + b) + c = a + (b + c)$$
$$(a \times b) \times c = a \times (b \times c)$$

3. Commutative properties:

$$a + b = b + a$$
$$a \times b = b \times a$$

4. Identity numbers:

$$a + 0 = a = 0 + a$$
$$a \times 1 = a = 1 \times a$$

5. Distributive property

$$a \times (b + c) = (a \times b) + (a \times c)$$
$$(b + c) \times a = (b \times a) + (c \times a)$$

Suppose we are to find the sum of $(+8)$ and (-2). We might observe a pattern as we know it from the addition of whole numbers.

$$(+8) + (+3) = (+11)$$
$$(+8) + (+2) = (+10)$$
$$(+8) + (+1) = (+9)$$
$$(+8) + 0 = (+8)$$
$$(+8) + (-1)?$$
$$(+8) + (-2)?$$

We notice that as we add $(+1)$ less each time, the sum decreases by $(+1)$ each time. If this pattern continues, then $(+8) + (-1)$ should be $(+7)$ and $(+8) + (-2)$ should be $(+6)$.

For a more logical justification of this sum, study the following:

$$(+8) + (-2) = (+6) + (+2) + (-2) \qquad \text{Rename } (+8)$$
$$= (+6) + [(+2) + (-2)] \qquad \text{Associative properties } +$$
$$= (+6) + 0 \qquad \text{Additive inverses}$$
$$= (+6) \qquad \text{Identity number}$$

Exercise 3.2

Find each sum by using the method illustrated above:

1. $(+7) + (-3)$	2. $(+12) + (-4)$	3. $(+13) + (-8)$
4. $(+26) + (-12)$	5. $(+46) + (-24)$	6. $(+87) + (-52)$
7. $(-8) + (+11)$	8. $(-5) + (+17)$	9. $(-9) + (+30)$

Now let us investigate the case where both addends are negative integers, such as in $(-2) + (-5)$. Since the set of integers is closed under addition, we know that this sum is an integer. Again, we might observe a pattern.

$$(+2) + (-5) = (-3)$$
$$(+1) + (-5) = (-4)$$
$$0 + (-5) = (-5)$$
$$(-1) + (-5) = ?$$
$$(-2) + (-5) = ?$$

If this pattern continues, then $(-1) + (-5)$ should be (-6) and $(-2) + (-5)$ should be (-7).

In deriving the sum of (-2) and (-5), let us consider the following:

$(+7) + (-2) + (-5)$	
$= (+5) + (+2) + (-2) + (-5)$	Rename $(+7)$
$= (+5) + (-5) + (+2) + (-2)$	Commutative and
	Associative properties $+$
$= 0 + 0$	Additive inverses
$= 0$	Addition

Since $(+7) + (-2) + (-5) = 0$, then $(-2) + (-5)$ must be the additive inverse of $(+7)$. That is, $(-2) + (-5) = (-7)$.

Exercise 3.3

Find each sum:

1. $(-2) + (-3)$
2. $(-7) + (-5)$
3. $(-8) + (-4)$
4. $(-10) + (-3)$
5. $(-2) + (-17)$
6. $(-21) + (-14)$
7. $(-35) + (-22)$
8. $(-18) + (-41)$
9. $(-72) + (-56)$
10. $(-105) + (-17)$

Finally, let us consider a sum of the type $(+8) + (-13)$.

$(+8) + (-13) = (+8) + (-8) + (-5)$	Rename (-13)
$= [(+8) + (-8)] + (-5)$	Associative properties
$= 0 + (-5)$	Additive inverses
$= (-5)$	Identity number

Exercise 3.4

Find each sum:

1. $(+4) + (-7)$
2. $(-14) + (+5)$
3. $(-21) + (+15)$
4. $(+8) + (-11)$
5. $(+52) + (-31)$
6. $(+7) + (+5)$
7. $(-7) + (+5)$
8. $(+7) + (-5)$
9. $(-7) + (-5)$
10. $(+121) + (-17)$
11. $(+82) + (+75)$
12. $(-17) + (-61)$
13. $(-44) + (+27)$
14. $(-44) + (-27)$

THE MEANING OF THE '−' OPERATION

We have already noticed that the set of whole numbers is not closed under subtraction. For example, $7 - 15$ and $3 - 9$ do not name whole numbers. An important reason for inventing the set of integers is that we desire a set of numbers that is closed under '−' which we continue to call subtraction.

Let us examine some subtractions in the set of whole numbers and some additions in the set of integers.

$$5 - 3 = 2; \quad (+5) + (-3) = (+2).$$
$$15 - 8 = 7; \quad (+15) + (-8) = (+7).$$
$$8 - 5 = 3; \quad (+8) + (-5) = (+3).$$

Apparently there is a close connexion between each addition, since the results are similar. Also, except for some of the signs, the same numerals are involved.

We still want addition and subtraction to be inverse operations for the set of integers. Recall that the difference between the whole numbers 5 and 3, denoted by $5 - 3$, is some number that when added to 3 yields a sum of 5.

$$3 + (5 - 3) = 5$$

Suppose we are to find the difference of two integers, such as $(+7) - (-3)$. Using our previous interpretation of subtraction, $(+7) - (-3)$ names some number, call it n, such that when added to (-3) yields a sum of $(+7)$

$$n + (-3) = (+7)$$

Since we know how to add integers, the value of n is easily determined.

$$n + (-3) = (+7)$$
and $$(+10) + (-3) = (+7)$$

Hence, $(+7) - (-3) = (+10)$ because $(+10) + (-3) = (+7)$.

Other such examples are:

(i) $(+8) - (-5) = (+13)$ because $(+13) + (-5) = (+8)$
(ii) $(+3) - (-4) = (+7)$ because $(+7) + (-4) = (+3)$

Now compare the following examples to (i) and (ii) above:

$$8 + 5 = 13 \text{ because } 13 - 5 = 8$$
$$3 + 4 = 7 \text{ because } 7 - 4 = 3$$

The result of subtracting (-5) from $(+8)$ is the same as adding $(+5)$ to $(+8)$. The result of adding (-5) to $(+13)$ is the same as subtracting $(+5)$ from $(+13)$. Similar relationships can be seen in the other examples. These relationships lead us to the following definition of subtraction.

Definition 3.4

The difference between any two integers a and b, denoted by $a - b$, is the integer $a + (-b)$.

In other words, to subtract an integer we can add the opposite (additive inverse) of the integer.

For all integers a and b, $a - b = a + (-b)$.

For example:

$$(+5) - (+7) = (+5) + (-7) = (-2)$$
$$(+13) - (-8) = (+13) + (+8) = (+21)$$
$$(-15) - (+7) = (-15) + (-7) = (-22)$$
$$(-9) - (-6) = (-9) + (+6) = (-3)$$

Exercise 3.5

Find each difference:

1. $(+7) - (-4)$ 2. $(-8) - (+12)$ 3. $(+11) - (+7)$
4. $(+8) - (+17)$ 5. $(-13) - (+8)$ 6. $(-13) - (-8)$
7. $(+13) - (+8)$ 8. $(+13) - (-8)$ 9. $(-14) - (+17)$
10. $(+72) - (+185)$ 11. $(+72) - (-185)$ 12. $0 - (+12)$

MULTIPLICATION ON A NUMBER LINE

As stated earlier, we require multiplication of integers to obey the same properties that multiplication of whole numbers obeys. Before using these properties in discovering how to multiply integers, let us use the number line to gain some understanding of multiplication of integers by whole numbers.

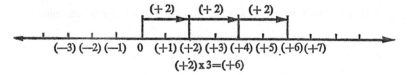

Since multiplication is to be commutative, $3 \times (-2) = (-2) \times 3$. We need only investigate one of these products. It is difficult to give meaning to $3 \times (-2)$ in terms of addition, since using 3 as an addend (-2) times is somewhat elusive. So let us show $(-2) \times 3$ as $(-2) + (-2) + (-2)$ on a number line.

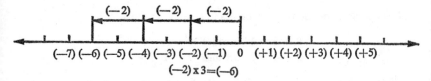

MULTIPLICATION OF INTEGERS

Now let us use the commutative and associative properties of multiplication, the distributive property of multiplication over addition (see page 44), and additive inverses to establish rules for multiplying integers.

A number line proved to be very helpful in forming initial ideas for the addition of integers. We then looked for patterns in the addition of whole numbers in order to establish a similar and consistent pattern for the addition of integers. Here we have just given a meaning to the multiplication of an integer by a whole number, therefore we will look for a pattern of results to suggest the result of multiplying two integers.

So far we have $3 \times 2 = 6$ $(+2) \times 3 = (+6)$ to suggest
$$(+3) \times (+2) = (+6)$$
and likewise $4 \times 7 = 28$ $(+7) \times 4 = (+28)$ to suggest
$$(+4) \times (+7) = (+28)$$

We need not consider further the case where both factors are positive integers, since this is similar to both factors being whole numbers. Let us begin by considering the case where either factor is a negative integer.

Study the following pattern:

$$(+5) \times (+3) = (+15)$$
$$(+5) \times (+2) = (+10)$$
$$(+5) \times (+1) = (+5)$$
$$(+5) \times 0 = 0$$
$$(+5) \times (-1) = ?$$
$$(+5) \times (-2) = ?$$

We notice that as the second factor decreases by $(+1)$ each time, the product decreases by $(+5)$ each time, i.e. the final result moves a distance of 5 units to the left on the number line. If this pattern continues, then

$$(+5) \times (-1) = 0 - (+5) = (-5)$$
$$(+5) \times (-2) = (-10)$$

In deriving the product

$$(+5) \times (-2)$$

let us consider the following:

$$(+2) + (-2) = 0 \qquad \text{Additive inverses}$$
$$(+5) \times [(+2) + (-2)] = 0 \qquad \text{Multiplied by zero}$$
$$[(+5) \times (+2)] + [(+5) \times (-2)] = 0 \qquad \text{Distributive property}$$
$$(+10) + [(+5) \times (-2)] = 0 \qquad \text{Multiplication}$$

Notice that the following number $(+5) \times (-2)$ is added to $(+10)$ and the sum is 0. Hence $(+5) \times (-2)$ must be the additive inverse of $(+10)$. That is $(+5) \times (-2) = (-10)$. This is in agreement with our idea of multiplication in terms of repeated addition.

$$(-2) \times (+5) = (-2) + (-2) + (-2) + (-2) + (-2) = (-10)$$

Also, by the commutative property of multiplication, we know that

$$(-2) \times (+5) = (+5) \times (-2) = (-10)$$

Exercise 3.6

Find each product.

1. $(+4) \times (-3)$
2. $(+8) \times (-2)$
3. $(-4) \times (+6)$
4. $(-5) \times (+7)$
5. $(+12) \times (-10)$
6. $(-12) \times (+10)$
7. $(+26) \times (-4)$
8. $(+21) \times (-13)$
9. $(+321) \times (-7)$
10. $(-5) \times (+205)$
11. $(-250) \times (+3)$
12. $(-126) \times (+23)$

Only one more case remains to be investigated—finding the product of two negative integers.

Knowing how to find the product of a positive integer and a negative integer, let us observe the following pattern:

$$(+3) \times (-4) = (-12)$$
$$(+2) \times (-4) = (-8)$$
$$(+1) \times (-4) = (-4)$$
$$0 \times (-4) = 0$$
$$(-1) \times (-4) = ?$$
$$(-2) \times (-4) = ?$$
$$(-3) \times (-4) = ?$$

As the first factor decreases by $(+1)$ each time, the product increases by $(+4)$ each time. If this pattern continues then $(-1) \times (-4) = (+4)$, $(-2) \times (-4) = (+8)$, and $(-3) \times (-4) = (+12)$.

In deriving the product

$$(-3) \times (-4)$$

let us consider the following:

$$(+4) + (-4) = 0 \qquad \text{Additive inverses}$$
$$(-3) \times [(+4) + (-4)] = 0 \qquad \text{Multiplied by zero}$$
$$[(-3) \times (+4)] + [(-3) \times (-4)] = 0 \qquad \text{Distributive properties}$$
$$(-12) + [(-3) \times (-4)] = 0 \qquad \text{Multiplication}$$

Notice that the number $(-3) \times (-4)$ is added to (-12) and the sum is 0. Hence, $(-3) \times (-4)$ must be the additive inverse of (-12). That is, $(-3) \times (-4) = (+12)$.

You have no doubt noticed a very close relationship between the multiplication of integers and the multiplication of whole numbers. The numerical computation is the same, but the sign or signs used to denote negative integers must be observed very closely in order to state the correct product numeral. We can conclude the following about multiplying integers:

1. The product of two positive integers is a positive integer.
2. The product of a positive integer and a negative integer is a negative integer.
3. The product of two negative integers is a positive integer.

Exercise 3.7

Find each product:

1. $(-3) \times (-5)$
2. $(-7) \times (-4)$
3. $(-7) \times (+4)$
4. $(+7) \times (-4)$
5. $(+7) \times (+4)$
6. $(-12) \times (-10)$
7. $(+12) \times (-10)$
8. $(-12) \times (+10)$
9. $(+12) \times (+10)$
10. $(-15) \times (-7)$
11. $(+122) \times (-3)$
12. $(-122) \times (-3)$
13. $(+402) \times (-4)$
14. $(-315) \times (-2)$

PROPERTY OF NEGATIVE ONE

Study the following:

$$(-1) \times (+7) = (-7), \quad (-1) \times (-7) = (+7)$$
$$(+15) \times (-1) = (-15), \quad (-13) \times (-1) = (+13)$$

We know that the number one is the identity number for multiplication. But what happens when we multiply by (-1)? Instead of the product being identical to the other factor, the product is the negative (or opposite) of the other factor.

$$a \times 1 = 1 \times a = a$$
$$a \times (-1) = (-1) \times a = (-a)$$

We call this property the property of negative one, or more concisely, the property of (-1).

We can use this property to verify some properties of integers, and in particular, some of the notions we already have about multiplication of integers. In the following examples, a, b, and c represent integers.

$$
\begin{aligned}
(-a) \times b &= [(-1) \times a] \times b & &\text{Properties of } (-1) \\
&= (-1) \times (a \times b) & &\text{Associative properties } \times \\
&= (-[a \times b]) \\
(-[a + b]) &= (-1) \times (a + b) & &\text{Properties of } (-1) \\
&= [(-1) \times a] + [(-1) \times b] & &\text{Distributive properties} \\
&= (-a) + (-b) & &\text{Properties of } (-1) \\
(-a) \times (-b) &= [(-1) \times a] \times [(-1) \times b] & &\text{Properties of } (-1) \\
&= [(-1) \times (-1)] \times (a \times b) & &\text{Commutative and} \\
& & &\text{Associative properties } \times \\
&= (+1) \times (a \times b) & &\text{Multiplication} \\
&= a \times b & &\text{Identity number } \times \\
a \times (b - c) &= a \times [b + (-c)] & &\text{Definition of subtraction} \\
&= (a \times b) + [a \times (-c)] & &\text{Distributive properties} \\
&= (a \times b) + [-(a \times c)] & &\text{Multiplication} \\
&= (a \times b) - (a \times c) & &\text{Definition of subtraction}
\end{aligned}
$$

Exercise 3.8

Find each product:

1. $0 \times (-52)$
2. $(-18) \times (-3)$
3. $82 \times (-13)$
4. 126×0
5. $315 \times (-1)$
6. $(-14) \times 0$
7. $(-1) \times 624$
8. $(-36) \times 3$
9. $(-12) \times (-12)$
10. $(-150) \times 6$
11. $(-15) \times (-6)$
12. $(-1) \times (-1)$

DIVISION OF INTEGERS

Every division can be stated as a multiplication by using the idea of multiplication and division as inverse operations. Recall that division of whole numbers was defined so that if $a \times b = c$ and $b \neq 0$, then $a = c \div b$.

Then division of whole numbers can also be defined so that if

$$c \div b = a \text{ and } b \neq 0, \text{ then } a \times b = c \tag{i}$$

Using the same definition for division of integers, we have the following:

$$(+20) \div (+4) = (+5) \text{ because } (+5) \times (+4) = (+20)$$
$$(+18) \div (+3) = (+6) \text{ because } (+6) \times (+3) = (+18)$$
$$(+250) \div (+25) = (+10) \text{ because } (+10) \times (+25) = (+250)$$

Now let us investigate the case where the divisor is a negative integer.

$$(+24) \div (-4) = q \text{ so } q \times (-4) = (+24) \text{ from (i) above.}$$

Then we ask ourselves, 'What number multiplied by (-4) yields a product of $(+24)$?' Obviously, $q = (-6)$. Since $(-6) \times (-4) = (+24)$, then $(+24) \div (-4) = (-6)$.

Suppose the dividend is a negative integer and the divisor is a positive integer, as in the following example:

$$(-32) \div (+8) = q \text{ so } q \times (+8) = (-32) \text{ from (i) above.}$$

Ask yourself, 'What number multiplied by $(+8)$ yields a product of (-32)?' From knowing how to multiply integers, we determine that $q = (-4)$. Since $(-4) \times (+8) = (-32)$, then $(-32) \div (+8) = (-4)$.

Thus far, we can conclude the following about division of integers:

1. The quotient of two positive integers is a positive number. It may or may not be an integer since the set of integers is not closed under division, e.g. $(+24) \div (+5)$ does not yield an integer for the result.

2. The quotient of a positive integer and a negative integer is a negative number. (It may or may not be an integer.) E.g. $(+24) \div (-5)$ does not yield an integer for the result.

Exercise 3.9

Find each quotient:

1. $(+10) \div (-2)$	2. $(-16) \div (+8)$	3. $(+72) \div (-9)$
4. $(+45) \div (-5)$	5. $(-63) \div (+7)$	6. $(+81) \div (-9)$
7. $(+75) \div (+15)$	8. $(+100) \div (+4)$	9. $(+100) \div (-4)$
10. $(-100) \div (+4)$	11. $(+256) \div (-8)$	12. $(-168) \div (+14)$

It may be the case that both the dividend and the divisor are negative integers.

$$(-18) \div (-3) = q \text{ so } q \times (-3) = (-18)$$

What number when multiplied by (-3) yields a product of (-18)? Again, by multiplication we know that $q = (+6)$.

Since $(+6) \times (-3) = (-18)$, then $(-18) \div (-3) = (+6)$.

Now we can conclude that the quotient of two negative integers is a positive number. (It may or may not be an integer.)

Exercise 3.10

1. $(-34) \div (-2)$	2. $(-19) \div (-19)$	3. $(-19) \div (+19)$
4. $(-21) \div (-3)$	5. $(-56) \div (-8)$	6. $(+132) \div (-11)$
7. $(+132) \div (+11)$	8. $(-132) \div (+11)$	9. $(-132) \div (-11)$
10. $(+13) \div (-1)$	11. $(-13) \div (-1)$	12. $(+99) \div (+11)$
13. $(-99) \div (+11)$	14. $(-99) \div (-11)$	

Earlier in this chapter we referred to the two uses of the symbols $+$ and $-$ (page 54), and in order to avoid possible confusion the integers were introduced with the aid of brackets, e.g. $(+3)$, (-2), etc. This was also done to ensure that we made some distinction between the positive integers and the whole numbers.

Now the set of integers is closed with respect to subtraction. Thus $(+4) - (+9) = (-5)$ states the correct result as an integer. Furthermore, it is clear which symbol $-$ stands for the operation of subtraction and which stands for 'negative'.

Unfortunately this distinction is not always made clear or even pointed out, consequently, the example just quoted is written as

$$4 - 9 = -5$$

Now if we are dealing with whole numbers the answer is meaningless. If, on the other hand, we are dealing with integers, how are we to distinguish between negative nine (a number) and subtract nine (an operation on a number)?

Let us suppose it means subtract nine, then (because -5 looks like -9) the answer must mean subtract five, which is an operation on a number (and not the same as a number). If it is a negative nine, what are we supposed to do with it since it is just written side by side with 4; in which case how do we get negative five?

With experience and practice we begin to understand the intended meanings, but only provided we have a clear understanding of the basic principles at the outset which we hope the reader will have gained in this chapter.

For practice in working from books in which the distinction above is not clear, try the next exercise.

Exercise 3.11

Determine the number n in each case:

1. $3 - 7 = n$
2. $13 - 22 = n$
3. $n - 9 = 4$
4. $-9 + n = 4$
5. $6 - n = 7$
6. $-n + 6 = 7$
7. $-8 + 9 = n$
8. $9 - 8 = n$
9. $-3 \times 7 = n$
10. $7 \times -3 = n$
11. $8 \div -2 = n$
12. $-2 \div 8 = n$

CHAPTER FOUR

SOLVING EQUATIONS AND PROBLEMS
RELATION SYMBOLS

By using mathematical symbols we are constantly building a language. In many respects it is more concise than the English language. Because of its brevity we must have a thorough understanding of the meaning of the symbols used in writing a number sentence. Study the meaning of each symbol given below.

$=$ is equal to,

$\neq$ is not equal to,

$<$ is less than,

$>$ is greater than.

These symbols serve as verbs in number sentences. Since they show how two numbers are related, they are called **relation symbols.**

The sentence $5 + 3 = 8$ is true, since $5 + 3$ and 8 are two names for the same number. But the sentence $7 - 3 = 8$ is false, since $7 - 3$ and 8 do not name the same number.

What relation symbol can we write between $7 - 3$ and 8 so that a true sentence is formed?

Since $7 - 3$ is not equal to 8, we see that $7 - 3 \neq 8$ is a true statement.

The statement $5 - 2 < 4 + 3$ is true, since $5 - 2$ or 3 is less than $4 + 3$ or 7. But the statement $8 + 5 < 9 - 6$ is false, since $8 + 5$ or 13 is greater than $9 - 6$ or 3.

The statement $17 - 5 < 12$ is false, since $17 - 5$ or 12 is equal to 12. Hence, we can change the relation symbol $<$ to the relation symbol $=$ and form the true statement $17 - 5 = 12$.

Exercise 4.1
Write T after each true statement below, and write F after each false statement.

1. $7 + 6 = 15$
2. $14 < 17 + 3$
3. $8 + 2 = 10$
4. $15 - 5 > 6$
5. $23 > 29 + 1$
6. $7 + 6 < 11 - 3$
7. $6 + 9 = 23 - 8$
8. $(+9) + (-1) = (+3) - (-5)$
9. $32 - 2 = 2 \times 15$
10. $3 \times 4 < 5 \times 2$

GROUPING SYMBOLS

Punctuation marks are used in any language to make clear what we want to say. Study how punctuation marks change the meaning of the first unpunctuated sentence.

Bob said Betty is honest.

Bob said, 'Betty is honest.'

'Bob,' said Betty, 'is honest.'

Punctuation marks are just as important in number sentences as they are in English sentences.

Study the following expression. What number does it name?

$$7 \times 2 + 5$$

Without being told by a symbol or some other means, we do not know whether to do the multiplication or the addition first.

If we multiply first:

$$7 \times 2 + 5 = 14 + 5 = 19$$

If we add first:

$$7 \times 2 + 5 = 7 \times 7 = 49$$

To avoid the confusion of such an expression naming two different numbers, let us use brackets ( ) to indicate which operation is to be done first.

$$(7 \times 2) + 5 = 14 + 5 = 19$$
$$7 \times (2 + 5) = 7 \times 7 = 49$$

When part of the sentence is enclosed within brackets think of that part as naming but one number. Think of (7×2) as naming 14 and think of $(2 + 5)$ as naming 7.

It is commonly agreed that if more than one operation, or all of the operations, are indicated in the same expression, without the use of brackets we multiply and divide first, then add and subtract.

$$5 + 3 \times 4 \text{ means } 5 + (3 \times 4) = 17$$
$$7 - 6 \div 2 \text{ means } 7 - (6 \div 2) = 4$$
$$7 + 3 \times 8 - 5 \text{ means } 7 + (3 \times 8) - 5 = 26$$
$$(+6) + (+3) \times (+2) \text{ means } (+6) + [(+3) \times (+2)] =$$
$$(+6) + (+6) = (+12)$$
$$(+7) + (+3) \times (+8) - (-5) \text{ means } (+7) + [(+3) \times (+8)] - (-5) =$$
$$(+7) + (+24) - (-5) = (+36)$$

In case only addition and subtraction are indicated (or only multiplication and division), we will perform the operations in the order indicated from left to right.

$$18 + 6 - 9 \text{ means } (18 + 6) - 9 = 15$$
$$6 \times 4 \div 3 \text{ means } (6 \times 4) \div 3 = 8$$
$$6 \div 3 \times 4 \text{ means } (6 \div 3) \times 4 = 8$$

It is not always necessary to write the multiplication sign. Multiplication is indicated in each of the following:

$$6(5 + 4) \text{ means } 6 \times (5 + 4) = 54$$
$$9(15) \text{ means } 9 \times 15 = 135$$
$$12n \text{ means } 12 \times n$$
$$(n + 2)(7 - 5) \text{ means } (n + 2) \times (7 - 5) = 2n + 4$$

Study how the brackets are handled in the following sentences:

$$(16 \div 2)(7 + 6) = 8 \times 13 = 104$$
$$(13 - 4) + (12 \div 3) = 9 + 4 = 13$$

Sometimes the sentence becomes so complicated that we need more than one set of brackets. Instead of two sets of ( ), let us use [ ] for the second set. In such cases we handle the **innermost** groupings first.

$$[4 \times (3 + 2)] - 8 = [4 \times 5] - 8$$
$$= 20 - 8$$
$$= 12$$
$$60 + [(8 \div 2) \times (4 + 3)] = 60 + [4 \times 7]$$
$$= 60 + 28$$
$$= 88$$

Exercise 4.2

What number is named by each of the following?

1. $(12 - 8) \times 7$ 2. $9 \times (32 \div 8)$ 3. $(11 - 6) + 12$
4. $(4 \times 2)(20 \div 5)$ 5. $28 \div (10 - 3)$ 6. $6 + (7 - 2) + 8$
7. $40 \div [(18 \div 3) + 2]$ 8. $[4 + (7 - 2)](9 - 3)$
9. $[14 \div (6 + 1)] \div 2$ 10. $8 - [7(5 - 3) - 6]$

Write brackets in each of the following expressions so that it will name the number indicated after it:

11. $30 - 12 \div 3 \times 2$ Number: 3   12. $30 - 12 \div 3 \times 2$ Number: 52
13. $30 - 12 \div 3 \times 2$ Number: 12   14. $30 - 12 \div 3 \times 2$ Number: 28
15. $30 - 12 \div 3 \times 2$ Number: 22

NUMBER SENTENCES

The symbols used in writing a number sentence are members of the following sets:

Number symbols or numerals:

$$0, 5, 72, 119, \ldots$$

Operation symbols:

$$+, -, \times, \div, \ldots$$

Relation symbols:

$$=, \neq, <, >, \ldots$$

Grouping symbols:

$$( ), [ ], \{ \}, \ldots$$

Placeholder symbols or variables:

$$n, x, y, t, \ldots$$

To write a number sentence we write a relation symbol between two different combinations of the other symbols.

A very important property of a number sentence which does not contain a placeholder symbol is that it is either true or false, but not both.

For example, the following number sentences are classified as true or false:

True	False
$4 + 7 = 19 - 8$	$6 + 9 = 17 - 8$
$5 \times 3 < 14 + 6$	$27 - 5 < 2 \times 9$
$48 \div 16 \neq 6$	$5 + 4 > 5 \times 4$

Exercise 4.3

Write T after each true sentence below and write F after each false sentence.

1. $5(3 + 4) = 21 + 14$
2. $(8 \div 4) + 7 < 4 \times 6$
3. $3 \times 4 > 20 - 8$
4. $6 + 8 \neq 15 \div 3$
5. $3(4 + 2) = (3 \times 4) + 2$
6. $(3 + 4)(5 - 5) < 5$
7. $(+2)[(+3) + (+4)] < (-3)$
8. $(+3) + (-9) < (+1)$

OPEN SENTENCES

In previous exercises we determined whether certain sentences were true or false. Now let us examine the following sentences to learn more about when a number sentence is true, when it is false, and when we are unable to determine which it is.

1. U Thant was elected Secretary General to the United Nations.
2. Florence Nightingale was elected Secretary General to the United Nations.
3. He was elected Secretary General to the United Nations.

You know that sentence 1 is true, and a little knowledge of history will show that sentence 2 is false. But what about sentence 3?

Until the word *He* is replaced by the name of a person, we are unable to tell whether sentence 3 is true or false.

Now consider the following number sentences:

$$7 + 6 = 13$$
$$9 - 5 = 27$$
$$8 + n = 15$$

Certainly, $7 + 6 = 13$ is true and $9 - 5 = 27$ is false, but we cannot tell whether $8 + n = 15$ is true or false until n is replaced by a numeral (7 in this case).

Definition 4.1

Mathematical sentences that contain letters (or some other symbols for placeholders) to be replaced by numerals, and are neither true nor false, are called *open* sentences.

Examples of open sentences are given below:

$$n - 15 = 7, \quad 3x + 2 < 12$$
$$14t = 64, \quad 26(y - 3) > 47$$

Exercise 4.4

After each sentence below write T if it is true, F if it is false, and O if it is open:

1. $(5 + 3) \div k = 2$
2. $(+4) - (-6) = (+10)$
3. $4y + (18 \div 6) < 4$
4. $3 + (-8) = 17 - (2 \times 10)$
5. $72 > (24 \div x) + 56$
6. $(16 \div 4) + 13 < (3 \times 4) + 5$

REPLACEMENT SET

Consider the replacements for the pronoun *she* in the following sentence.

She was a great musician.

Certainly it would be sensible to replace *she* with the name of a person. It would not be sensible to replace *she* with the name of a state, a building, a ship, and so on. If we want the resulting sentence to be meaningful we must know the set from which we can select the replacements for the pronoun *she*.

This idea also underlies an open sentence. We must know which set of numbers we are allowed to use when replacing a placeholder symbol or variable by a numeral.

Definition 4.2

The set of numbers whose names are to be used as replacements for a variable is called the *replacement set.*

SOLUTION SET

Use {1, 2, 3, 4, 5, 6} as the replacement set for x in the following open sentence:

$$3 + x < 7$$

We can replace x by each of the numbers 1, 2, 3, 4, 5, and 6 to determine which of them make the resulting sentence true.

True	False
$3 + 1 < 7$	$3 + 4 < 7$
$3 + 2 < 7$	$3 + 5 < 7$
$3 + 3 < 7$	$3 + 6 < 7$

We see that only the numbers 1, 2, and 3 can replace x to make the resulting sentence true. Hence, {1, 2, 3} is called the *solution set* for $3 + x < 7$.

Definition 4.3

The set of replacements for the variable in an open sentence that make the resulting sentence true is called the *solution set*. Each member of the solution set is called a *solution* or a *root* of the open sentence.

In the previous example, {1, 2, 3} is the solution set, and the roots of the open sentence are 1, 2, and 3.

Study the following examples of finding a solution set:

Example 1

Find the solution set of $n + 8 = 17$ if the replacement set is the set of whole numbers. We know that $9 + 8 = 17$, so 9 is a solution and belongs in the solution set. If n represents any whole number other than 9 the sentence is false. Hence, the only root is 9 and the solution set is {9}.

Example 2

Find the solution set of $7 - y = 10$ if the replacement set is the set of whole numbers. Since 0 is the least of the whole numbers and $7 - 0 = 7$, we know that y must represent a number less than zero. There is no such whole number, so the solution set is ϕ. This does not mean that there is no solution set. It merely means that there is no solution in the set of whole numbers.

Example 3

Find the solution set of $7 - y = 10$ if the replacement set is the set of integers. Since $7 - (-3) = 10$, we know that (-3) is a root. We could try other integer replacements for y to convince ourselves that this is the only root. Hence, the solution set is $\{(-3)\}$.

Exercise 4.5

Using $\{(-5), (-4), (-3), (-2), (-1), 0, (+1), (+2), (+3), (+4), (+5)\}$ as the replacement set, find the solution set of each of these open sentences:

1. $n + (+7) = (+9)$
2. $3x + (+2) = (-7)$
3. $(+3) + k < (+1)$
4. $(+7) - r < 0$
5. $(n \div 2) + (+3) = (+5)$
6. $4t - (+7) < (+10)$
7. $(+21) - n = (+21)$
8. $13 \times k = (-13)$
9. $3((+2) + n) = 0$
10. $4[(+15) \div n)] = (-12)$

EQUATIONS

Those number sentences that state that two expressions are names for the same number are called **equations**. The following are examples of an equation:

$$7 + 6 = 13, \quad n = 9$$
$$5 + 2 = 76, \quad 4(5 + r) = 28.$$

We see that an equation might be true, such as $7 + 6 = 13$, or it might be false, such as $5 + 2 = 76$, or it might be an open sentence, such as $4(5 + r) = 28$.

If an equation is not an open sentence all we need do is determine whether it is true or false. Hence, we are primarily concerned with those equations that are open sentences. When no confusion is possible let us refer to such open sentences as equations.

Definition 4.4

To *solve an equation* means to find its solution set.

Throughout this chapter consider the replacement set to be the **set of integers for all equations**, unless directed otherwise.

In an equation such as $n = (+5)$ or $k = (-56)$ the solution set is obvious. We can guess to find the solution set of an equation such as $n + (+3) = (+5)$ or $5t = (+15)$.

Study how the following equation might be solved mentally.

Solve $$3n + (+5) = (+11).$$

By previous agreement, the replacement set for n is the set of integers. Ask yourself: what number added to $(+5)$ gives a sum of $(+11)$? Since $(+6) + (+5) = (+11)$, then $3n = (+6)$. Then ask yourself: 3 times what number gives a product of $(+6)$. Since $3 \times (+2) = (+6)$, then $n = (+2)$. The solution set is $\{(+2)\}$. Check your answer by replacing n by $(+2)$ in the original equation $\{3n + (+5) = (+11)\}$ and compute to see if this replacement makes the resulting sentence true.

When we used the letter n as a placeholder for a whole number there was very little difficulty associated with finding that number as a replacement for n in equations such as $2n + 3 = 11$. We simply realized that since $8 + 3 = 11$ and $2 \times 4 = 8$, then n must be 4.

Now consider $2n + 3 = 1$. If we are looking for n in the set of whole numbers we discover that there is no value which will satisfy our equation. On the other hand, if we look for n in the set of integers it looks as though (-1) will do. Substitution of this value yields $2(-1) + 3 = 1$, which is really meaningless because of the confusion of $+3$ with $(+3)$ and 1 with $(+1)$. It would, however, have been a solution to

$$2n + (+3) = (+1)$$
because $$2(-1) + (+3) = (+1).$$

But do observe that we had to give a meaning to $2(-1)$ as being the same as $(+2)(-1)$ and (-2) earlier in our work. What meaning could we give to $-1(+2)$? Well, we know that subtracting this integer is the same as adding its opposite, which is (-2).

$$\therefore \quad -1(+2) = +1(-2); \text{ Similarly, } -5(+2) = +5(-2) \text{ and so on.}$$

We summarize these results in the following examples, even though they have been examined in the preceding chapter.

$$(+8) \times (+2) = 8(+2) = (+16)$$
$$(-8) \times (+2) = (-8)(+2) = -8(+2) = (-16)$$
$$(+8) \times (-2) = (+8)(-2) = (-16)$$

$(-8) \times (-2) = (-8)[(+1) - (+3)]$	Rename (-2)
$= (-8)(+1) - (-8)(+3)$	Distributive properties
$= (-8) - (-24)$	
$= (-8) + (+24)$	Additive inverse
$= (+16)$	

If we now consider w, x, y as placeholders for integers, then we know that since integers satisfy the distributive law of multiplication over addition we may write:

1. $$2(w + x + y) = 2w + 2x + 2y$$
or $$3(w - x - y) = 3w - 3x - 3y$$

2.
$$2(x + y) + 3(y + w) + 4(x - y - w)$$
$$= 2x + 2y + 3y + 3w + 4x - 4y - 4w$$
$$= 2x + 4x + 2y + 3y - 4y + 3w - 4w$$
$$= 6x + 1y - 1w$$

But suppose we try to simplify $6x - 14x$. In whole numbers this is meaningless. What is the meaning of the result for integers? There is a strong temptation to write

$$6x - 14x = -8x$$

but what meaning can we give $-8x$? The difficulty is really due to the mixed use of whole numbers 6 and 8 and the integer x.

$-x$ is the subtraction of the integer x, which is the same as adding its inverse, i.e. $+(-x)$. Where we write $(-x)$ as the inverse of integer x.

$$\therefore \quad -8x = +8(-x)$$

But since $8(x) + 8(-x) = 8[(x) + (-x)] = 0$, then $-8x$ means the integer which is opposite (or the inverse) of $8x$.

Example 1

We know that
$$-8x = (-16)$$
$$-8(+2) = (-16)$$
$$\therefore \quad x = (+2)$$

Example 2

We know that
$$-4x = (+16)$$
$$-4(-4) = (-4)(-4) = (+16)$$
$$\therefore \quad x = (-4)$$

Example 3

We know that
$$7x = (-42)$$
$$7(-6) = (-42)$$
$$\therefore \quad x = (-6)$$

The following exercise 4.6 is written in the form to be found in the more traditional books of mathematics.

(*a*) There is no distinction between whole numbers and integers.
(*b*) There is no distinction between the negative number and the subtracted number.

Try to interpret the question correctly and then solve the equation.

Exercise 4.6

Solve these equations:

1. $n + 5 = 12$ 2. $5k = 35$ 3. $36 \div y = 4$
4. $16 + x = 10$ 5. $7 + 3t = 19$ 6. $4r + 20 = 0$
7. $x(12 - 8) = 28$ 8. $8k \div 2 = -16$ 9. $x(8 - 12) = 28$
10. $5(1 - 2n) = 35$

ADDITION PROPERTY OF EQUATIONS

As we attempt to solve more complicated equations we become aware that a more systematic or logical procedure is needed. That is, we should like to develop a sequence of reasoning for restating an equation until it becomes simple enough for us to solve mentally. However, each step in the reasoning process should be based on the properties of numbers, the properties of the operations, or on the properties of equations that we shall assume.

Suppose we are to solve $r + (+5) = (+17)$. We already know that $r = (+12)$, since $(+12) + (+5) = (+17)$, but let us examine more closely how to arrive at such a root.

Can we somehow restate the equation so that only r remains on one side of the equal sign?

Let us begin by thinking of $r + (+5)$. We can undo the adding of $(+5)$ by adding its additive inverse, which is (-5).

$$[r + (+5)] + (-5) = r + [(+5) + (-5)] \qquad \text{Associative properties } +$$
$$= r + 0 \qquad \text{Additive inverses}$$
$$= r \qquad \text{Identity number } +$$

Since $r + (+5) = (+17)$ means that $r + (+5)$ and $(+17)$ are two names for the same number, we must also add (-5) to $(+17)$ in order that the two new expressions still name the same number. Hence, we could solve the equations as illustrated below:

$$r + (+5) = (+17)$$
$$[r + (+5)] + (-5) = (+17) + (-5)$$
$$r + (+5) + (-5) = (+12) \qquad \text{Associative properties } +$$
$$r + 0 = (+12) \qquad \text{Additive inverses}$$
$$r = (+12) \qquad \text{Identity number } +$$

The first step in the above solution uses what we call the addition property of equations.

For all integers a, b, and c,
if $a = b$, then $a + c = b + c$

In other words, we can add the same number to both sides of an equation. Of course, we know that adding a negative integer is equivalent to subtracting its opposite. Hence, we do not need such a property for subtracting.

Study the following examples:

Example 1

Solve $(+17) = n - (+8)$

$$(+17) + (+8) = [n - (+8)] + (+8) \qquad \text{Additive properties of equations}$$
$$(+25) = [n + (-8)] + (+8) \qquad \text{Addition}$$
$$(+25) = n + [(-8) + (+8)] \qquad \text{Associative properties } +$$
$$(+25) = n + 0 \qquad \text{Additive inverses}$$
$$(+25) = n \qquad \text{Identity number } +$$

Example 2

Solve $(+12) + x = (+31)$

Since addition is commutative, we can add (-12) on the right or on the left of $(+12) + x$ and $(+31)$.

$$(+12) + x = (+31)$$
$$(-12) + [(+12) + x)] = (-12) + (+31) \quad \text{Additive properties of equations}$$
$$[(-12) + (+12)] + x = (+19) \quad \text{Associative properties}$$
$$0 + x = (+19) \quad \text{Additive inverses}$$
$$x = (+19) \quad \text{Identity number } +$$

Exercise 4.7

As in Exercise 4.6, the following questions are written in traditional form. Interpret the question and solve the equations.

1. $k + 7 = 21$
2. $29 + x = 36$
3. $y - 12 = 43$
4. $8 + r = -9$
5. $-5 + n = 72$
6. $t + 6 = -18$
7. $17 = r + 6$
8. $n - (-3) = 11$
9. $-7 = r + 15$
10. $76 + n = 76$

MULTIPLICATION PROPERTY OF EQUATIONS

We can denote division by several symbols. In an equation such as $n \div 3 = 17$ it is convenient to state

$$n \div 3 \text{ as } \frac{n}{3}.$$

Again to restate the equation so that it can be solved mentally we should like to restate the equation so that only n remains on one side of the equal sign.

We can think of undoing the dividing by 3, by multiplying by 3, since multiplication and division are inverse operations.

$$n \div 3 = 17$$
$$(n \div 3) \times 3 = 17 \times 3$$
$$n = 51$$

another way of writing this is:

$$\frac{n}{3} = 17$$

$$\frac{n}{3} \times 3 = 17 \times 3$$

$$\frac{n \times 3}{3} = 51$$

$$n \times \frac{3}{3} = 51$$

$$n \times 1 = 51$$

$$n = 51$$

In the first step of this solution we used the **multiplication property of equations**.

For all integers a, b, and c, if $a = b$, then $a \times c = b \times c$. Study how this property is used in solving the following equation:

$$(-14) = \frac{a}{4}$$

$$(-14)(4) = \frac{a}{4} \times 4$$

$$(-56) = \frac{a \times 4}{4}$$

$$(-56) = a \times \frac{4}{4}$$

$$(-56) = a \times 1$$

$$(-56) = a$$

Fractions occurred in both of the preceding examples. You are probably familiar with the operations on these fractional numbers. These operations will be fully explained in the chapter dealing with rational numbers.

Exercise 4.8

Solve the following equations:

1. $\frac{c}{5} = 8$ 2. $\frac{x}{(-3)} = 9$ 3. $21 = \frac{n}{10}$

4. $0 = \frac{t}{3}$ 5. $\frac{n}{(-4)} = (-7)$ 6. $\frac{x}{12} = (-6)$

7. $25 = \frac{c}{3-4}$ 8. $\frac{a}{(-9)} = 1$ 9. $\frac{a}{2+5} = 8$

10. $\frac{t}{4(-5)} = (-8)$ 11. $3(-6) = \frac{n}{5}$ 12. $\frac{r}{(-1)} = 1$

DIVISION PROPERTY OF EQUATIONS

Consider solving the equation $4n = 32$. In this case n is multiplied by 4. Since multiplication and division are inverse operations, we can undo multiplying by 4 by dividing by 4. So we might divide both $4n$ and 32 by 4, as shown in the following example:

$$4n = 32$$

$$\frac{4n}{4} = \frac{32}{4}$$

$$\frac{4}{4} \times n = 8$$

$$1 \times n = 8$$

$$n = 8$$

In this case we have used the division property of equations. For all integers a, b, and c, where $c \neq 0$, if

$$a = b, \text{ then } \frac{a}{c} = \frac{b}{c}.$$

We might ask why we have a division property and no subtraction property. Once we have invented the set of rational (fractional) numbers, we can show that the division property is no longer needed. Until that time we will use this property of equations. Study how this property is used in the following examples:

Example 1

Solve $5n = 35$

$$5n = 35$$
$$\frac{5n}{5} = \frac{35}{5} \quad \text{Division property of equations}$$
$$\frac{5}{5} \times n = 7$$
$$1 \times n = 7$$
$$n = 7$$

Example 2

Solve $(-8)c = 56$

$$(-8)c = 56$$
$$\frac{(-8)c}{(-8)} = \frac{56}{(-8)} \quad \text{Division property of equations}$$
$$\frac{(-8)}{(-8)} \times c = (-7)$$
$$1 \times c = (-7)$$
$$c = (-7)$$

Exercise 4.9

Solve these equations expressed in traditional form:

1. $4n = 36$
2. $42 = -6r$
3. $-7k = 63$
4. $-5x = -75$
5. $-k = 47$
6. $72 = 8x$
7. $-52 = -a$
8. $t(5 + 2) = 91$
9. $n(5 - 9) = -24$
10. $x(21 \times 3) = 0$

SOLVING EQUATIONS

As we attempt to solve more and more complicated equations we may need to use more than one of the properties of equations or use the same property more than once.

In solving an equation such as $5t + (+6) = (+21)$ it is generally advisable to use the property that is most convenient for changing the expression $5t + (+6)$ to $5t$ first. Then use the property for changing $5t$ to t. Study the following example:

$$5t + (+6) = (+21)$$
$$[5t + (+6)] + (-6) = (+21) + (-6) \quad \text{Addition property of equations}$$
$$5t + (+6) + (-6) = (+21) + (-6) \quad \text{Associative property}$$
$$5t + 0 = (+15) \quad \text{Addition}$$
$$5t = (+15) \quad \text{Identity number}$$
$$\frac{5t}{5} = \frac{(+15)}{5} \quad \text{Division property of equations}$$
$$t = (+3) \quad \text{Division}$$

Then we can check our work by replacing t by $(+3)$ in the original equation.

$$5t + (+6) = (+21)$$
$$5(+3) + (+6) = (+21)$$
$$(+15) + (+6) = (+21)$$
$$(+21) = (+21)$$

Since we have shown that $5(+3) + (+6)$ and $(+21)$ name the same number, we know that $(+3)$ is a root of the equation.

Since we know that adding a number and its additive inverse is equivalent to adding zero, we can combine some of the steps when writing the previous solution. Also, we may make some of the calculations mentally in the process. However, the example shows the thinking steps necessary in solving the equation.

As you study the following examples, think of the reason or reasons for each step.

Example 1

Solve $\frac{n}{(+4)} + (-13) = (+3)$

$$\frac{n}{(+4)} + (-13) + (+13) = (+3) + (+13)$$
$$\frac{n}{(+4)} = (+16)$$
$$\frac{n}{(+4)} \times (+4) = (+16) \times (+4)$$
$$n = (+64)$$

Example 2

Solve $\frac{x - 5}{4} = 6$ (Traditional form)

$$\frac{x - 5}{4} = 6$$
$$\frac{x - 5}{4} \times 4 = 6 \times 4$$
$$x - 5 = 24$$
$$(x - 5) + 5 = 24 + 5$$
$$x = 29$$

Example 3

Solve $\frac{3t + 7}{2} = 26$ (Traditional form)

$$\frac{3t + 7}{2} = 26$$
$$\frac{3t + 7}{2} \times 2 = 26 \times 2$$
$$3t + 7 = 52$$
$$(3t + 7) + (-7) = 52 + (-7)$$
$$3t = 45$$
$$\frac{3t}{3} = \frac{45}{3}$$
$$t = 15$$

Example 4

Solve $5(2t - 14) = 60$ (Traditional form)

Solution 1

$$5(2t - 14) = 60$$
$$\frac{5(2t - 14)}{5} = \frac{60}{5}$$
$$2t - 14 = 12$$
$$(2t - 14) + 14 = 12 + 14$$
$$2t = 26$$
$$\frac{2t}{2} = \frac{26}{2}$$
$$t = 13$$

Solution 2

$$5(2t - 14) = 60$$
$$10t - 70 = 60$$
$$(10t - 70) + 70 = 60 + 70$$
$$10t = 130$$
$$\frac{10t}{10} = \frac{130}{10}$$
$$t = 13$$

Exercise 4.10

Solve these equations in traditional form. Check your answers by replacing the variable in the original equation with the root or solution you have found.

1. $3n + 4 = 19$

2. $14 = \frac{a}{2} - 6$

3. $-26 = 1 - 3x$

4. $15 + 3t = 0$

5. $3t + 5 = 29$

6. $\frac{r}{4} + 8 = 7$

7. $4(k + 3) = 0$

8. $\frac{3n}{4} = 9$

9. $\frac{2c}{5} + 6 = 10$

10. $\frac{n + 7}{5} = 16$

11. $\frac{5n}{3} - 13 = 2$

12. $12 = 11t - 10$

13. $4 = \frac{3n - 4}{8}$

14. $7t - 18 = 73$

15. $3(4n - 1) = 21$

16. $\frac{n + 14}{3} = 20$

MORE ABOUT SOLVING EQUATIONS

If a variable occurs more than once in the same equation it must be replaced by the same numeral in both instances. For example, if either of the letters k in $5k + 2 = 7 + k$ is replaced by the numeral 3, then the other must also be replaced by 3.

Before solving equations where the same variable occurs more than once let us investigate some expressions of this type.

What is a simpler name for $2y + 5y$? Our first guess would probably be $7y$. We could test our guess for a few replacements of y to see if $2y + 5y = 7y$. For example:

If we replace y by 3, then

$$2y + 5y = 2(3) + 5(3) = 6 + 15 = 21$$
$$7y = 7(3) = 21.$$

If we replace y by (-5), then

$$2y + 5y = 2(-5) + 5(-5) = (-10) + (-25)$$
$$= (-35)$$
$$7y = 7(-5) = (-35).$$

At least for these two replacements of y the sentence $2y + 5y = 7y$ is true. Since it is impossible to test this equation for all values of y, let us use the properties of numbers and their operations to verify that $2y + 5y = 7y$.

$$
\begin{aligned}
2y + 5y &= (2 \times y) + (5 \times y) \\
&= (2 + 5) \times y \quad \text{Distributive property} \\
&= 7 \times y \quad\quad\ \text{Addition} \\
&= 7y
\end{aligned}
$$

We can also use the distributive property to show that $9x - 5x = 4x$.

$$
\begin{aligned}
9x - 5x &= (9 - 5)x \\
&= 4x
\end{aligned}
$$

Since a placeholder or variable names a number, we can treat it just as we do a numeral when solving an equation, as shown in the following:

Example 1
Solve $9t = 40 + t$ (Traditional form)
$$
\begin{aligned}
9t &= 40 + t \\
9t + (-t) &= (40 + t) + (-t) \\
8t &= 40 \\
\frac{8t}{8} &= \frac{40}{8} \\
t &= 5
\end{aligned}
$$

Example 2
Solve $5(2n - 3) = 21 + n$ (Traditional form)
$$
\begin{aligned}
5(2n - 3) &= 21 + n \\
10n - 15 &= 21 + n \\
(10n - 15) + (-n) &= (21 + n) + (-n) \\
(-n) + [10n + (-15)] &= 21 + 0 \\
[(-n) + 10n] + (-15) &= 21 \\
9n + (-15) &= 21 \\
[9n + (-15)] + 15 &= 21 + 15 \\
9n &= 36 \\
\frac{9n}{9} &= \frac{36}{9} \\
n &= 4
\end{aligned}
$$

Exercise 4.11

Solve these equations in traditional form:

1. $7k + 8 = 11k$ 2. $8n = 14 + n$ 3. $4t - 5 = t + 1$
4. $7r + 3r = 130$ 5. $17x - 11x = 42$ 6. $c + 3(5 + c) = 23$
7. $a + 14 = 5(a - 2)$ 8. $2t + 18 = t + 6$ 9. $3(2t - 18) = 11 + t$
10. $15 - t = 2(6 + t)$

TRANSLATING ENGLISH PHRASES

One of the most important skills in problem solving is the ability to translate a problem stated in the English language into the language of mathematics. That is, we want to write an open sentence that says essentially the same thing as a 'story problem'.

We know from our study of the English language that sentences may contain phrases. Before translating sentences, let us investigate what we shall call *open phrases*, such as $k + 5$.

Suppose we want to express John's age 6 years ago and we do not know what John's age is now.

We might think like this:

Number of years in John's age now —— n
Number of years in John's age 6 years ago —— $n - 6$

Suppose Bob's age is 5 years more than 3 times his sister's age. How can we express Bob's age?

Number of years in his sister's age —— s
3 times the number of years in his sister's age —— $3s$
5 years more than 3 times the number of years in his sister's age —— $3s + 5$

In making the translation from English to mathematics, we first choose some letter to use as the variable. Then decide which operation or operations say essentially the same thing as the English words.

Exercise 4.12

Translate the following English phrases into open phrases. Use the letter n for the variable in each open phrase.

1. Seven more than some number.
2. Three less than 2 times some number.
3. The sum of a number and twice the number.
4. Mary's age 8 years from now.
5. The number of elements in n pairs and $(7 - n)$ triples.
6. A man's age is 9 years greater than 2 times his son's age.
7. The sum of 3 times some number and 4 times the number.
8. Five more than twice the number of cars Jim has.
9. The number of feet in the perimeter of a square.
10. Bob's score on a test if he answered 3 problems incorrectly.

TRANSLATING ENGLISH SENTENCES

We usually describe a problem situation in the English language. Some of the English sentences can be translated into mathematical sentences and others cannot be so translated. For example, the sentence 'The rose is red' does not lend itself to a mathematical translation.

Consider the sentence,

John is 34 years old.

We can easily translate this into the language of mathematics; it is just as meaningful if stated as follows:

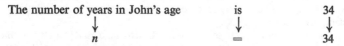

Now consider the sentence,

Six years ago John was 28 years old.

This is just as meaningful if stated as follows:

The number of years in John's age

        6 years ago   was   28

           x =   28

Notice that both of these sentences were rewritten in order to emphasize that the variable, in this case n and x represent a number.

Exercise 4.13

Translate each of these English sentences into open sentences. Use a different letter for the variable in each open sentence:

1. If he spends £3 he will have £4 left.
2. The product of some number and 12 is 32.
3. 3 times some number is 7 greater than 8.
4. Alice received 51 votes, which is 7 more votes than George received.
5. When a certain number is divided by 5 the quotient is (-9).

SOLVING PROBLEMS

There is no one set of rules for solving problems, nor is there only one way to apply mathematics to the physical world. However, some suggestions can be made for solving problems.

    (*a*) Study the problem carefully and think about the situation in terms of which operation or operations to use and what open sentence you might write for the problem.
    (*b*) Translate the problem into an open sentence.
    (*c*) Form the equation.
    (*d*) Use the root or roots of the equation to answer the problem.

Study how each of the following problems are translated into an open sentence and how the root of the open sentence is used to answer the problem.

Example 1

The George Washington Bridge has two end spans of the same length and a centre span that is 1125 metres long. The overall length of the bridge is 1500 metres. How long is each end span?

Let f = the number of metres in the length of each end span.

$$2f + 1125 = 1500$$
$$(2f + 1125) + (-1125) = 1500 + (-1125)$$
$$2f = 375$$
$$\frac{2f}{2} = \frac{375}{2}$$
$$f = 187 \cdot 5$$

Each end span is 187·5 metres long.

Example 2

Jim and Bill were the only candidates for secretary. Jim received 52 more votes than Bill. If 264 votes were cast, how many votes did each receive?

Let e = the number of votes for Bill
$e + 52$ = the number of votes for Jim

$$e + (e + 52) = 264$$
$$(e + e) + 52 = 264$$
$$2e + 52 = 264$$
$$(2e + 52) + (-52) = 264 + (-52)$$
$$2e = 212$$
$$\frac{2e}{2} = \frac{212}{2}$$
$$e = 106$$

Bill received 106 votes.
Jim received $e + 52$ or 158 votes.

Example 3

Jean's age is 7 years more than twice her sister's age. If Jean is 19 years old, how old is her sister?

Let s = the number of years in her sister's age

$$2s + 7 = 19$$
$$(2s + 7) + (-7) = 19 + (-7)$$
$$2s = 12$$
$$s = 6$$

Her sister is 6 years old.

Exercise 4.14

Solve these problems:

1. The sum of two times a certain number and 6 is 22. What is the number?
2. One number is 3 more than a second number. Their sum is 67. What are the two numbers?
3. Ted has a mass 9 kg less than Roger. Their combined mass is 103 kg. What is the mass of each boy?
4. A rope 26 metres long is cut into 2 pieces so that one piece is 8 metres longer than the other.
   How long is the shorter piece of rope?
5. A rectangle is 12 metres long. Its perimeter is 44 metres. How wide is the rectangle?
6. The sum of a number and 4 times the same number is 75. What is the number?
7. The difference between a number and 5 times the same number is 32. What is the number?
(*Hint:* There are two possible answers—one positive and one negative.)

CHAPTER FIVE

RATIONAL NUMBERS

A NEED FOR NEW NUMBERS

As a result of inventing the integers we obtained a set of numbers which was closed under the operation of subtraction, but we were still unable to provide an answer within this set for problems such as $3n = 8$.

If we attempt to solve this equation our work may appear as follows:

$$3n = 8$$
$$\frac{3n}{3} = \frac{8}{3}$$
$$n = \frac{8}{3}$$

But this is not an integer, so we therefore need to invent some more numbers.

The symbol $\frac{8}{3}$ means two things:

(i) The result of 8 divided by 3.
(ii) It is a numeral for a number.

(i) implies the mathematical operation of division, (ii) is merely a numeral or name for a *rational number*.

Definition 5.1

A **rational number** is any number of the form $\frac{a}{b}$ where a and b are integers and $b \neq 0$. The integer a is called the *numerator* and the integer b is called the *denominator*.

You remember that the integers were given as . . ., (-4), (-3), (-2), (-1), 0, $(+1)$, $(+2)$, $(+3)$, . . ., and a distinction was made between the whole numbers $0, 1, 2, 3, \ldots$ and the positive integers $0, (+1), (+2), (+3), \ldots$ By maintaining this distinction we were able to avoid the confusion between, say, the operations of subtraction and the idea of the negative of a number, and thereby it was hoped that a proper understanding would be gained of the basic fundamentals of Algebra.

Having obtained this understanding, we were able to interpret the intended meaning of the traditional type of problem, and one or two exercises were given in this procedure in the last chapter. This is very necessary, because it will probably take fifty years or so before all books in Arithmetic, Book-keeping, Engineering, Science, etc., are brought up to date.

In the meantime we have to accept the following meanings:

The whole numbers are taken as positive integers.
The integers are written as $\ldots$, -4, -3, -2, -1, 0, 1, 2, 3, 4, $\ldots$

-3 is the same meaning as $-(+3)$ or is a version of (-3) without the brackets;

$+3$ is the same meaning as $+(+3)$ or is a version of $(+3)$ without brackets;

3 is the same meaning as $(+3)$.

After a time, of course, we all know what the required meaning is, but the confusion makes it very difficult for the beginner. Thus, $4 \times -3 = -12$ means $(+4) \times (-3) = (-12)$; $-4 + 3 = (+3) + (-4) = (-1)$ and so on.

Now the following fractions name rational numbers.

$$\frac{+1}{+2}, \frac{-4}{+3}, \frac{+6}{+2}, \frac{0}{+5}, \frac{+27}{+5}, \frac{+100}{+25}$$

The third numeral also appears to name 3 and the last names 4. Does this mean that all integers may also be considered as rational numbers? The answer is 'yes', because the integers may be written as

$$\ldots, \frac{-2}{+1}, \frac{-1}{+1}, \frac{0}{+1}, \frac{+1}{+1}, \frac{+2}{+1}, \frac{+3}{+1}, \ldots$$

Summing up therefore, this means that in practice we are choosing to ignore the mathematical differences between the whole numbers, integers, and rationals, and also the differences between the various operations of addition and multiplication which are defined on each separate set of numbers. We ignore these differences in practical work simply because it is too inconvenient to do otherwise.

Looking forward to Chapter Six, we see that the set of natural numbers with the usual operations of addition and multiplication, which are defined to yield results consistent with our physical experience, is isomorphic to the set of positive integers with its operations of 'addition' and 'multiplication'. It is also isomorphic to the set of rational numbers of the form $\frac{(+n)}{(+1)}$, with its operations of 'addition and multiplication', so that results such as

$$4 + 3 = 7, \quad 4 \times 3 = 12$$

may be interpreted as being true for each system.

Exercise 5.1

Solve the following equations by using the set of rational numbers as the replacement set.

1. $4x = 17$
2. $-3n = 11$
3. $8t = 9$
4. $7r = 28$
5. $\dfrac{3a}{5} = 4$
6. $2n + 7 = 12$
7. $5c - 4 = 26$
8. $\dfrac{5r}{7} = -15$
9. $-3a + 2 = 29$
10. $15 + 4n = -17$

OTHER NAMES FOR RATIONAL NUMBERS

From our knowledge of division of whole numbers we know that $\frac{4}{2} = 2$ and $\frac{6}{2} = 3$. That is, $\frac{4}{2}$ and 2 are two names for the same number and $\frac{6}{2}$ and 3 are two names for the same number. Then $\frac{5}{2}$ must name a number between 2 and 3.

Since $\frac{5}{2}$ means $5 \div 2$, we can find another name for $\frac{5}{2}$ as follows:

$$
\begin{aligned}
\frac{5}{2} = 5 \div 2 &= (4 + 1) \div 2 \\
&= (4 \div 2) + (1 \div 2) \quad \text{Distributive property} \\
&= 2 + \tfrac{1}{2}
\end{aligned}
$$

A short way to write $2 + \frac{1}{2}$ is $2\frac{1}{2}$. Hence, $\frac{5}{2} = 2\frac{1}{2}$, which means that $\frac{5}{2}$ and $2\frac{1}{2}$ are two names for the same number. Each of these names is as acceptable as the other. However, each has its advantages and disadvantages.

For convenience of reference, let us refer to a numeral such as $2\frac{1}{2}$ as a 'mixed numeral'.

A person probably gets a better idea of the quantity if we say $5\frac{3}{8}$ litres than if we say $\frac{43}{8}$ litres. However, for most computational purposes the name $\frac{43}{8}$ is more convenient.

Exercise 5.2

Name each of the following by mixed numerals:

1. $\frac{8}{5}$
2. $\frac{41}{10}$
3. $\frac{51}{7}$
4. $\frac{62}{7}$
5. $\frac{38}{9}$
6. $\frac{85}{9}$
7. $\frac{136}{11}$
8. $\frac{236}{25}$
9. $\frac{421}{50}$

THE RATIONAL NUMBER LINE

A number line is a very useful tool for picturing some of our ideas about numbers and operations. Recall that we established a one-to-one correspondence between the set of integers and some selected points on a number line.

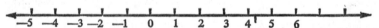

Obviously there are many points on the line which are not matched with these numbers. Some of these points (but not all of them) can be matched one-to-one with the rational numbers. Let us consider that segment of the number line from 0 to 1. The point midway between 0 and 1 is matched with the rational number $\frac{1}{2}$.

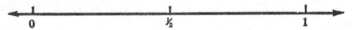

Then we could lay off the segment of length $\frac{1}{2}$ to the left and to the right to locate those points to be matched with the rational numbers $\ldots, \dfrac{-3}{2}, \dfrac{-2}{2}, \dfrac{-1}{2}, \dfrac{0}{2}, \dfrac{1}{2}, \dfrac{2}{2}, \dfrac{3}{2}, \ldots$

Those points that separate the segments from 0 to 1 into three parts of equal length can be matched with the rational numbers $\frac{1}{3}$ and $\frac{2}{3}$.

Then we could lay off the segment of length $\frac{1}{3}$ to the left and to the right to locate those points to be matched with rational numbers $\ldots$, $\dfrac{-4}{3}, \dfrac{-3}{3}, \dfrac{-2}{3}, \dfrac{-1}{3}, \dfrac{0}{3}, \dfrac{1}{3}, \dfrac{2}{3}, \dfrac{3}{3}, \dfrac{4}{3}$.

Similarly, we could locate points to be matched with any rational number. We might be tempted to say this time that such a one-to-one correspondence would involve every point on the number line, but this is far from true. There still remain many unmatched points on the number line. Later in this chapter we shall discuss those numbers that we match with these points.

MULTIPLICATION OF RATIONAL NUMBERS

Just as we did when we extended the set of whole numbers to the set of integers, we want the operations on the rational numbers to obey the same properties that the operations on the integers obey. In other words, let us assume both commutative properties, both associative properties, the distributive property, 0 as the identity number of addition, and 1 as the identity number of multiplication for the set of rational numbers.

When you first learned about fractions you probably interpreted the fraction $\frac{1}{4}$ to indicate that some object had been separated into 4 parts of equal size, and you were considering 1 of these parts.

These intuitive ideas also give us a clue to the product of $\frac{1}{4}$ and $\frac{1}{2}$.

Think of separating something into 4 parts of equal size. Then separate one of these parts into 2 parts of equal size. What part of the object is one of these latter pieces? Obviously $\frac{1}{2}$ of $\frac{1}{4}$ is $\frac{1}{8}$, or $\frac{1}{4} \times \frac{1}{2} = \frac{1}{8}$.

By using other fractions, we can find that $\frac{1}{2} \times \frac{1}{5} = \frac{1}{10}$, that $\frac{1}{3} \times \frac{1}{5} = \frac{1}{15}$, and so on. But do notice that we have decided that this shall be the meaning of the multiplication, i.e. we are agreeing to interpret $\frac{1}{2} \times \frac{1}{5}$ as meaning one-fifth of one-half.

Therefore we define the following multiplication:

$$\frac{1}{a} \times \frac{1}{n} = \frac{1 \times 1}{a \times n} = \frac{1}{an}$$

for all rational numbers $\dfrac{1}{a}$ and $\dfrac{1}{n}$, since it is consistent with practical experience.

Exercise 5.3

Find each product:

1. $\frac{1}{3} \times \frac{1}{4}$ 2. $\frac{1}{6} \times \frac{1}{2}$ 3. $\frac{1}{7} \times \frac{1}{3}$ 4. $\frac{1}{8} \times \frac{1}{5}$
5. $\frac{1}{2} \times \frac{1}{9}$ 6. $\frac{1}{10} \times \frac{1}{3}$ 7. $\frac{1}{12} \times \frac{1}{5}$ 8. $\frac{1}{13} \times \frac{1}{3}$
9. $\frac{1}{4} \times \frac{1}{17}$

We can get a more general notion of multiplication of rational numbers by investigating the number line.

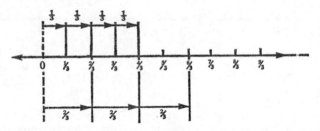

The arrows above the number line show $\frac{1}{3} + \frac{1}{3} + \frac{1}{3} + \frac{1}{3} = \frac{4}{3}$. From our previous idea of multiplication of integers, we may also interpret $\frac{1}{3} + \frac{1}{3} + \frac{1}{3} + \frac{1}{3}$ as $\frac{1}{3} \times 4$.

Then it appears that

$$\frac{1}{3} \times 4 = \frac{4}{3} = \frac{1 \times 4}{3}$$

The arrows below the number line show $\frac{2}{3} + \frac{2}{3} + \frac{2}{3} = \frac{6}{3}$.

In terms of multiplication it appears that

$$\frac{2}{3} \times 3 = \frac{2 \times 3}{3} = \frac{6}{3}$$

The first example reveals an important way of renaming any rational number. Let us begin with $\frac{4}{3}$ and work through the example.

$$\frac{4}{3} = \frac{4 \times 1}{3} = 4 \times \frac{1}{3}$$

Let us assume that this is true for all rational numbers.

For every rational number $\frac{a}{b}$,

$$\frac{a}{b} = \frac{a \times 1}{b} = a \times \frac{1}{b}$$

Now let us use these ideas to find the product of $\frac{2}{3}$ and $\frac{4}{5}$.

$$
\begin{aligned}
\frac{2}{3} \times \frac{4}{5} &= (2 \times \tfrac{1}{3})(4 \times \tfrac{1}{5}) \\
&= 2 \times (\tfrac{1}{3} \times 4) \times \tfrac{1}{5} \qquad \text{Associative property} \\
&= 2 \times (4 \times \tfrac{1}{3}) \times \tfrac{1}{5} \qquad \text{Commutative property} \\
&= (2 \times 4)(\tfrac{1}{3} \times \tfrac{1}{5}) \qquad \text{Associative property} \\
&= 8 \times \tfrac{1}{15} \\
&= \frac{8 \times 1}{15} = \frac{8}{15}
\end{aligned}
$$

Then we notice that $8 = 2 \times 4$ and $15 = 3 \times 5$.

We could have multiplied as follows:

$$\frac{2}{3} \times \frac{4}{5} = \frac{2 \times 4}{3 \times 5} = \frac{8}{15}$$

From other such examples, we are led to the following definition of multiplication of rational numbers.

Definition 5.2

For all rational numbers $\dfrac{a}{n}$ and $\dfrac{r}{s}$,

$$\frac{a}{n} \times \frac{r}{s} = \frac{a \times r}{n \times s} = \frac{ar}{ns}$$

The multiplication of rational numbers is usually written in a shortened form.

For example, to evaluate $\qquad \dfrac{72}{11} \times \dfrac{121}{48}$,

instead of writing this as

$$\frac{3 \times 24 \times 11 \times 11}{11 \times 2 \times 24} = \frac{11}{11} \times \frac{24}{24} \times \frac{11 \times 3}{2} = 1 \times 1 \times \frac{11 \times 3}{2} = \frac{33}{2}$$

we effect the division straight away as $\dfrac{\overset{3}{72}}{\underset{1}{11}} \times \dfrac{\overset{11}{121}}{\underset{2}{48}}$.

Exercise 5.4

Find each product:

1. $\frac{3}{7} \times \frac{4}{5}$ 2. $\frac{2}{3} \times \frac{5}{7}$ 3. $\frac{4}{9} \times \frac{10}{7}$ 4. $\frac{0}{7} \times \frac{5}{6}$

5. $\frac{5}{13} \times \frac{4}{3}$ 6. $\frac{4}{5} \times \frac{2}{9}$ 7. $\frac{8}{13} \times \frac{2}{3}$ 8. $\frac{1}{2} \times \frac{9}{8}$

9. $\frac{22}{13} \times \frac{19}{33} \times \frac{26}{5}$ 10. $\frac{42}{19} \times \frac{38}{12} \times \frac{3}{14}$ 11. $\frac{17}{64} \times \frac{16}{38} \times \frac{57}{51}$ 12. $\frac{11}{63} \times \frac{7}{55} \times \frac{81}{27}$

RENAMING RATIONAL NUMBERS

We have already assumed the number 1 to be the identity number of multiplication. For the set of integers we discovered that $4 \div 4 = 1$, $3 \div 3 = 1$, and so on, for any non-zero number. Since $4 \div 4$ can be written as $\frac{4}{4}$, we can conclude this:

For any non-zero integer n, the rational number $\dfrac{n}{n} = 1$.

Consider the following multiplication, when 1 is named as a rational number of the form $\dfrac{n}{n}$.

$$\frac{2}{3} \times 1 = \frac{2}{3} \times \frac{2}{2} = \frac{2 \times 2}{3 \times 2} = \frac{4}{6}$$

$$\frac{2}{3} \times 1 = \frac{2}{3} \times \frac{3}{3} = \frac{2 \times 3}{3 \times 3} = \frac{6}{9}$$

Now consider the following:

$$\frac{8}{12} = \frac{4 \times 2}{4 \times 3} = \frac{4}{4} \times \frac{2}{3} = 1 \times \frac{2}{3} = \frac{2}{3}$$

$$\frac{18}{30} = \frac{6 \times 3}{6 \times 5} = \frac{6}{6} \times \frac{3}{5} = 1 \times \frac{3}{5} = \frac{3}{5}$$

From these examples we can make the assumption that for any rational number $\frac{a}{c}$ and for any non-zero integer r,

$$\frac{a}{c} = \frac{ra}{rc}$$

In other words, both the numerator and denominator of a rational number can be multiplied or divided by the same non-zero integer without changing the rational number. All we do in the process is write another name for the number. This is the idea behind what is commonly called 'reducing fractions' to lowest terms when there is no whole number, other than 1, that will divide both the numerator and the denominator.

Example 1

$$\frac{4}{5} \times \frac{3}{8} = \frac{4 \times 3}{5 \times 8} = \frac{12}{40} = \frac{4 \times 3}{4 \times 10} = \frac{3}{10}$$

Example 2

$$\frac{4}{9} \times \frac{3}{14} = \frac{4 \times 3}{9 \times 14}$$
$$= \frac{2 \times 2 \times 3}{3 \times 3 \times 2 \times 7}$$
$$= \frac{2 \times 3 \times 2}{2 \times 3 \times 3 \times 7}$$
$$= \frac{2 \times 3}{2 \times 3} \times \frac{2}{3 \times 7}$$
$$= 1 \times \frac{2}{21}$$
$$= \frac{2}{21}$$

Exercise 5.5

Find each product. State the answer in the lowest terms.

1. $\frac{5}{16} \times \frac{4}{5}$

2. $\frac{3}{4} \times \frac{8}{15}$

3. $\frac{15}{16} \times \frac{20}{21}$

4. $\frac{15}{2} \times \frac{8}{5}$

5. $\frac{2}{3} \times \frac{3}{4} \times \frac{5}{6}$

6. $\frac{5}{8} \times \frac{4}{7} \times \frac{21}{25}$

7. $\frac{3}{2} \times \frac{4}{9} \times \frac{5}{6}$

8. $\frac{7}{10} \times \frac{5}{8} \times \frac{6}{21}$

MIXED NUMERALS IN MULTIPLICATION

One or more of the factors might be named by a mixed numeral. Our previous definition of multiplication covers such cases.

$$\begin{aligned} 5\tfrac{1}{2} \times \frac{4}{7} &= \frac{11}{2} \times \frac{4}{7} \\ &= \frac{11}{2} \times \frac{4}{7} \\ &= \frac{2 \times 2 \times 11}{2 \times 7} \\ &= \frac{2}{2} \times \frac{2 \times 11}{7} \\ &= 1 \times \frac{22}{7} \\ &= \frac{22}{7} \end{aligned}$$

All we need do is change the mixed numeral to a fraction and proceed as usual.

Exercise 5.6

Find each product. State the answer in lowest terms.

1. $4\tfrac{1}{2} \times \tfrac{4}{5}$
2. $5\tfrac{2}{3} \times \tfrac{3}{2}$
3. $3\tfrac{1}{5} \times \tfrac{1}{8}$
4. $4\tfrac{1}{3} \times 12$
5. $\tfrac{6}{7} \times 3\tfrac{1}{2} \times 4$
6. $4\tfrac{1}{2} \times \tfrac{5}{3} \times \tfrac{4}{5}$
7. $5\tfrac{1}{2} \times 6\tfrac{2}{5} \times 3\tfrac{2}{11}$
8. $\tfrac{3}{2} \times 1\tfrac{5}{7} \times \tfrac{7}{8}$

RECIPROCALS OR MULTIPLICATIVE INVERSES

We have seen that the identity number of multiplication of rational numbers is 1. Now we wish to investigate the possibility of two rational numbers having a product of 1.

Study these examples:

$$\frac{4}{3} \times \frac{3}{4} = \frac{4 \times 3}{3 \times 4} = \frac{12}{12} = 1$$

$$\frac{5}{6} \times \frac{6}{5} = \frac{5 \times 6}{6 \times 5} = \frac{30}{30} = 1$$

In general, for any rational number $\frac{r}{s}$ we can find a rational number $\frac{s}{r}$, so that

$$\frac{r}{s} \times \frac{s}{r} = 1$$

Definition 5.3

If the product of two rational numbers is 1, then the two rational numbers are called the **reciprocals** or the **multiplicative inverses** of each other.

Since $\frac{3}{4} \times \frac{4}{3} = 1$, $\frac{3}{4}$ and $\frac{4}{3}$ are each the reciprocal or the multiplicative inverse of the other. Since 5 can be named as $\frac{5}{1}$, then $\frac{5}{1} \times \frac{1}{5} = 1$ and 5 and $\frac{1}{5}$ are reciprocals of each other.

Does every integer have a reciprocal? Since $(-3) \times (-\frac{1}{3}) = 1$, we see that negative integers have reciprocals. What number could n represent so that $0 \times n = 1$? There is no such number n, and we conclude that zero has no reciprocal.

Exercise 5.7

State the reciprocal of each of the following numbers:

1. $\frac{5}{8}$ 2. 7 3. $-\frac{4}{9}$ 4. $\frac{7}{12}$ 5. $\frac{9}{31}$
6. $-\frac{15}{37}$ 7. $3\frac{1}{2}$ 8. -9 9. $-6\frac{5}{7}$

DIVISION OF RATIONAL NUMBERS

Since multiplication and division are inverse operations, we can state every division as a multiplication. Suppose we are to solve the equation $\frac{5}{7} \div \frac{2}{3} = n$.

$\frac{5}{7} \div \frac{2}{3} = n$
$\frac{5}{7} = n \times \frac{2}{3}$ Inverse operations
$\frac{5}{7} \times 1 = n \times \frac{2}{3}$ Identity number $\times$
$\frac{5}{7} \times (\frac{3}{2} \times \frac{2}{3}) = n \times \frac{2}{3}$ State 1 as product of $\frac{2}{3}$ and its reciprocal.
$(\frac{5}{7} \times \frac{3}{2}) \times \frac{2}{3} = n \times \frac{2}{3}$ Associative property $\times$

Since the second factor on both sides of the equation is $\frac{2}{3}$, we conclude that

$$\frac{5}{7} \times \frac{3}{2} = n.$$

Replacing n in the original equation by this expression we have

$$\frac{5}{7} \div \frac{2}{3} = \frac{5}{7} \times \frac{3}{2}$$

Dividing by a rational number is essentially the same as multiplying by its reciprocal (or multiplicative inverse).

Definition 5.4

For all rational numbers $\dfrac{a}{n}$ and $\dfrac{r}{s}$,

$$\frac{a}{n} \div \frac{r}{s} = \frac{a}{n} \times \frac{s}{r}$$

Hence, to perform division of rational numbers, merely restate the division as a multiplication and proceed according to the rules for multiplying rational numbers.

Exercise 5.8

Find each quotient. State your answer in lowest terms.

1. $\frac{8}{9} \div \frac{4}{9}$ 2. $\frac{4}{5} \div \frac{8}{15}$ 3. $\frac{5}{6} \div 4$ 4. $\frac{8}{11} \div \frac{1}{2}$ 5. $\frac{2}{3} \div \frac{3}{8}$
6. $\frac{10}{11} \div \frac{2}{5}$ 7. $15 \div 3\frac{1}{3}$ 8. $1\frac{1}{6} \div \frac{1}{9}$ 9. $\frac{9}{10} \div \frac{27}{50}$ 10. $2\frac{2}{3} \div \frac{3}{8}$
11. $4\frac{1}{5} \div \frac{4}{3}$ 12. $1\frac{1}{5} \div 24$ 13. $8\frac{1}{3} \div 1\frac{2}{3}$ 14. $5\frac{5}{9} \div 8\frac{1}{3}$ 15. $3\frac{1}{3} \div 2\frac{1}{2}$

ADDITION OF RATIONAL NUMBERS

By using the properties of operations and what we have learned about multiplication, the addition of rational numbers is easy when the denominators are the same.

$$\frac{5}{9} + \frac{2}{9} = (5 \times \frac{1}{9}) + (2 \times \frac{1}{9})$$
$$= (5 + 2) \times \frac{1}{9}$$
$$= 7 \times \frac{1}{9}$$
$$= \frac{7}{9}$$

A comparison of $\frac{5}{9} + \frac{2}{9}$ and $\frac{7}{9}$ reveals that we might have added the numerators and used the same denominator.

$$\frac{5}{9} + \frac{2}{9} = \frac{5+2}{9} = \frac{7}{9}$$

In general, for rational numbers $\frac{a}{n}$ and $\frac{c}{n}$,

$$\frac{a}{n} + \frac{c}{n} = \frac{a+c}{n}$$

Furthermore, knowing that addition is associative, we can state that

for all rational numbers $\frac{a}{n}, \frac{c}{n},$ and $\frac{e}{n},$

$$\frac{a}{n} + \frac{c}{n} + \frac{e}{n} = \frac{a+c+e}{n}$$

Exercise 5.9

Find each sum. State the answer in lowest terms.

1. $\frac{2}{3} + \frac{1}{3}$ 2. $\frac{5}{11} + \frac{6}{11}$ 3. $\frac{3}{8} + \frac{5}{8}$ 4. $\frac{2}{13} + \frac{5}{13}$
5. $\frac{4}{15} + \frac{2}{15} + \frac{4}{15}$ 6. $\frac{3}{20} + \frac{7}{20} + \frac{5}{20}$ 7. $\frac{1}{9} + \frac{2}{9} + \frac{3}{9}$ 8. $\frac{4}{25} + \frac{3}{25} + \frac{8}{25}$
9. $2\frac{1}{3} + \frac{2}{3}$ 10. $7\frac{1}{3} + \frac{1}{3}$ 11. $1\frac{3}{7} + \frac{2}{7}$ 12. $3\frac{1}{5} + 4\frac{2}{5}$

Suppose we are to find the sum of $\frac{4}{5}$ and $\frac{2}{3}$. In this case the denominators are not the same. But we know how to rename these numbers so that the denominators become the same. Then we could proceed as in the previous exercises.

$$\frac{4}{5} + \frac{2}{3} = \left(\frac{4}{5} \times 1\right) + \left(\frac{2}{3} \times 1\right)$$
$$= \left(\frac{4}{5} \times \frac{3}{3}\right) + \left(\frac{2}{3} \times \frac{5}{5}\right)$$
$$= \frac{4 \times 3}{15} + \frac{5 \times 2}{15}$$
$$= \frac{12}{15} + \frac{10}{15}$$
$$= \frac{12 + 10}{15}$$
$$= \frac{22}{15}$$

But how did we know to choose $\frac{3}{3}$ and $\frac{5}{5}$ as names for 1? Notice that $\frac{4}{5}$ was multiplied by $\frac{3}{3}$ and that 3 is the denominator of the other factor.

Is this also true for $\frac{2}{3} \times \frac{4}{5}$? This procedure may not yield the least possible denominator common to both rational numbers.

Definition 5.5

For all rational numbers $\dfrac{a}{n}$ and $\dfrac{r}{s}$,

$$\frac{a}{n} + \frac{r}{s} = \frac{as + nr}{ns}$$

For example, let us find the sum of $\frac{7}{8}$ and $\frac{2}{5}$.

$$
\begin{aligned}
\frac{7}{8} + \frac{2}{5} &= \frac{(7 \times 5) + (8 \times 2)}{8 \times 5} \\
&= \frac{35 + 16}{40} \\
&= \frac{51}{40}
\end{aligned}
$$

Exercise 5.10

Find each sum. State the answer in lowest terms.

1. $\frac{2}{3} + \frac{1}{7}$ 2. $\frac{2}{9} + \frac{1}{3}$ 3. $\frac{2}{5} + \frac{1}{2}$ 4. $\frac{2}{3} + \frac{3}{4}$
5. $\frac{2}{9} + \frac{3}{7}$ 6. $\frac{5}{24} + \frac{1}{8}$ 7. $\frac{1}{4} + \frac{2}{9}$ 8. $\frac{3}{11} + \frac{2}{7}$
9. $\frac{1}{2} + \frac{1}{8} + \frac{1}{9}$ 10. $\frac{2}{5} + \frac{1}{4} + \frac{1}{3}$ 11. $\frac{1}{12} + \frac{1}{5} + \frac{3}{8}$ 12. $\frac{1}{2} + \frac{1}{3} + \frac{1}{4}$

MIXED NUMERALS IN ADDITION

It is usually easier to add rational numbers named by mixed numerals in vertical form rather than horizontal form. However, we shall use the horizontal form and the properties we know to verify that the vertical arrangement is valid. Suppose we are to find the sum of $5\frac{2}{3}$ and $3\frac{1}{2}$.

$$
\begin{aligned}
5\tfrac{2}{3} + 3\tfrac{1}{2} &= (5 + \tfrac{2}{3}) + (3 + \tfrac{1}{2}) \\
&= 5 + (\tfrac{2}{3} + 3) + \tfrac{1}{2} \\
&= 5 + (3 + \tfrac{2}{3}) + \tfrac{1}{2} \\
&= (5 + 3) + (\tfrac{2}{3} + \tfrac{1}{2})
\end{aligned}
$$

At this point in the solution we notice that we are to find the sum of two whole numbers, the sum of two rational numbers, and then to add these sums. Hence we can state it like this:

$$
\begin{array}{ll}
5\tfrac{2}{3} & \text{and think of:} \quad 5 + \tfrac{2}{3} \\
+3\tfrac{1}{2} & \hphantom{\text{and think of:} \quad} 3 + \tfrac{1}{2} \\
\overline{\hphantom{5\tfrac{2}{3}}} & \hphantom{\text{and think of:}} \overline{\hphantom{(5 + 3) + \tfrac{2}{3} + \tfrac{1}{2}}} \\
& \hphantom{\text{and think of:} \quad} (5 + 3) + \tfrac{2}{3} + \tfrac{1}{2}
\end{array}
$$

We already know how to find the required sums in this example.

$$5\tfrac{2}{3} \quad \text{and think of:} \quad 5\tfrac{4}{6}$$
$$+3\tfrac{1}{2} \qquad\qquad\qquad +3\tfrac{3}{6}$$
$$\overline{} \qquad\qquad\qquad \overline{8\tfrac{7}{6}}$$

Since $\tfrac{7}{6} = \tfrac{6}{6} + \tfrac{1}{6} = 1 + \tfrac{1}{6}$, we can restate the sum like this:

$$8\tfrac{7}{6} = 8 + \tfrac{7}{6} = 8 + (1 + \tfrac{1}{6}) = (8 + 1) + \tfrac{1}{6} = 9 + \tfrac{1}{6} = 9\tfrac{1}{6}$$

We say that a mixed numeral is in lowest terms if the fractional part names a number less than one, and if the fractional part is in lowest terms.

Exercise 5.11

Find each sum. State the answer in lowest terms.

1. $6\tfrac{3}{8}$
   $+5\tfrac{1}{8}$

2. $18\tfrac{3}{4}$
   $+7\tfrac{1}{2}$

3. $12\tfrac{2}{3}$
   $5\tfrac{1}{4}$
   $+3\tfrac{1}{2}$

4. $28\tfrac{9}{10}$
   $+43\tfrac{4}{5}$

5. $27\tfrac{3}{5}$
   $+6\tfrac{5}{8}$

6. $9\tfrac{7}{8}$
   $13\tfrac{1}{4}$
   $+52\tfrac{1}{2}$

7. $215\tfrac{3}{4}$
   $+82\tfrac{1}{6}$

8. $752\tfrac{3}{7}$
   $+54\tfrac{1}{2}$

9. $115\tfrac{1}{6}$
   $607\tfrac{2}{3}$
   $+79\tfrac{1}{4}$

SUBTRACTION OF RATIONAL NUMBERS

Most of this chapter thus far has used primarily the positive integers, and hence positive rational numbers, since they receive the greater attention in arithmetic. However, the definitions for multiplication, division, and addition of rational numbers apply as well to the negative rational numbers.

In the set of integers we defined subtraction in this way: for all integers a and b, $a - b$ means $a + (-b)$. That is, to subtract a number, we added its opposite or additive inverse. This same idea applies to subtraction of rational numbers. For example:

$$\tfrac{5}{8} - \tfrac{3}{8} = \tfrac{5}{8} + (-\tfrac{3}{8})$$

Before completing this example, let us investigate how we might rename $-\tfrac{3}{8}$. By the property of -1, $-\tfrac{3}{8} = (-1) \times \tfrac{3}{8}$. But we already know that -1 can be named by $\dfrac{-1}{1}$ or $\dfrac{1}{-1}$. Hence, we can use either of these in place of -1, as shown below.

$$-\frac{3}{8} = (-1) \times \frac{3}{8} \qquad\qquad\qquad -\frac{3}{8} = (-1) \times \frac{3}{8}$$

$$= \frac{-1}{1} \times \frac{3}{8} \qquad\qquad\qquad\qquad = \frac{1}{-1} \times \frac{3}{8}$$

$$= \frac{(-1) \times 3}{1 \times 8} \qquad\qquad\qquad\qquad = \frac{1 \times 3}{(-1) \times 8}$$

$$= \frac{-3}{8} \qquad\qquad\qquad\qquad\qquad = \frac{3}{-8}$$

Hence, $-\dfrac{3}{8} = \dfrac{-3}{8} = \dfrac{3}{-8}.$

In fact, for any rational number $-\dfrac{a}{n}$,

$$-\frac{a}{n} = \frac{-a}{n} = \frac{a}{-n}$$

By using this idea we can always rename a rational number, so that the sign preceding the fraction is $+$, which we have agreed will be implied when no sign is written.

Now let us return to $\frac{5}{8} - \frac{3}{8}$

$$\begin{aligned}\frac{5}{8} - \frac{3}{8} &= \frac{5}{8} + \left(-\frac{3}{8}\right) \\ &= \frac{5}{8} + \frac{-3}{8} \\ &= \frac{5 + (-3)}{8} \\ &= \frac{2}{8} \text{ or } \frac{1}{4}\end{aligned}$$

We can define subtraction of rational numbers by applying the above idea to the definition of addition of rational numbers.

Definition 5.6

For all rational numbers $\dfrac{a}{n}$ and $\dfrac{r}{s}$

$$\frac{a}{n} - \frac{r}{s} = \frac{as - nr}{ns}$$

Study how this definition is used to find the following difference:

$$\begin{aligned}3\frac{4}{5} - \frac{2}{3} &= \frac{19}{5} - \frac{2}{3} \\ &= \frac{(19 \times 3) - (5 \times 2)}{5 \times 3} \\ &= \frac{57 - 10}{15} \\ &= \tfrac{47}{15} \text{ or } 3\tfrac{2}{15}\end{aligned}$$

Exercise 5.12

Find each difference. State the answer in lowest terms.

1. $\frac{7}{9} - \frac{3}{9}$ 2. $\frac{7}{15} - \frac{2}{15}$ 3. $5 - \frac{3}{5}$ 4. $3\frac{7}{10} - \frac{3}{5}$ 5. $\frac{8}{11} - \frac{2}{3}$

6. $\frac{3}{5} - \frac{2}{7}$ 7. $\frac{7}{9} - \frac{2}{3}$ 8. $\frac{12}{25} - \frac{2}{5}$ 9. $\frac{7}{12} - \frac{5}{16}$ 10. $\frac{7}{8} - \frac{4}{5}$

11. $7\frac{1}{6} - 4\frac{3}{4}$ 12. $8\frac{2}{5} - 6\frac{3}{8}$ 13. $45\frac{5}{9} - 37\frac{5}{6}$ 14. $86\frac{4}{7} - 55\frac{7}{8}$ 15. $47\frac{4}{9} - 36\frac{3}{5}$

DECIMALS

Whenever the denominator is a power of ten or whenever the rational number can be renamed so the denominator is a power of ten, the rational number can easily be named as a **decimal numeral** or simply a **decimal**.

Study how each of the following rational numbers is named as a decimal.

$$\frac{27}{100} = 0.27$$
$$\frac{3}{25} = \frac{4}{4} \times \frac{3}{25} = \frac{12}{100} = 0.12$$
$$3\frac{2}{5} = \frac{17}{5} = \frac{2}{2} \times \frac{17}{5} = \frac{34}{10} = 3.4$$
$$\frac{3}{250} = \frac{4}{4} \times \frac{3}{250} = \frac{12}{1000} = 0.012$$

The digits to the right of the decimal point (·) name the numerator, and the number of such digits indicates the power of ten which is the denominator. For example, 0·547 denotes a numerator of 547 and a denominator of 10^3 or 1 000. Study how these decimals are rewritten as fractions:

$$0.57 = \frac{57}{100}$$
$$0.32 = \frac{32}{100} = \frac{4 \times 8}{4 \times 25} = \frac{8}{25}$$
$$5.13 = 5 + 0.13 = 5 + \frac{13}{100} = 5\frac{13}{100}$$

Exercise 5.13

Write each of the following as a decimal:

1. $\frac{47}{100}$ 2. $\frac{4}{5}$ 3. $\frac{17}{25}$ 4. $\frac{3}{4}$ 5. $\frac{28}{250}$ 6. $\frac{13}{20}$

Write each of the following as a fraction or a mixed numeral:

7. 12·7   8. 0·0037   9. 0·125   10. 0·6   11. 0·034   12. 0·0475

DECIMAL NUMERATION

Base-ten numeration is also called decimal numeration. The scheme of place value is shown by diagram:

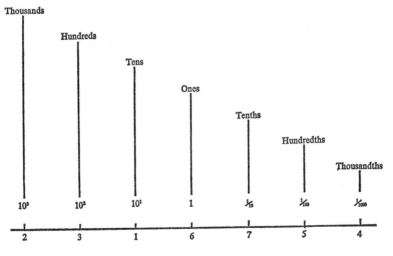

The numeral just shown is read *two thousand three hundred sixteen and seven hundred fifty-four thousandths*. Notice that the digits to the left of the decimal point are read just as for a whole number; the decimal point is read *and*; and the digits to the right of the decimal point are read just as for a whole number and followed by the place-value name of the rightmost digit.

DECIMALS IN ADDITION AND SUBTRACTION

Before discussing how to compute with decimals, it is important to realize that we are still performing operations on rational numbers. The properties, such as commutative and associative properties, are properties of the operations. Hence, we have nothing new to learn about these operations; we only learn the rules for using the decimal numerals when the operations are to be performed.

By using expanded notation for some decimals, the usual pattern for writing decimals for addition or subtraction becomes evident. Suppose we are to find the sum of 34·12 and 47·35? To make the addition easy we might write these numerals another way and add those numbers that have the same place value.

$$34\cdot12 = 30 + 4 + \frac{1}{10} + \frac{2}{100}$$
$$+47\cdot35 = 40 + 7 + \frac{3}{10} + \frac{5}{100}$$
$$70 + 11 + \frac{4}{10} + \frac{7}{100} = 81\cdot47$$

In this scheme we notice that the numerals for the tens (30 and 40) are aligned vertically, the numerals for the ones (4 and 7) are aligned vertically, and the same is true for the tenths and hundredths.

When writing decimals for addition or subtraction such an alignment becomes nothing more than aligning the decimal points. Then we add or subtract just as if the numerals named whole numbers. Then write the decimal point in the numeral for the sum or difference below those used to name the numbers in the addition or subtraction.

Addition Examples

34·12	5·907	312·5907
+47·35	+0·218	+ 36·4725
81·47	6·125	349·0632

Subtraction Examples

36·73	7·459	472·3652
−19·52	−0·536	− 83·0764
17·21	6·923	389·2888

Exercise 5.14
Find each sum:

1.   36·49
    +18·72

2.   3·595
    0·608
   +7·172

3.   248·6
    + 35·7

4. 0·4321
    0·7006
    +0·3895
    ────

5. 7·26105
    +5·72964
    ────

6. 0·501247
    0·000986
    +0·720058
    ────

Find each difference:

7. 63·35
    −19·75
    ────

8. 926·3
    −756·4
    ────

9. 80·05
    −26·16
    ────

10. 0·308
    −0·269
    ────

11. 4·7092
    −0·8135
    ────

12. 0·51246
    −0·36189
    ────

DECIMALS IN MULTIPLICATION

Our definition of multiplication of rational numbers stated that we find the product of the numerators and the product of the denominators. Since decimals are shortened names for rational numbers, we should expect the same definition to apply.

$$0·3 \times 0·23 = \frac{3}{10} \times \frac{23}{100} = \frac{69}{1000} = 0·069$$

Since 0·3 denotes the numerator 3 and the denominator 10, and since 0·23 denotes the numerator 23 and the denominator 100, we can find the product of the numerators as we do with fractions, and then place the decimal point so that the indicated denominator is the product of 10 and 100. We can state this multiplication as follows:

$$
\begin{array}{r}
0·23 \\
\times 0·3 \\
\hline
0·069 \\
\hline
\end{array}
$$

To find the product of 3·201 and 0·7, think: the numerator will be 3 201 × 7, and the decimal point should be placed so that it indicates a denominator $10^3 \times 10$ or 10 000. In other words, there should be 4 digits to the right of the decimal point.

$$
\begin{array}{r}
3·201 \\
\times 0·7 \\
\hline
2·2407 \\
\hline
\end{array}
$$

How many digits are to the right of the decimal point in 3·201? In 0·7? In 2·2407? Do you see a relationship between these? The number of digits to the right of the decimal point in the numeral for the product is the sum of the numbers of digits to the right of the decimal points in the two factors.

Exercise 5.15

Find each product:

1. 3·4
    ×0·8
    ⎯

2. 0·76
    × 5
    ⎯

3. 84·3
    × 2·4
    ⎯

4. 5·12
    ×0·37
    ⎯

5. 76·29
    ×0·34
    ⎯

6. 315·6
    ×0·215
    ⎯

DECIMALS IN DIVISION

Suppose you are to find the quotient when 35 is divided by 0·7?

Since
$$\frac{0·35}{0·7} = \frac{0·35}{0·7} \times 1,$$

we can rename 1 so that the product has a denominator that is a whole number. In this case we name 1 as $\frac{10}{10}$, since $0·7 \times 10 = 7$, which is a whole number.

$$\frac{0·35}{0·7} = \frac{0·35}{0·7} \times 1 = \frac{0·35}{0·7} \times \frac{10}{10} = \frac{3·5}{7}$$

Since this can be done for any rational number, we need only concern ourselves with the case where the divisor is a whole number.

$$\frac{3·5}{7} = 3·5 \div 7 = \frac{35}{10} \div 7 = \frac{35}{10} \times \frac{1}{7}$$
$$= \frac{35 \times 1}{10 \times 7} = \frac{35 \times 1}{7 \times 10} = \frac{35}{7} \times \frac{1}{10}$$

The final product above shows that we can divide as if both numerator and denominator were whole numbers if we multiply the quotient by $\frac{1}{10}$, $\frac{1}{100}$, $\frac{1}{1000}$, and so on, as appropriate.

Another example will help clarify this idea.

Suppose we divide 0·875 by 0·25? In this case we think:

$$\frac{0·875}{0·25} = \frac{0·875}{0·25} \times \frac{100}{100} = \frac{87·5}{25}$$

This step is equivalent to multiplying both the dividend and the divisor by 100. In the usual manner this would appear as:

$$0·25\overline{)0·875} \quad \text{becomes} \quad 25\overline{)87·5}$$

Then we divide as if both divisor and dividend were whole numbers.

$$
\begin{array}{r}
3·5 \\
25\overline{)87·5} \\
75 \\
\hline
125 \\
125 \\
\hline
\end{array}
$$

But how do we know whether to multiply this quotient by $\frac{1}{10}$, or $\frac{1}{100}$, and so on? If 87·5 were written as a fraction, what is the denominator? How many decimal places are to the right of the decimal point in 87·5? Is there a relationship between these? Hence, we must multiply by $\frac{1}{10}$, and the quotient is $35 \times \frac{1}{10}$ or 3·5.

$$\begin{array}{r} 3 \cdot 5 \\ \hline 25\overline{)87 \cdot 5} \end{array}$$

Perhaps you can discover a short cut for placing the decimal point in the numeral for the quotient.

Exercise 5.16

Find each quotient:

1. 86·95 ÷ 4·7
2. 1·008 ÷ 3·6
3. 6·7332 ÷ 0·124
4. 3·087 ÷ 0·21

TERMINATING DECIMALS

Now that we have used decimals to name numbers some interesting questions arise. Can every rational number be named by a decimal? Does every decimal name a rational number?

It is easy to change from a fraction to a decimal, since the fraction notation is also considered to denote division.

The following show how to change the fractions $\frac{1}{8}$, $\frac{17}{20}$, and $\frac{4}{25}$ to decimals.

$$\begin{array}{r} 0 \cdot 125 \\ \hline 8\overline{)1 \cdot 000} \\ 8 \\ \hline 20 \\ 16 \\ \hline 40 \\ 40 \\ \hline 0 \end{array} \qquad \begin{array}{r} 0 \cdot 85 \\ \hline 20\overline{)17 \cdot 00} \\ 160 \\ \hline 100 \\ 100 \\ \hline 0 \end{array} \qquad \begin{array}{r} 0 \cdot 16 \\ \hline 25\overline{)4 \cdot 00} \\ 25 \\ \hline 150 \\ 150 \\ \hline 0 \end{array}$$

We say that each of these quotients is named by a **terminating decimal**, since we reach a remainder of zero in each case. Notice, however, that each divisor (8, 20, 25) is a factor of 100 or 1 000, and hence the rational numbers $\frac{1}{8}$, $\frac{17}{20}$, and $\frac{4}{25}$ could be multiplied by the identity number of multiplication in the form $\frac{n}{n}$ so that the denominator is either 100 or 1 000. Therefore, we have been discussing only special kinds of fractions.

Recall that a decimal indicates a denominator that is a power of 10. Since the factors of 10 are 2 and 5 (not including 1 and 10 itself), any rational number whose denominator has factors of only 2s and 5s, or both 2s and 5s, can be named as a terminating decimal. If we have a rational number such as

$$\frac{9}{40} = \frac{9}{2 \times 2 \times 2 \times 5}$$

we can make its denominator a power of 10 by pairing each 2 with a 5 (as long as they last) and then supplying any 2^s or 5^s needed to complete the pairings. The needed 2^s and 5^s are supplied by multiplying both numerator and denominator by the same number.

$$\frac{9}{40} = \frac{9}{2 \times 2 \times 2 \times 5} \times \frac{5 \times 5}{5 \times 5}$$

$$= \frac{9 \times 5 \times 5}{2 \times 2 \times 2 \times 5 \times 5 \times 5}$$

$$= \frac{9 \times 25}{(2 \times 5)(2 \times 5)(2 \times 5)}$$

$$= \frac{225}{10 \times 10 \times 10} \text{ or } \frac{225}{1000} \text{ or } 0.225$$

Hence, we need only inspect the denominator to determine whether or not a rational number can be named by a terminating decimal. For example, $\frac{7}{16}$ can be named by a terminating decimal since $16 = 2 \times 2 \times 2 \times 2$.

Exercise 5.17

Use division to name each of these as a terminating decimal.

1. $\frac{7}{20}$ 2. $\frac{7}{8}$ 3. $\frac{11}{40}$ 4. $\frac{19}{50}$ 5. $2\frac{3}{4}$ 6. $58\frac{9}{25}$

REPEATING DECIMALS

Now let us consider a rational number, such as $\frac{4}{11}$, where the denominator has factors other than 2 or 5.

$$
\begin{array}{r}
0.36 \\
11{\overline{)4.0000}} \\
33 \\
\overline{} \\
70 \\
66 \\
\overline{} \\
4
\end{array}
$$

Can you tell without further division the next two digits of the quotient? Does the decimal naming the quotient ever terminate? A decimal such as 0·3636 . . . is called a **repeating decimal** or a **periodic decimal**, since the digits 36 continue to repeat. In order to name such a decimal more concisely, let us draw a bar over the sequence or period of digits that repeats.

$$\tfrac{4}{11} = 0.36\overline{36} \text{ or } 0.\overline{36}$$
$$\tfrac{1}{3} = 0.33\overline{3} \text{ or } 0.3\overline{3} \text{ or } 0.\overline{3}$$

The number of times we repeat the period of digits before drawing a bar over it is immaterial.

One repetition is sufficient.

Think about naming $\frac{5}{7}$ as a decimal. What are the possible remainders when the divisor is 7? (0, 1, 2, 3, 4, 5, 6.) This means that after annexing 0^s

to the dividend numeral, a remainder must repeat in at most 7-division steps. If a remainder of zero occurs, the decimal terminates. As soon as a remainder repeats, we can stop dividing, since we know what the repeating decimal is.

Thus, we see that every rational number can be named by either a terminating or a repeating decimal.

PERCENTAGES

On page 92 we obtained a method for renaming a rational number as $\frac{a}{c} = \frac{ra}{rc}$ ($r \neq 0$).

Now suppose that $\frac{a}{c} = 0\cdot 3$, then it follows that we may write $0\cdot 3 = \frac{r}{r} \times 0\cdot 3$, and if we choose $1 = \frac{100}{100}$ we obtain $0\cdot 3 = \frac{100 \times 0\cdot 3}{100} = \frac{30}{100}$.

When the denominator of a rational number is 100 we may refer to the number as a percentage. Thus $\frac{30}{100}$ is called 30 per cent, $\frac{45}{100}$ is called 45 per cent and, as a final example, $\frac{170}{100}$ is called 170 per cent. We have an abbreviation for the words per cent, it is the sign %. The above examples would be written 30%, 45%, and 170% respectively.

Note: per cent may be thought of as per hundred, i.e. $\frac{23}{100}$ is clearly 23 per hundred or just 23%.

Example 1

Express $\frac{4}{11}$ as a percentage.

We have just obtained $\frac{4}{11} = 0\cdot\overline{36}$.

Now $\frac{4}{11} \times \frac{100}{100} = \frac{36\cdot\overline{36}}{100}$ which we may call $36\cdot\overline{36}$ per cent.

Example 2

Express $\frac{1}{3}$ as a percentage. We have $\frac{1}{3} = 0\cdot\overline{3}$. Therefore $\frac{1}{3} = \frac{33\cdot\overline{3}}{100}$, which we may call $33\cdot\overline{3}$ per cent (or $33\frac{1}{3}\%$).

Exercise 5.18

Use a division to name each of these as a repeating decimal:

1. $\frac{5}{6}$ 2. $\frac{1}{9}$ 3. $\frac{2}{3}$ 4. $\frac{5}{33}$ 5. $\frac{11}{14}$ 6. $\frac{4}{7}$
7. Express the rational numbers of Questions 1, 2, 3, 4, Exercise 5.17 as per cents.

CHANGING DECIMALS TO FRACTIONS

There is no problem in changing a terminating decimal to a fraction. Merely reading the decimal tells us the fraction. For example, we read $0\cdot 213$ as two hundred and thirteen thousandths and write the fraction $\frac{213}{1000}$. Since this is already in lowest terms, we leave it as it is.

How can we change a repeating decimal to a fraction? For example, let us change $0\cdot\overline{23}$ to a fraction. Think of $0\cdot\overline{23}$ as naming some rational number k.

$$k = 0\cdot\overline{23}$$

Then $\qquad\qquad 100k = 23\cdot\overline{23}$

If we should subtract k from $100k$ we should eliminate the decimal and need work only with whole numbers.

$$100k = 23 \cdot \overline{23}$$
$$k = 0 \cdot \overline{23}$$
$$\overline{99k = 23 \cdot 00}$$
$$k = \tfrac{23}{99}$$

Since $\tfrac{23}{99}$ is already in lowest terms, we conclude that $0 \cdot \overline{23} = \tfrac{23}{99}$

Another example might be to name $0 \cdot \overline{618}$ as a fraction. Since the digits repeat in a period of 3 digits, we will multiply by 1 000 in this case.

Let $t = 0 \cdot \overline{618}$
Then $1\,000t = 618 \cdot \overline{618}$
$$t = 0 \cdot \overline{618}$$
$$\overline{999t = 618 \cdot 00}$$
$$t = \tfrac{618}{999} \text{ or } \tfrac{206}{333}$$
We conclude that $0 \cdot \overline{618} = \tfrac{206}{333}$

Thus, we see that every terminating or repeating decimal can be changed to a fraction.

Can you think of a decimal that never repeats nor terminates? How about $0 \cdot 010110111 \ldots$ or $0 \cdot 343343334 \ldots$?

These decimals name numbers that are *not* rational. Such numbers are called **irrational numbers**. You may have already worked with some irrational numbers, such as $\sqrt{2}$, $\sqrt{5}$, and π. The union of the set of rational numbers and the set of irrational numbers is the set of what we call *real* numbers.

Exercise 5.19

Change each of these decimals to a fraction:

1. $0 \cdot 35\overline{35}$
2. $5 \cdot \overline{66}$
3. $0 \cdot 39\overline{6}$
4. $0 \cdot \overline{6315}$
5. $0 \cdot 73\overline{3}$
6. $2 \cdot 26\overline{12}$

Calculate the following:

7. $0 \cdot \overline{2} + 0 \cdot \overline{2}$
8. $0 \cdot 3\overline{6} + 0 \cdot 2\overline{1}$
9. $0 \cdot 4\overline{8} + 0 \cdot 5\overline{2}$
10. $1 \cdot \overline{3} + 2 \cdot \overline{6}$

Explain how you would calculate the next two results.

11. $0 \cdot \overline{3} \times 0 \cdot \overline{3}$
12. $0 \cdot \overline{1} \times 0 \cdot 47$

DENSITY OF RATIONAL NUMBERS

Certainly there are no integers between 2 and 3 or between 0 and -1. We can summarize this by saying that there is no integer between two consecutive integers. This enables us to say that two such integers *are* consecutive.

We might ask: is there a rational number between any two given rational numbers? For example, is there a rational number between $\tfrac{1}{4}$ and $\tfrac{3}{4}$? In this case we can easily see that $\tfrac{2}{4}$ or $\tfrac{1}{2}$ is between these two rational numbers.

Then we might ask if there is a rational number between $\tfrac{1}{4}$ and $\tfrac{1}{2}$. Since $\tfrac{1}{4} = \tfrac{2}{8}$ and $\tfrac{1}{2} = \tfrac{4}{8}$ this question is the same as asking if there is a rational number between $\frac{2}{8}$ and $\frac{4}{8}$. Now it is more obvious that $\frac{2}{8} < \frac{3}{8} < \frac{4}{8}$ (which is another way of saying that $\frac{3}{8}$ is between $\frac{2}{8}$ and $\frac{4}{8}$).

We could continue this line of questioning and become quite convinced that between any two rational numbers there is always a third rational number. But let us investigate a general case on a number line. Let a and c represent any two rational numbers such that $a < c$.

Now let us locate the point for $a + c$.

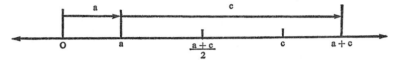

Next, let us find the point midway between the points for 0 and $a + c$, as shown above by $\dfrac{a + c}{2}$.

Is this point between the points for a and c? This process could be used for any two rational numbers. The rational number $\dfrac{a + c}{2}$ is the average of the two given rational numbers a and c. Hence, to find a rational number between two given rational numbers, find the average of the two given numbers. Thus we may conclude:

Between any two rational numbers there is always a third rational number. We say that the rational numbers are **dense**. Even so, there are gaps between the rational numbers. The numbers which occupy these gaps are called **irrationals**.

The set R, of rational numbers is:

Closed for $+$, $\times$, $-$, $\div$

Commutative for $+$ and $\times$ but not for $-$ and $\div$

Associative for $+$ and $\times$ but not for $-$ and $\div$

Distributive for $\times$ over $+$, i.e. $\frac{4}{9} \times (\frac{7}{6} + \frac{8}{11}) = (\frac{4}{9} \times \frac{7}{6}) + (\frac{4}{9} \times \frac{8}{11})$

for $\times$ over $-$, i.e. $\frac{4}{9} \times (\frac{7}{6} - \frac{8}{11}) = (\frac{4}{9} \times \frac{7}{6}) - (\frac{4}{9} \times \frac{8}{11})$

From the } for $\div$ over $+$, i.e. $(\frac{7}{6} + \frac{8}{11}) \div \frac{4}{9} = (\frac{7}{6} \div \frac{4}{9}) + (\frac{8}{11} \div \frac{4}{9})$

right only $\}$ for $\div$ over $-$, i.e. $(\frac{7}{6} - \frac{8}{11}) \div \frac{4}{9} = (\frac{7}{6} \div \frac{4}{9}) - (\frac{8}{11} \div \frac{4}{9})$

The set R has the identity element 1 for $\times$ and 0 for $+$.

Each rational number $\dfrac{a}{b}$ has an inverse for $\times$, i.e. $\dfrac{b}{a}$ the multiplicative inverse. $(a \neq o)$.

Each rational number $\dfrac{a}{b}$ has an inverse for $+$, i.e. $-\dfrac{a}{b}$ the additive inverse.

These are the laws which govern the calculations we perform with our basic rational number system and constitute what we call a **mathematical system**.

In general, a mathematical system is described by:

(i) a set of objects (not necessarily numbers);
(ii) defining one or more operations for these objects of the set;
(iii) the basic properties of the operations (this includes reference to the identity number, etc.).

We have been discussing systems with an infinite number of elements. In the next chapter we will consider finite systems commonly called modular or finite arithmetics.

CHAPTER SIX

FINITE ARITHMETICS

CLOCK ARITHMETIC

We can illustrate the construction of a small mathematical system or arithmetic by considering the one-handed five-hour clock of Fig. 6.1. Recall that a mathematical system is a set for which there is a binary operation defined on the set. The system will use the set of numbers {0, 1, 2, 3, 4}, and the operations to be considered are defined as follows:

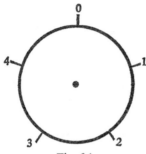

Fig. 6.1

1. **Addition** is to be carried out by rotating the hand clockwise through the number of places concerned. Thus, 2 added to 3 will mean: starting from 0, move the hand clockwise through two places and then another three places. We could invent a new symbol to represent the rotation, for example $\circlearrowright$, so that the sum we have just done could be written $2 \circlearrowright 3$, but why bother when the usual symbol $+$ will do providing we keep its new meaning in mind. The result of the addition will be given by the final position of the hand.

In this case the hand ends up at 0. We represent this equivalence by the following statements:

$$2 + 3 = 0$$

Further examples are $1 + 4 = 0$; $2 + 4 = 1$; $3 + 4 = 2$; $4 + 4 = 3$, etc.

Now it is clear that since we have such a small system we can offer a complete definition of addition by listing all possible results in the form of the table Fig. 6.2. Since all the results belong to the set {0, 1, 2, 3, 4}, then the set is closed with respect to addition.

We locate the result of $2 + 3$ at the intersection of the row labelled 2 and the column labelled 3. Likewise the result of $3 + 2$ will be found at the intersection of the row labelled 3 and the column labelled 2. In each case the result is 0. The addition as defined here is commutative, and this is evident from the table by noticing that the upper shaded half is the reflection of the

109

lower half in the dotted diagonal. More precisely, if *a* and *b* are any two
rows and columns of a square table, then the operation it represents is
commutative if the result at the intersection of row *a* and column *b* is the
same as the result at the intersection of row *b* and column *a*. We have just
shown this for *a* = 2, *b* = 3. Try some other values for *a* and *b*.

It is easy to see from Fig. 6.2 that addition is associative simply by examining all the possible combinations of (*a* + *b*) + *c* and *a* + (*b* + *c*), but this,

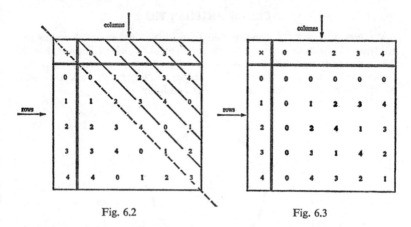

Fig. 6.2   Fig. 6.3

of course, takes a great deal of time. However, we know from previous work
that whole numbers are associative with respect to addition, and since we are
adding whole numbers, and then only interpreting the result as a clockwise
rotation, it follows that addition is associative in clock arithmetic.

2. Multiplication. Consider the following:

$$4 + 4 + 4$$
$$1 + 1 + 1 + 1$$
$$2 + 2 + 2$$

These are all examples of continued addition of the same number which we
have already seen is equivalent to multiplication. Consequently we may
write these as 4 × 3, 1 × 4, 2 × 3. The result is given by the number to which
the hand points after starting from 0 and passing through 3 + 3 + 3 + 3
places, or 1 + 1 + 1 + 1 places, and so on. Thus 4 × 3 = 2, 1 × 4 = 4,
2 × 3 = 1. The set is closed with respect to multiplication because all the
results are members of the set {0, 1, 2, 3, 4}.

Since multiplication of whole numbers is commutative and associative
then so is multiplication in our clock arithmetic.

Exercise 6.1

Use the tables of Fig. 6.2 and Fig. 6.3 to calculate the following:

1. 3 × 3   2. 3 × 4   3. 2 + 4 + 3   4. (3 × 4) + 2
5. (4 × 4) + 4   6. 2 × (2 + 1)   7. (2 × 2) + (2 × 1)

8. $4 \times (3 + 2)$ 9. $(4 \times 3) + (4 \times 2)$ 10. $2 \times (3 + 4)$
11. $(2 \times 3) + (2 \times 4)$

12. Compare the questions (6, 7); (8, 9); (10, 11). These are examples of which law?

ADDITIVE INVERSES AND SUBTRACTION

As in the case of whole numbers, 0 is the identity element for addition, and since we have such a small system, we can list every one of the additive possibilities of combining elements of the set with 0.

$$4 + 0 = 0 + 4 = 4$$
$$3 + 0 = 0 + 3 = 3$$
$$2 + 0 = 0 + 2 = 2$$
$$1 + 0 = 0 + 1 = 1$$
$$0 + 0 \qquad = 0$$

This system has properties which the set of whole numbers does not possess. For example, every element of this system has an **additive inverse**.

Remember that the additive inverse of a number when added to that number yields the identity element with respect to addition. Let us represent the additive inverse of x by x^*. Therefore $x + x^* = 0 = x^* + x$. Examining the addition table of Fig. 6.2, we see that $2^* = 3$, meaning that the additive inverse of 2 is 3, or $2 + 3 = 0$. What is 3^*? Well, just looking at the last equation tells us that $3^* = 2$.

Now since each element of our system has an additive inverse, it follows that we should be able to perform the inverse operation to addition; in other words, subtraction.

Definition 6.1

In clock arithmetic $a - b$ is that number c such that $a = b + c$.

Consider the problem $2 - 4$. We are looking for a number c such that $2 = 4 + c$, i.e. which number must we add to 4 to give the result 2?

From the addition table we see that $c = 3$, therefore $2 - 4 = 3$. If we combine the operations of addition and subtraction into the same problem we agree to evaluate the bracketed parts first. If no brackets are present, then we read from left to right in the usual manner (as already mentioned on page 69).

Example 1

$$
\begin{aligned}
1 - (4 + 3) + 2 &= 1 - 2 + 2 \quad && (4 + 3) = 2 \\
&= 4 + 2 \quad && 1 - 2 = 4 \\
&= 1 \quad && \text{or} \; -2 + 2 = 0
\end{aligned}
$$

Example 2

$$
\begin{aligned}
1 - (1 - 3) - 4 &= 1 - 3 - 4 \quad && (1 - 3) = 3 \\
&= 3 - 4 \quad && (1 - 3) = 3 \\
&= 4
\end{aligned}
$$

Example 3

$$
\begin{aligned}
(2 - 4) + (1 - 4) &= 3 + (1 - 4) \quad && (2 - 4) = 3 \\
&= 3 + 2 \quad && (1 - 4) = 2 \\
&= 0
\end{aligned}
$$

Exercise 6.2

Insert the missing number:

1. $4^* = -$ 2. $1^* = -$ 3. $3 - 4 = -$
4. $2 - 3 = -$ 5. $1 - 4 = -$ 6. $3 + 2 - 4 = -$
7. $1 + 2 - 4 = -$ 8. $(4 - 2) - 3 = -$ 9. $(3 - 4) - (1 - 3) = -$
10. $2^* - 3 + 4^* = -$ 11. $2^* + 3^* = -$ 12. $4^* - 1^* = -$

MULTIPLICATIVE INVERSES AND DIVISION

The identity element for multiplication is 1, a fact which we can show in the full list of results, where we see that

$$4 \times 1 = 1 \times 4 = 4$$
$$3 \times 1 = 1 \times 3 = 3$$
$$2 \times 1 = 1 \times 2 = 2$$
$$1 \times 1 = 1 \times 1 = 1$$
$$0 \times 1 = 1 \times 0 = 0$$

Multiplication by the identity element 1 leaves each element of the set unchanged. In general, the existence of the identity element is indicated by the table having a row and a column identical with the headings. In Fig. 6.3 the row and column labelled 1 is the same as the headings, so that 1 is the identity element, while in Fig. 6.2 the row and column labelled 0 applies.

What we want to know is, do the elements of our clock arithmetic have multiplicative inverses? For example, which number multiplied by 3 will yield 1? Examination of the table of Fig. 6.3 reveals that $2 \times 3 = 1$ and $3 \times 2 = 1$, so that 2 is the multiplicative inverse of 3.

Definition 6.2

The *multiplicative inverse* of an element a is that element which when multiplied by a yields the multiplicative identity.

The inverse of a will be written a^{-1} (read, 'a inverse'). Thus

$$a \times a^{-1} = a^{-1} \times a = 1.$$

From the table we see that $3^{-1} = 2$ and $2^{-1} = 3$, so that $3^{-1} \times 3 = 2 \times 3 = 1$, etc.

What are 1^{-1} and 4^{-1}? Notice that all the elements have an inverse except 0. It must not be taken for granted that *all* elements in a clock arithmetic have multiplicative inverses, as we shall see later on.

Now we define division as the inverse operation to multiplication. For example, by $2 \div 3$ we mean 2×3^{-1}, i.e. dividing by an element of our set is the same as multiplying by its inverse, *provided*, of course, that the inverse *exists*.

Definition 6.3

If a, b are two numbers or elements of clock arithmetic, then a divided by b written $a \div b$ is the same as $a \times b^{-1}$ where $b \neq 0$ and b^{-1} exists.

Example 1

$$2 \div 3 = 2 \times 3^{-1} = 2 \times 2 = 4$$

Example 2

$$1 \div 4 = 1 \times 4^{-1} = 1 \times 4 = 4$$

Example 3

$$3 \div 2 = 3 \times 2^{-1} = 3 \times 3 = 4$$

Exercise 6.3

Use Tables 6.2 and 6.3 to evaluate the following:

1. 1^{-1}	2. $2 \div 4$	3. $4 \div 2$
4. $1 \div 3$	5. $2^{-1} \times 3^{-1}$	6. $2^{-1} \times 4^{-1}$
7. $1 \div 2$	8. $3 \times (1 \div 3)$	9. $2 \times (1 \div 2)$
10. $(1 + 3)^{-1}$	11. $2 - 3^{-1}$	12. $(2 + 3)^{-1}$

MODULAR ARITHMETIC

You may have noticed that once the tables of Figs. 6.2 and 6.3 are given or completed there is no longer any need to examine the original clock. This is because the tables completely define the mathematical system, by showing the structure of its results, and it is the structure of the system that we are interested in.

We all use a twelve-hour clock or a twenty-four-hour clock, so that a more familiar arithmetic could be constructed based on 12 or 24. Indeed, any clock, provided it has at least two places on its face, will provide us with a mathematical system.

Systems of this kind are called **modular arithmetics**. For the clock we have just discussed, the number 5 is called the **modulus** and the system itself is called **arithmetic modulo 5**, written in abbreviated form as **arithmetic (mod 5)**.

If we wish to express any whole number as a number in arithmetic modulo 5 we divide that number by 5 and the remainder is the representation of the number in arithmetic modulo 5.

Thus, $23 = 3$ (mod 5) because $23 = 5 \times 4 + 3$ and $37 = 2$ (mod 5) because $37 = 5 \times 7 + 2$.

But note especially that the clock arithmetic we have been discussing is not quite the same as arithmetic (mod 5), because in the clock arithmetic we dealt only with numbers or elements of the set {0, 1, 2, 3, 4}, and such numbers as 23 and 37 were therefore meaningless.

To indicate that 32 and 27 have the same remainder when divided by 5 we write $32 \equiv 27$ (mod 5). This last phrase is read as '32 is congruent to 27, mod 5'.

We could also say that $32 \equiv 22$ (mod 5), or $32 \equiv 2$ (mod 5), so that an equation such as $x \equiv 2$ (mod 5) with the whole numbers for replacement set has the solution set

$$\{2, 7, 12, 17, 22, 27, \ldots\}$$

To gain a little more experience in this work let us consider the addition and multiplication tables for arithmetic (mod 6).

The results are located in exactly the same manner as for Figs. 6.2 and 6.3. We complete the table in the following manner:

$$4 \times 5 = 20 \equiv 2 \ (\text{mod } 6)$$

The 2 is entered in the intersection of row 4 and column 5. $5 \times 4 = 20 \equiv 2$ (mod 6), this 2 being entered in the intersection of row 5 and column 4. Similarly for all the other results. Three entries have been omitted from each table. See if you can insert the required results. (Answers at the end of the next exercise.) From row 2 in Fig. 6.5 we notice that the identity element of multiplication, namely 1, doesn't occur. Therefore 2 does not have a multiplicative inverse, i.e. there is no 2^{-1} such that $2 \times 2^{-1} = 1$.

It also follows that $3 \div 2$ is meaningless because we recall that the definition of division gives $3 \div 2$ as being the same as 3×2^{-1} if 2^{-1} exists, and

+	0	1	2	3	4	5
0	0	1	2	3	4	5
1	1	2	3	4	5	0
2	2	3	4	5	0	1
3	3	4	5		1	2
4	4	5	0	1	2	
5	5	0	1	2		4

Fig. 6.4

×	0	1	2	3	4	5
0	0	0	0	0	0	0
1	0	1	2	3	4	5
2	0	2	4	0		4
3	0	3	0	3	0	
4	0	4		0	4	2
5	0	5	4	3	2	1

Fig. 6.5

we have just seen that 2^{-1} doesn't exist. Now does this mean that division is not an operation? Before reading on have a look back at the definition of an operation and division on pages 15 and 112 respectively.

You will see that the definition depends upon the **uniqueness** of the result **where it exists**. Well, here the results of division are unique (i.e. only one result for each division) where they exist, so division is an operation. In fact the definition 6.3 is different from definition 2.5 in order that division may be preserved as an operation.

We have already met a similar situation with the operation of subtraction of whole numbers. The results were unique when they existed, but there were any number of occasions when the results didn't exist in the set of whole numbers, for example $4 - 7$, $6 - 19$ were meaningless in the set of whole numbers, but, these cases were ruled out by the definition 2.3 on page 36.

Exercise 6.4

Using the tables of Fig. 6.4 and Fig. 6.5, answer the following questions:

1. Is the set of numbers closed under the operations of addition and multiplication? How do you obtain this answer?

2. Is division an operation in arithmetic (mod 6)?

3. Which element of the set is the additive identity and which is the multiplicative identity?

4. Which elements of the set have additive inverses?

5. Which elements of the set have multiplicative inverses?

6. Simplify the following:

(*a*) 3 × (4 + 2) (*b*) 2 × (5 + 4) (*c*) 4 × (2 + 3)

7. Does your answer to Question 6 indicate that multiplication is distributive over addition?

8. Verify, by simplifying each side separately, that

(*a*) 3 − 5 = 3 + 5* (*b*) 2 − 4 = 2 + 4* (*c*) 5 − 1 = 5 + 1*

9. Simplify the following (where possible)

(*a*) 2 ÷ 2 (*b*) 4 ÷ 3 (*c*) 3 ÷ 5 (*d*) 2 ÷ 4 (*e*) 4 ÷ 5

10. Show that $(2 \div 5) + (1 \div 5) = 3$

'FRACTIONS' IN FINITE ARITHMETIC

We shall let the symbol $\frac{a}{b}$ represent $a \div b$ or $a \times b^{-1}$. We have already used such expressions as $2 \div 3$, which in arithmetic (mod 5) gave us the result of 4, but in arithmetic (mod 6) was meaningless. We now encounter another strange situation. From our experience with the integers we saw that we could write $\frac{1}{5}$ as $\frac{2}{10}$ because

$$\tfrac{1}{5} = 1 \times \tfrac{1}{5} = \tfrac{2}{2} \times \tfrac{1}{5} = \tfrac{2}{10}$$

What happens if we do this in arithmetic (mod 6)?

$$\frac{1}{5} = \frac{2}{2} \times \frac{1}{5} = \frac{2 \times 1}{2 \times 5}$$

$$= \frac{2}{4} \text{ since } 2 \times 5 = 4 \text{ (mod 6)}$$

$$= \frac{1}{2}$$

This is meaningless, so there must be something wrong here, because we know that $\frac{1}{5} = 5^{-1} = 5$ (mod 6) simply by looking at the table.

Where is the mistake? Well, notice the step $1 \times \frac{1}{5}$ written as $\frac{2}{2} \times \frac{1}{5}$. This means $2 \times 2^{-1} \times 5^{-1}$ so with 2^{-1} meaningless the mistake was clearly made at this stage.

Studying situations like this in Finite Arithmetics enables us to appreciate the wider range of results in the Arithmetic of the integers. More important still, these situations enable us to bring into the open those rules and definitions for the integers which everyday use has encouraged us to take so much for granted, so much so that we eventually forget that the rules and definitions need to be stated at all.

Let us examine this last situation in arithmetic (mod 5) with the aid of the table in Figs. 6.2 and 6.3.

Here we have $\frac{1}{2} = 3$, $\frac{1}{3} = 2$, $\frac{1}{4} = 4$, and as previously noted, each element, except 0, has a multiplicative inverse. Suppose we examine $\frac{2}{4}$? Is it possible to divide numerator and denominator by 2 so as to write $\frac{2}{4} = \frac{1}{2}$? This amounts to writing

$$\frac{2}{4} = \frac{2 \times 1}{2 \times 2} = 1 \times \frac{1}{2}$$

which is perfectly all right (mod 5) because $\frac{2}{2}$ exists and is equal to 1, i.e. perfectly all right because 2^{-1} exists.

Since $\qquad \frac{2}{4} = 2 \div 4 = 2 \times 4^{-1} = 2 \times 4 = 3$ (mod 5)

and $\qquad \frac{1}{2} = 2^{-1} = 3$ (mod 5)

the correctness of our result is confirmed.

Suppose we now try reversing the procedure by considering $\frac{3}{3} \times \frac{1}{2} = \frac{1}{2}$ (mod 5)? This appears to be $\dfrac{3 \times 1}{3 \times 2} = \dfrac{3}{1} = 3$, which, of course, is 2^{-1} and another correct result confirmed.

Now let us consider $\frac{2}{3} \times \frac{1}{4}$. Is this the same as $\dfrac{2 \times 1}{3 \times 4}$ (mod 5)? We know it cannot be true for (mod 6) because $\frac{1}{3}$ doesn't exist.

$$\frac{2}{3} = 4 \text{ (mod 5)}$$
$$\frac{1}{4} = 4 \text{ (mod 5)}$$
$$\therefore \quad \frac{2}{3} \times \frac{1}{4} = 4 \times 4 = 1 \text{ (mod 5)}$$
$$\frac{2 \times 1}{3 \times 4} = \frac{2}{2} = 1 \text{ (mod 5)}$$

Correct again!

If we calculate all the possible products of this kind we shall eventually see that if x, y, z, w are any elements of the set $\{0, 1, 2, 3, 4\}$ in arithmetic (mod 5), then

$$\frac{x}{y} \times \frac{z}{w} = \frac{x \times z}{y \times w}$$

provided that y and w are not 0.

It is clear that for this rule to hold in *any* arithmetic, all the elements except 0 must have multiplicative inverses. In the arithmetic of the integers we usually take this rule for granted or as intuitively obvious. Our study above shows this to be far from obvious.

Now consider the addition of fractions in arithmetic (mod 5), i.e. $\frac{1}{2} + \frac{3}{4}$ (mod 5).

Treating each fraction separately, this problem is rewritten

$$3 + 2 = 0 \quad \text{because} \quad \frac{1}{2} = 2^{-1} = 3 \quad \text{and} \quad \frac{3}{4} = 3 \times 4^{-1} = 2$$

What we really want to know is, can we add fractions in arithmetic (mod 5) in the same manner as the integers? Thus

$$\frac{1}{2} + \frac{3}{4} = \frac{2 + 3}{4} = \frac{5}{4} = 0 \tag{A}$$

We see that the result previously obtained is now confirmed. Since the method (A) simply consists of establishing a common denominator for both fractions by the multiplication of $\frac{1}{2}$ by $\frac{2}{2}$, we knew it would work in any case, because we have just deduced the result $\frac{x}{y} \times \frac{z}{w} = \frac{x \times z}{y \times w}$ for arithmetic (mod 5) or any arithmetic in which all elements (except 0) have unique multiplicative inverses. Let us try the same addition in arithmetic (mod 6) $\frac{1}{2} + \frac{3}{4} = ?$ Since neither $\frac{1}{2}$ nor $\frac{3}{4}$ exists in arithmetic (mod 6), small wonder we write '?'!

The reader may have been puzzled by the idea of replacement sets for solving equations on page 72. We shall now see why we obtain our solutions in this manner by considering the following examples:

Example 1

Solve (if possible) the equation $2x = 4$ in arithmetic mod 6 with replacement set {0, 1, 2, 3, 4, 5}. Our first inclination, based on previous experience, is to divide both sides by 2 to simplify the equation to the form $x = 2$, and this would be taken as the solution. But if we re-examine what we have done in a little more detail we see that we suggested

$$2x \div 2 = 4 \div 2 \quad \textit{in arithmetic mod 6}$$

and this means $2x \times 2^{-1} = 4 \times 2^{-1}$, and since 2^{-1} doesn't exist in arithmetic (mod 6), it follows that our manipulation of this equation is incorrect.

So what we really search for is a number with which to *replace* x so that $2x$ or $x + x = 4$. Examination of the table in Fig. 6.5 reveals in the row labelled 2 that x may be 2 or 5 and the solution set is {2, 5}, a result which we confirm by replacing x by 2 and 5 respectively. We have seen therefore how we may obtain the solutions by employing the replacement when the usual method of solution fails completely.

Example 2

Solve, if possible, the equation $3x = 3$ in arithmetic (mod 6). The replacement set is {0, 1, 2, 3, 4, 5}.

Examining the table of Fig. 6.5, we see that $3 \times 1 = 3, 3 \times 3 = 3, 3 \times 5 = 3$. Therefore we have the solution set as {1, 3, 5} for this equation. Notice once again that the division of the equation by 3 is not permitted because 3^{-1} doesn't exist in arithmetic (mod 6).

Example 3

Solve (if possible) the equation $x^x = 1$ (mod 6), using {0, 1, 2, 3, 4, 5} as replacement set.

Replacing x as suggested, we must try $0^0, 1^1, 2^2, 3^3, 4^4, 5^5$. 0^0 is meaningless since it is not defined.

$1^1 = 1$ $2^2 = 2 \times 2 = 4$
$3^3 = 3 \times 3 \times 3 = 3$ $4^4 = 4 \times 4 \times 4 \times 4 = 4$
$5^5 = 5 \times 5 \times 5 \times 5 \times 5 = 5$
The solution set is {1}.

Example 4

Solve (if possible) $\frac{x}{2} = 3$ in arithmetic (mod 6)

The left-hand side of this equation is $x \times 2^{-1}$, but in arithmetic (mod 6) we know that 2^{-1} doesn't exist, consequently the equation is meaningless and no solution is possible.

118 *New Mathematics Made Simple*

Example 5

Solve (if possible) $4x + 1 = 3$ (mod 6) with replacement set {0, 1, 2, 3, 4, 5}.

We know that the additive inverse of 1 exists, namely 5 (i.e. $1^* = 5$) because $(1 + 1^* = 0)$.

$$\therefore \quad 4x + 1 + 1^* = 3 + 1^*$$
$$4x = 2$$

Examining the table of Fig. 6.5, and in particular the row labelled 4, we see that $4 \times 2 = 2$ and $4 \times 5 = 2$.

The solution set is therefore {2, 5}.

Exercise 6.5

Simplify the following Numbers 1–5 in both arithmetic (mod 5) and (mod 6):

1. $\frac{1}{3} + \frac{1}{4}$ 2. $\frac{2}{3} + \frac{3}{4}$ 3. $\frac{1}{5} + 2$ 4. $\frac{1}{3} - \frac{1}{4}$ 5. $\frac{4}{3} - \frac{1}{4}$

6. Construct addition and multiplication tables for arithmetic (mod 4), then answer the following questions:

(a) Is the set {0, 1, 2, 3} closed under addition and multiplication for arithmetic (mod 4)?

(b) Which element is the additive identity and what is the additive inverse of each element?

(c) Is there a multiplicative identity?

(d) Which elements have multiplicative inverses, and what are they?

(e) Is division an operation in arithmetic (mod 4)?

Before attempting the next questions make certain you have obtained the correct tables for arithmetic (mod 4) by checking your answer for Question 6.

7. Which of the following are meaningless in arithmetic (mod 4)?

(a) 2^* (b) 3^{-1} (c) 2^{-1} (d) $\frac{3}{2}$ (e) $\frac{1}{3}$

8. The replacement set for the following equations is {0, 1, 2, 3}. Obtain the solution sets for x (mod 4) in each case.

(a) $2x = 2$ (b) $3x = 1$ (c) $2x + 1 = 3$ (d) $x - 3 = 3$ (e) $x^x = 1$

OTHER FINITE SYSTEMS

The systems we have discussed so far have all had numbers as elements and the familiar operations of addition and multiplication. We now examine the possibility of having a system without numbers, together with operations other than those already discussed. Let us start by examining the table of Fig. 6.6. This is a table of results for the operation of $\otimes$ which we have yet to define.

Just by looking at the table we see that:

(a) The set of elements being discussed is T = {I, H, V, R}.

(b) The operation is $\otimes$.

(c) The set T is closed under the operation of $\otimes$.

(d) The element I is the identity element for T and the operation $\otimes$ because the row and column labelled I is identical to the headings.

(e) Since I occurs in each row or column then each element has an inverse with respect to the operation $\otimes$.

(f) The operation of $\otimes$ is commutative, since the lower part is a reflection of the upper part in the diagonal.

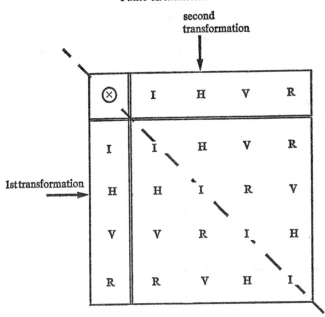

Fig. 6.6

All this we are able to state by just looking at the table, and yet we have not even bothered to give any meaning to ⊗ or I, H, V, R. But we can give *one* set of meanings to the elements of the table as follows:

Consider a rectangular piece of paper with one corner removed and having two axes drawn in the paper as illustrated by the dotted lines of Fig. 6.7. We remove the corner simply to aid identification of the position of the rectangle at any time. The axes, parallel to the edges, pass through centre O.

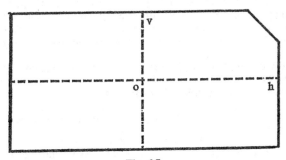

Fig. 6.7

We shall call each element of T a transformation of this rectangle from one position to another in Fig. 6.8. Four different positions will be considered, and these are numbered for convenience in the separate locations of Fig. 6.8.

The elements H, V, R, I are rotations defined as follows:

H = half a complete turn about the horizontal axis. Thus the application of H will change the position of the rectangle from position 1 to position 4, and vice versa, or from position 2 to position 3, and vice versa.

V = half a complete turn about the vertical axis *v*. Thus the application of V will change the position of the rectangle from position 1 to position 2, and vice versa, or from position 3 to position 4, or vice versa.

R = half a complete anti-clockwise turn about an axis perpendicular to the paper and passing through O. Thus, the application of R will change the

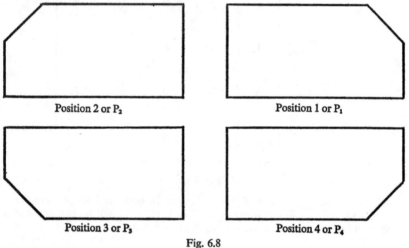

Fig. 6.8

position of the rectangle from P_2 to P_4, and vice versa, or P_1 to P_3, and vice versa.

I = leaves the rectangle unmoved. That is, I means 'leave the rectangle where it is'. The operation ⊗ will be called 'circle multiplication' for obvious reasons, although any word would do. We interpret the result of combining the elements as that single element which could produce the same final position of the rectangle so that *H* ⊗ *V means 'starting from P perform H first, then perform V second'*.

Now H first transforms P_1 into P_4, or we say P_1 is mapped into P_4, and write $P_1 \rightarrow P_4$.

Then V will transform P_4 into P_3 or $P_4 \rightarrow P_3$.

So, H ⊗ V means $P_1 \rightarrow P_4 \rightarrow P_3$ or simply $P_1 \rightarrow P_3$. We could have reached P_3 by using the single element R so that the meaning of

$$H \otimes V \text{ is } R \quad \text{or} \quad H \otimes V = R.$$

We can easily see that $V \otimes H = R$, starting from any P.

Furthermore, we could evaluate all the results such as H ⊗ R ⊗ V to establish that the operation is associative as well. But the important thing to realize is that once we have the table of results we may discard the original physical model of the paper rectangle, just as we discarded the clock face earlier in the chapter. For example, we know that I is the identity element, so that if we want the inverse of H (which we shall call H^{-1}) we look for a solution to $H^{-1} \otimes H = I$ in the row labelled H and immediately see that $H = H^{-1}$, in other words, H is its own inverse.

We shall now deal with one or two everyday uses of finite arithmetic or rather the ways in which people already use finite arithmetic without bothering about its rules.

Example 1

Christmas day 1967 falls on a Monday. On which day of the week is Christmas day 1968?

Now Christmas day 1968 is 366 days after Christmas day 1967 because 1968 is a leap year. Taking Monday as zero day of the week, we require the smallest number congruent to 366 (mod. 7). But $366 \equiv 2$ (mod 7), so that Christmas day 1968 falls on a Wednesday.

Example 2

If today is Sunday 9 July 1967, what date is the second Wednesday in October 1967?

Taking today as zero day there are 22 days left in July, 31 days in August, and 30 days in September. This gives a total of 83 days to the last day in September. But $83 \equiv 6$ (mod 7), and since we are working in arithmetic (mod 7) with Sunday as zero day, it follows that day 6 is a Saturday. Therefore 1 October, being a Sunday, the next two Wednesdays are 4 and 11 October.

The second Wednesday is therefore 11 October.

Example 3

A Scots night-watchman works three nights on, followed by three nights off, throughout the year. Sunday 8 October is the first of three nights off. Will he be free on New Year's Eve?

The watchman works a cycle of 6 nights. Three off duty, followed by three nights on duty.

Taking 8 October as zero night, we see that he is off duty on day numbers which are congruent to 0, 1, 2 (mod 6). There are 23 nights left in October, 30 nights in November, and 31 nights in December, so that on 8 October, New Year's Eve is 84 nights away.

But $84 \equiv 0$ (mod 6). He will be off duty. Furthermore, $84 \equiv 0$ (mod 7), which shows that New Year's Eve is also on a Sunday.

Exercise 6.6

1. Using the table of Fig. 6.6, simplify the following:

(*a*) H $\otimes$ R   (*b*) R $\otimes$ V   (*c*) H $\otimes$ R $\otimes$ V
(*d*) V $\otimes$ H $\otimes$ R   (*e*) R $\otimes$ R $\otimes$ H   (*f*) H $\otimes$ H $\otimes$ V

2. Which element is the inverse of: (*a*) V; (*b*) R?

3. Is $\otimes$ an operation?

4. Figs. 6.9 and 6.10 show the union ($\cup$) and intersection ($\cap$) tables for the set of elements M = {A, B, C, D}.

    (*a*) Is the set closed with respect to $\cup$ and $\cap$?
    (*b*) Are $\cup$ and $\cap$, operations in the set M?
    (*c*) Are $\cup$ and $\cap$ commutative?
    (*d*) Which is the identity element for $\cup$?
    (*e*) Which is the identity element for $\cap$?
    (*f*) Which elements have inverses for $\cup$ and what are these inverses?
    (*g*) Which elements have inverses for $\cap$ and what are the inverses?

U	A	B	C	D
A	A	A	A	A
B	A	B	A	B
C	A	A	C	C
D	A	B	C	D

Fig. 6.9

∩	A	B	C	D
A	A	B	C	D
B	B	B	D	D
C	C	D	C	D
D	D	D	D	D

Fig. 6.10

5. Fig. 6.11 shows a piece of paper cut in the form of a rhombus with a piece cut from one side in order to identify the final positions more easily.

Obtain the circular multiplication table for the set of transformations T = {I, H, V, R} defined in the example above for the rectangle of Fig. 6.7.

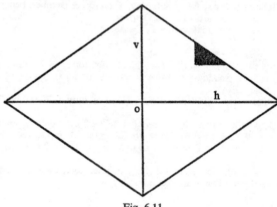

Fig. 6.11

6. If 14 August is a Monday, which day of the week is 9 October?

7. If Boxing Day one year falls on a Monday, on which day will it fall the next year if

    (i) the next year is a leap year?
    (ii) the next year is not a leap year?

8. Two friends do shift work throughout the 7-day week. One works 3 nights on and 3 nights off, the other works 2 nights on followed by 2 nights off. If they started work on exactly the same day, Monday 28 August, and the Rugby Test Match takes place on 4 November, will they both be able to go? Will they both be able to see the local match on 11 November?

GROUPS

The mathematical systems which have been discussed so far have arisen from a variety of situations, yet they have all had similar properties. For example, the set of transformations on the rectangle of Fig. 6.7 and the rhombus of Fig. 6.11 produced exactly the same table for the operation ⊗. All the addition tables for modular arithmetic had common properties, even though the tables were not the same size, It would be an advantage if we could discover the common properties of these mathematical systems and identify them under a common name. We use the name 'group' and give it the following definition:

Definition 6.4

A **group** is a set of elements, G, together with one operation, denoted by ○, which satisfy the following conditions:

(1) The set G is closed under the operation ○.

(2) The associative law holds in G so that if x, y, z are any elements of G then $x \bigcirc (y \bigcirc z) = (x \bigcirc y) \bigcirc z$.

(3) The identity element i for the operation ○ belongs to G such that if x is **any** element of G then $i \bigcirc x = x = x \bigcirc i$.

(4) Every element x of G has an inverse x^{-1} with respect to ○ such that $x \bigcirc x^{-1} = i = x^{-1} \bigcirc x$.

Let us examine Table Fig. 6.6 again. Does the set T = {I, H, V, R} together with the operation ⊗ constitute a group? On page 118 when the table

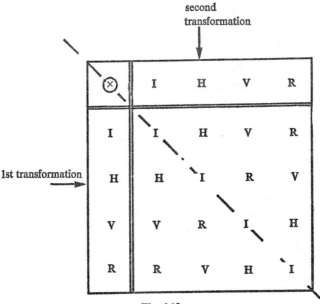

Fig. 6.12

was introduced we noticed the group properties (1), (3), (4) at once and we can quite easily evaluate all possible products to show group property (2) holds, so that the answer to our question is 'yes'.

Examination of the table of Fig. 6.9 in Exercise 6.6 (repeated below) reveals:

    1. Closure with respect to the operation ∪.
    2. Associativity with respect to the operation ∪ (by trying all possible combinations).
    3. An identity element, D.

Unfortunately only D has an inverse, so that we are unable to satisfy condition (4) for a group, and consequently the set {A, B, C, D} does not form a group under the operation ∪.

∪	A	B	C	D
A	A	A	A	A
B	A	B	A	B
C	A	A	C	C
D	A	B	C	D

Fig. 6.13

ISOMORPHISM

Consider the table of Fig. 6.14, which is the multiplication table in the set {1, −1}. We notice straight away that

    (1) the set is closed under the operation ×;
    (2) the set is associative under the operation ×;
    (3) the identity element is 1;
    (4) each element has an inverse, i.e. the inverse of 1 is 1 and the inverse of −1 is −1.

Consequently, the set {1, −1} is a group under the operation ×.
Now compare this with Fig. 6.15. There is a strong similarity between these two tables.

In fact, if we substitute 1 for I and -1 for H and change the operation from $\times$ to $\otimes$ we obtain exactly the same table. Consequently, we may say straight away that the set {I, H} forms a group under the operation of $\otimes$, although of course we could have deduced this by examining the table to see if it satisfied the four conditions for being a group in the first place. When

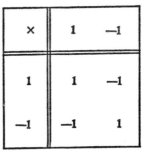

Fig. 6.14

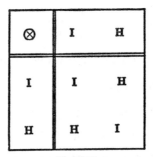

Fig. 6.15

this situation occurs we say that the two groups are **isomorphic**. More precisely we have

Definition 6.5

Two groups   G = {A, B, C, D, . . .} operation $\times$
                T = {P, Q, R, S, . . .} operation $\otimes$

are said to be *isomorphic* if a one-to-one correspondence can be established between their respective elements such that

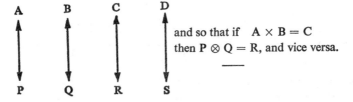

and so that if   A $\times$ B = C
then P $\otimes$ Q = R, and vice versa.

The addition table for arithmetic (modulo 4) was

+	0	1	2	3
0	0	1	2	3
1	1	2	3	0
2	2	3	0	1
3	3	0	1	2

Fig. 6.16

and the set {0, 1, 2, 3} forms a group under the operation of addition arithmetic (modulo 4).

Now consider the following table:

×	A	B	C	D
A	A	B	C	D
B	B	C	D	A
C	C	D	A	B
D	D	A	B	C

Fig. 6.17

Comparing Fig. 6.17 with Fig. 6.16 we see that we may establish a one-to-one correspondence between {A, B, C, D} and {0, 1, 2, 3}, so that if A × B = B, then $0 + 1 = 1$, etc. Consequently, not only does the set {A, B, C, D} form a group under $\times$ but the two groups are isomorphic.

We see therefore that though isomorphic groups may have different elements and different operations, they do have the same structure.

SUBGROUPS

We have already seen that the set {I, H} under operation $\otimes$ forms a group (see Fig. 6.15), and the set {I, H, V, R} under operation $\otimes$ also forms a group (see Fig. 6.12). In this case we call the group {I, H}, a **subgroup** of {I, H, V, R} under the operation of $\otimes$.

But do note that {I} is a group as well. However, as subgroups of {I, H, V, R} we regard {I} and {I, H, V, R} as **trivial** or **improper** subgroups, and consequently define a proper subgroup as being a subgroup other than:

(i) the entire group itself;
(ii) the group of the identity element only.

The subgroups are not always so easily found. For example, in Fig. 6.4 the set {0, 3} forms a subgroup. This is clearer when we detach this part of the table as seen in Fig. 6.18. Of course the one thing to always look for is the identity element, because this must belong to every subgroup.

+	0	3
0	0	3
3	3	0

Fig. 6.18

Exercise 6.7

1. In arithmetic (modulo 6) the set of elements {0, 1, 2, 3, 4, 5} forms a group under + (see Fig. 6.4). There are two proper subgroups of 2 and 3 elements respectively. Identify them.

2. In the table of Fig. 6.12, identify each proper subgroup and make a table for each one.

3. You should be able to obtain three proper subgroups in Question 2. Are they isomorphic?

4. Is the set {0, 1, 2, 3, 4} a group under + in arithmetic (modulo 5)?

5. Is the set {0, 1, 2, 3, 4} a group under $\times$ in arithmetic (modulo 5)? (See Fig. 6.3.)

6. Is the set {1, 2, 3, 4} a group under $\times$ in arithmetic (modulo 5)? If so make tables for all its proper subgroups.

7. The three tables (i), (ii), (iii) show the results of the operations of ①, ⊙, and ⊖ performed in the respective sets {A, B, C, D}; {P, Q, R, S}; {1, 2, 3, 4}.

(a) Does {A, B, C, D} form a group under the operation of ①?
(b) Does {P, Q, R, S} form a group under the operation of ⊙?
(c) Does {1, 2, 3, 4} form a group under the operation of ⊖?
(d) Give the identity elements for each table.
(e) Is there an isomorphism?
(f) Name the inverse of each element in (i) and (ii).
(g) Make a table for the proper subgroups in (i).

①	A	B	C	D
A	A	B	C	D
B	B	D	A	C
C	C	A	D	B
D	D	C	B	A

(i)

⊙	P	Q	R	S
P	P	Q	R	S
Q	Q	S	P	R
R	R	P	S	Q
S	S	R	Q	P

(ii)

Θ	1	2	3	4
1	4	3	2	1
2	1	4	3	2
3	2	1	4	3
4	3	2	1	4

(iii)

CHAPTER SEVEN

MULTIBASE ARITHMETIC

CONVERSION TO BASE TEN

Our everyday arithmetic is the decimal system based on the number ten. We therefore employ ten digits 0, 1, 2, 3, 4, 5, 6, 7, 8, 9. In order to represent larger numbers than 9; 99; 999, etc., we have to invent the idea of tens, hundreds, and thousands, and allocate place values to digits which take up different positions in the complete written number. For example, the number written 769 is taken to be seven hundreds, six tens, and nine units, and represents the result of $700 + 60 + 9$. If we employ the decimal point we agree that this will indicate the separation of the whole-number part from the fractional part in tenths, hundredths, and so on, so that a number written as $54 \cdot 67$ has the meaning $50 + 4 + \frac{6}{10} + \frac{7}{100}$. Clearly we may spread out in both directions from the point as the table of Fig. 7.1 shows.

tenthousands	thousands	hundreds	tens	units	tenths	hundredths	thousandths
10^4	10^3	10^2	10^1	10^0	10^{-1}	10^{-2}	10^{-3}
10000	1000	100	10	1	$\frac{1}{10}$	$\frac{1}{100}$	$\frac{1}{1000}$

Fig. 7.1

What is important at this moment is to realize the connexion between the adjacent columns. Starting from the unit column and proceeding left, we multiply by ten to enter each new column. Hence the columns are headed:

$$\begin{aligned} \text{tens} \quad & 10 & = 10^1 \\ \text{hundreds} \quad & 10 \times 10 & = 10^2 \\ \text{thousands} \quad & 10 \times 10 \times 10 & = 10^3 \\ \text{ten thousands} \quad & 10 \times 10 \times 10 \times 10 & = 10^4 \end{aligned}$$

If we now retrace our steps we should observe the pattern 10^4, 10^3, 10^2, 10^1, and be able to suggest the headings for the unit column as 10^0, the tenths column as 10^{-1}, and so on, as indicated in Fig. 7.1.

The base for all our numbers, which are called **denary numbers**, has so far been the number ten. Suppose now we were to change this base to three, say. The new numbers are called **ternary numbers**, and we have to restrict ourselves to using the digits 0, 1, 2 only, so the denary headings for our new representation table are as indicated in Fig. 7.2.

The two numbers which have been entered in Fig. 7.2 are numbers to base 3. Notice that the fractional part is represented in terms of thirds, ninths, twenty-sevenths, and so on.

Through constant practice, we feel more at home in the scale of ten, so that we shall probably want to translate numbers in other bases back into the denary system. In fact, this is a very good reason for studying other systems, since a conscious effort to translate will ensure that we shall gradually come to understand our denary system a little better.

twentysevens	nines	threes	units	thirds	ninths	twentysevenths
3^3	3^2	3^1	3^0	3^{-1}	3^{-2}	3^{-3}
27	9	3	1	$\frac{1}{3}$	$\frac{1}{9}$	$\frac{1}{27}$
1	2	1	0			
	2	0	2	2	1	

Fig. 7.2

In order to distinguish the numbers in different bases, we shall use a subscript notation. For example,

1. 44_{10} reads 'four four to base ten', i.e. the usual forty-four.
2. 21_3 reads 'two one to base three'. (*Note:* It is not read as twenty-one because this is the wording we use for the denary numbers.)
3. 37_8 reads 'three seven to base eight'.
4. 64_{13} reads 'six four to base thirteen'. (*Note:* The base subscript is always taken as a denary number.)

Looking again at the numbers entered in Fig. 7.2, the first number is 1210_3 read as 'one, two, one, zero to base three'.

To obtain the size of this number in the denary scale we look at the headings of each column to see that the number is equal to

$$(3^3 \times 1) + (3^2 \times 2) + (3^1 \times 1) + (3^0 \times 0)$$
$$= 27 + 18 + 3 + 0 \text{ (all digits denary)}$$
$$= 48$$

We could, of course, write this last result as 48_{10}.

Thus, $1210_3 = 48_{10}$

The second number entered into Fig. 7.2 was 202·21. Notice how we still use the point (·) to separate the whole number from the fractional part. We translate this number into the denary scale as before, simply by observing the headings of the columns of the table. Therefore,

$$202 \cdot 21_8 = (3^2 \times 2) + (3^1 \times 0) + (3^0 \times 2) + (3^{-1} \times 2) + (3^{-2} \times 1)$$
$$= (9 \times 2) + (3 \times 0) + (1 \times 2) + (\tfrac{1}{3} \times 2) + (\tfrac{1}{9} \times 1)$$
$$= 18 + 0 + 2 + \tfrac{2}{3} + \tfrac{1}{9}$$
$$= 20 + \tfrac{2}{3} + \tfrac{1}{9}$$
$$= 20 + \tfrac{6}{9} + \tfrac{1}{9}$$
$$= 20\tfrac{7}{9}$$
$$\therefore \quad 202 \cdot 21_8 = 20\tfrac{7}{9}{}_{10}$$

We already know that $\tfrac{1}{9}$ is represented by the recurring decimal $0 \cdot \dot{1}$, so that we may progress a little farther by writing

$$202 \cdot 21_8 = 20 \cdot \dot{7}_{10}$$

Suppose we now consider numbers to base eight. We start by constructing the reference table Fig. 7.3 in the now familiar manner.

8^4	8^3	8^2	8^1	8^0	8^{-1}	8^{-2}	8^{-3}
4096	512	64	8	1	$\tfrac{1}{8}$	$\tfrac{1}{64}$	$\tfrac{1}{512}$
		6	4	5	7		
2	1	7	6	5	4	3	

Fig. 7.3

The first number $645 \cdot 7_8 = (8^2 \times 6) + (8^1 \times 4) + (8^0 \times 5) + (8^{-1} \times 7)$
$$= 384 + 32 + 5 + \tfrac{7}{8}$$
$$= 421\tfrac{7}{8}$$
$$645 \cdot 7_8 = 421 \cdot 875_{10}$$

The second number of Fig. 7.3 is given by

$$21\,765 \cdot 43_8 = (8^4 \times 2) + (8^3 \times 1) + (8^2 \times 7) + (8^1 \times 6) + (8^0 \times 5) +$$
$$(8^{-1} \times 4) + (8^{-2} \times 3)$$
$$= 8\,192 + 512 + 448 + 48 + 5 + \tfrac{4}{8} + \tfrac{3}{64}$$
$$= 9\,205 + \tfrac{4}{8} + \tfrac{3}{64}$$
$$= 9\,205 + 0 \cdot 5 + 0 \cdot 046875$$
$$21\,765 \cdot 43_8 = 9\,205 \cdot 546875_{10}$$

By this time we are beginning to notice that the conversion of the fractional part is becoming a little tedious. What started out as a number to two fractional places (0.43_8) has now become a number to six fractional places (0.546875_{10}). It would appear at first sight that things might be easier if we concentrated on using a smaller base. Let us try the base of two. The numbers

2^4	2^3	2^2	2^1	2^0	2^{-1}	2^{-2}	2^{-3}	2^{-4}	2^{-5}
16	8	4	2	1	$\frac{1}{2}$	$\frac{1}{4}$	$\frac{1}{8}$	$\frac{1}{16}$	$\frac{1}{32}$
1	1	0	1	1	1				
1	0	0	0	1	0	1	1		

Fig. 7.4

to base two are called **binary numbers** and only involve the use of two digits 0 and 1. The reference table is Fig. 7.4, and the first number is given by the same method as before to be

$$11011 \cdot 1_2 = 16 + 8 + 0 + 2 + 1 + \tfrac{1}{2}$$
$$= 27 \cdot 5_{10}$$

and the second number is

$$10001 \cdot 011_2 = 16 + 0 + 0 + 0 + 1 + 0 + \tfrac{1}{4} + \tfrac{1}{8}$$
$$= 17\tfrac{3}{8}$$
$$= 17 \cdot 375_{10}$$

The conversion appears to be easier, but this is deceiving, because the numbers we have chosen are not very difficult. Suppose we had considered the $\frac{3}{64}$ of Fig. 7.3 in the binary scale? We need to write

$$\tfrac{3}{64} = \tfrac{1}{32} + \tfrac{1}{64}$$
$$= 0 \cdot 00001 + 0 \cdot 000001$$
$$= 0 \cdot 000011$$

So that $0 \cdot 3_8 = 0 \cdot 000011_2$, and we see that the difficulties of converting the fractional part of a number are not necessarily removed by choosing smaller bases.

Exercise 7.1

Convert the following numbers into denary numbers or fractions:

1. 12_3	2. 12_8	3. 21_3	4. 121_8
5. 27_8	6. 132_4	7. $112 \cdot 2_3$	8. $312 \cdot 3_4$
9. $13 \cdot 62_8$	10. $111 \cdot 11_2$	11. $413 \cdot 22_5$	12. $83 \cdot 17_9$

CONVERSION FROM BASE TEN

Let us now reverse the process and convert denary numbers into numbers with other different bases.

Consider converting 512_{10} to a number with base eight. Using the reference table of Fig. 7.3, we see that 512_{10} corresponds to 1 entered in the column headed 512 or 8^3. Therefore we have straight away that

$$512_{10} = 1000_8$$

Similarly
$$527_{10} = 512_{10} + 15_{10}$$
$$= 512_{10} + 8_{10} + 7_{10}$$
$$= 1017_8$$

This seems easy enough for the moment, now let us try to convert 1945_{10}.

Following the idea above, we remove 512 first and write

$$1945_{10} = 512_{10} + 1433_{10}$$

whereupon we see that this step may be repeated to yield

$$1945_{10} = 512_{10} + 512_{10} + 921_{10}$$

and then again $$1945_{10} = 512_{10} + 512_{10} + 512_{10} + 409_{10}$$

This, of course, was equivalent to dividing by 512 to get a quotient of 3 and a remainder of 409.

We now need to distribute 409 into the other columns, the first of which is headed 64.

Since $$409 = 64 \times 6 + 25$$
and $$25 = 8 \times 3 + 1$$

we are now able to write

$$1945_{10} = (512 \times 3) + (64 \times 6) + (8 \times 3) + 1$$
$$= (8^3 \times 3) + (8^2 \times 6) + (8^1 \times 3) + (8^0 \times 1)$$
$$= 3631_8$$

Unfortunately this is not a very convenient method because it involves division by large numbers such as 512 and 64.

It is better to return to the basic idea of finding the powers of eight in 1945_{10} simply by successive division by eight.

Consider this division:

Step (i) 8 | 1945

Step (ii) 8 | 243 remainder 1 $1945 = (8^1 \times 243) + (8^0 \times 1)$

Step (iii) 8 | 30 remainder 3 $1945 = (8^2 \times 30) + (8^1 \times 3) + (8^0 \times 1)$

Step (iv) 8 | 3 remainder 6 $1945 = (8^3 \times 3) + (8^2 \times 6) + (8^1 \times 3) +$
$$(8^0 \times 1)$$

0 remainder 3

Final result $$1945_{10} = 3631_8$$

Consider another conversion, 1945_{10} to base three.

The required division is as follows:

Information

Step				3^6	3^5	3^4	3^3	3^2	3^1	3^0
Step (i) 3	1 945									
Step (ii) 3	648 remainder 1								648	1
Step (iii) 3	216 remainder 0							216	0	1
Step (iv) 3	72 remainder 0						72	0	0	1
Step (v) 3	24 remainder 0					24	0	0	0	1
Step (vi) 3	8 remainder 0				8	0	0	0	0	1
Step (vii) 3	2 remainder 2		2	2	0	0	0	0	0	1
3	0 remainder 2		2	2	0	0	0	0	0	1

result, $1945_{10} = 2200001_3$. We can dispense with the information section and simply read the answer from the remainders.

We thus see that the conversion of any denary number using any base whatever is a simple process of successive division.

The next point to arise is how to convert a denary number which contains a fractional part.

To start with let us consider the fractional part in isolation by changing 0.53_{10} to a number in base three. If we are observant we ought to spot what to do by retracing the pattern of the last example.

Thus, to obtain the entry for the column headed 3^1 we divide by 3^1

to obtain the entry for the column headed 3^2 we divide by 3^2

to obtain the entry for the column headed 3^3 we divide by 3^3

the process being continued until each column contains a digit entry 0, 1, or 2.

Since the first column of the fractional part is 3^{-1} we divide by 3^{-1} then by 3^{-2} and so on. But dividing by 3^{-1} is the same as multiplying by the inverse of 3^{-1}, namely 3. Dividing by 3^{-2} is the same as multiplying by the inverse of 3^{-2}, namely 3^2 and so on. Perhaps the process will appear clearer from the next working:

Information

	0·53	3^{-1}	3^{-2}	3^{-3}	3^{-4}	3^{-5}
Step (i)	3					
	1⌋ ·59	1				
Step (ii)	3					
	1⌋ ·77	1	1			
Step (iii)	3					
	2⌋ ·3 1	1	1	2		
Step (iv)	3					
	0⌋ ·9 3	1	1	2	0	
Step (iv)	3					
	2⌋ ·7 9	1	1	2	0	2

Let us examine this a little more closely. We started with 0·53, and from this we extracted 3^{-1} as step (i) by writing

$$0·53 = \frac{3}{3} \times 0·53$$

$$= \frac{1·59}{3} = \frac{1 + 0·59}{3}$$

$$= \frac{1}{3} + \frac{0·59}{3} \quad \text{thus extracting the } \tfrac{1}{3}\text{'s (thirds)}$$

For step (ii) we extract $\frac{1}{9}$'s (ninths) from $\frac{0·59}{3}$ by writing

$$\frac{0·59}{3} = \frac{3}{3} \times \frac{0·59}{3}$$

$$= \frac{1·77}{9} = \frac{1 + 0·77}{9}$$

$$= \frac{1}{9} + \frac{0·77}{9}$$

For step (iii) we extract $\frac{1}{27}$'s from $\frac{0·77}{9}$ by writing

$$\frac{0·77}{9} = \frac{3}{3} \times \frac{0·77}{9}$$

$$= \frac{2·31}{27} = \frac{2 + 0·31}{27}$$

$$= \frac{2}{27} + \frac{0·31}{27}$$

Pausing at this stage to collect our thoughts, we can see that so far we have obtained

$$0·53 = \frac{1}{3} + \frac{1}{9} + \frac{2}{27} + \frac{0·31}{27}$$

and the remaining steps should now be clear.

Our result so far is $\qquad 0·53_{10} = 0·11202_3$ (5 places)

and we notice that we could carry on for some time yet.

Putting the last two results together, we are able to write

$$1945·53_{10} = 2200001·11202_3 \quad \text{(5 places)}$$

For a further example consider $77·65_{10}$ to be converted to base eight.

Taking the whole number part first we have

	8^2	8^1	8^0	8^{-1}	8^{-2}	8^{-3}	8^{-4}
Step (i) 8 | 77							
Step (ii) 8 |<u> 9 </u> remainder 5		9	5				
Step (iii) 8 |<u> 1 </u> remainder 1	1	1	5				
Step (iv) |<u> 0 </u> remainder 1	1	1	5				
For the fractional part							
0·65							
8							
Step (i) <u>5|</u> ·20				5			
8							
Step (ii) <u>1|</u> ·6				5	1		
8							
Step (iii) <u>4|</u> ·8				5	1	4	
8							
Step (iv) <u>6|</u> ·4				5	1	4	6

Combining the above results we have

$$77 \cdot 65_{10} = 115 \cdot 5146_8 \quad \text{(4 places)}$$

Exercise 7.2

Carry out the following conversions:

1. 127_{10} to base eight
2. 143_{10} to base three.
3. 97_{10} to base six.
4. 17_{10} to base two.
5. $0 \cdot 36_{10}$ to base eight (3 places).
6. $0 \cdot 41_{10}$ to base three (3 places).
7. $127 \cdot 36_{10}$ to base eight (3 places).
8. $143 \cdot 41_{10}$ to base three (3 places).
9. $17 \cdot 75_{10}$ to base two.
10. $17 \cdot 75_{10}$ to base four.

ROUNDING OFF

Consider the denary number 7·493. This is a number given to three places of decimals or simply three places after the point. Sometimes we are not interested in such accuracy, and consequently we give the number to fewer decimal places (i.e. round off). For example, to quote a denary number to two places we must examine the digit in the third place. If this is 5 or more, then we add 1 to the digit in the second place and then quote the result.

Thus   7·935   becomes   7·94   (2 places)

         9·297   becomes   9·30   (2 places)

         0·145   becomes   0·15   (2 places)

If the digit under inspection is less than 5 we quote the result without any addition.

Thus   7·943   becomes   7·94   (2 places)

         9·293   becomes   9·29   (2 places)

The same procedure applies to any place in the denary number.

Thus   4·2946   becomes   4·295   (3 places)

                        or   4·29   (2 places)

Notice that in obtaining this last result we quote from the original number and *not* from 4·295 (3 places).

4·2946   becomes   4·3   (1 place)

The problem now arises as to how we offer similar approximations for numbers in other bases.

Consider the number $2·414_8$ to be quoted (2 places).

Suppose we quote the result as $2·41_8$ (2 places)? This means we have lost $0·004_8$. Had we suggested $2·42_8$ (2 places), it would have meant a gain of $0·002_8$, so clearly this second suggestion is more accurate.

Suppose the number had been $2·513_6$? A suggested result of $2·51_6$ (2 places) has lost $0·003_6$ and a suggested result of $2·52_6$ (2 places) has gained $0·003_6$. Clearly we have to decide which result will be taken. So, to be consistent with our treatment of denary numbers, we choose $2·52_6$ (2 places).

Similarly, of course, for numbers to base eight $2·524_8$ becomes $2·53_8$ (2 places). In other words, if the digit under inspection is equal to or greater than half the base of the number we add 1 to the digit in the previous place.

Thus $2·3464_8$ becomes $2·347_8$ (3 places) or $2·4_8$ (1 place)

$3·22_4$ becomes $3·3_4$ (1 place)

$1·0101_2$ becomes $1·011_2$ (3 places) or $1·1_2$ (1 place)

$1·101_2$ becomes $1·11_2$ (2 places) or $1·1_2$ (1 place)

This has been easy to put into practice for bases which are an even number. But when the base is an odd number we have to inspect two digits rather than one.

For example, $0·245_9$ quoted as $0·2_9$ (1 place) means a loss of $0·045_9$, but quoted as $0·3_9$ (1 place) means a gain of $0·044_9$ and therefore a better approximation, because this quotation is nearer the original.

We therefore have the following rules:

For base nine if the next two digits form a number equal to or greater than 45 we add 1 to the digit in the previous place.

Thus $0·645_9$ becomes $0·7_9$ (1 place) and $0·65_9$ (2 places)

again $1·2736_9$ becomes $1·274_9$ (3 places)

or $1·27_9$ (2 places)

or $1·3_9$ (1 place)

For numbers to base seven if the next two digits form a number equal to or greater than 34.

For numbers in base five, if the next two digits form a number greater than or equal to 23.

For numbers in base three if the next two digits form a number greater than or equal to 12.

Thus $1 \cdot 112_3$ becomes $1 \cdot 2_3$ (1 place) but $1 \cdot 111_3$ becomes $1 \cdot 1_3$ (1 place)

$4 \cdot 123_5$ becomes $4 \cdot 2_5$ (1 place) but $4 \cdot 122_5$ becomes $4 \cdot 1_5$ (1 place)

$6 \cdot 534_7$ becomes $6 \cdot 6_7$ (1 place) but $6 \cdot 533_7$ becomes $6 \cdot 5_7$ (1 place)

$3 \cdot 245_9$ becomes $3 \cdot 3_9$ (1 place) but $3 \cdot 244_9$ becomes $3 \cdot 2_9$ (1 place)

Exercise 7.3

Quote the following numbers to the required places:

1. $2 \cdot 123_4$ (2, 1) places
2. $1 \cdot 1101_2$ (3, 2) places
3. $4 \cdot 654_8$ (2, 1) ,,
4. $4 \cdot 223_5$ (2, 1) ,,
5. $1 \cdot 212_3$ (2, 1) ,,
6. $2 \cdot 643_7$ (1) ,,
7. $4 \cdot 2445_9$ (2, 1) ,,
8. $1 \cdot 1223_5$ (2, 1) ,,
9. $5 \cdot 2334_7$ (2, 1) ,,
10. $5 \cdot 2334_6$ (2, 1) ,,

ADDITION

The operations of $+$, $-$, $\times$, $\div$ with numbers of other bases makes an interesting revision of basic ideas about the denary system.

Example 1

Consider adding 543_8 to 637_8

```
   6 3 7
   5 4 3
  ------
 1 4 0 2
   1 1
```

As always in addition we start with the first column on the right, and each time we acquire 8 in any column we represent it by carrying 1 in to the next column.

Step (i) Thus $7 + 3 = 10_{10} = 12_8$ we register 2 in the answer and carry the 1 in the usual manner;

Step (ii) $3 + 4 + 1 = 8_{10} = 10_8$ we register 0 in the answer and carry the 1 in to the next column;

Step (iii) $6 + 5 + 1 = 12_{10} = 14_8$ we register 4 in the answer and carry the 1 to the next column.

Our result is $543_8 + 637_8 = 1402_8$

Example 2

Find the sum of 645_7, 454_7, and 366_7.

Using the usual layout we write

```
   6 4 5
   4 5 4
   3 6 6
  ------
 2 1 3 1
   2 2 2
```

Step (i) $5 + 4 + 6 = 15_{10} = 21_7$. Register 1 and carry 2;

Step (ii) $4 + 5 + 6 + 2 = 17_{10} = 23_7$. Register 3 and carry 2;

Step (iii) $6 + 4 + 3 + 2 = 15_{10} = 21_7$. Register 1 and carry 2.

Our result is $645_7 + 454_7 + 366_7 = 2131_7$

SUBTRACTION

Study the following example:

Example 1

Determine $254_8 - 146_8$.

A vertical layout of the problem is

2	5	4
−1	4	6

Step (i) decompose the upper number to yield

2	5^4	14
−1	4	6
		6

We can now perform the first step $14 - 6 = 6$. (Remember we are working in base eight.)

Step (ii) complete the subtraction

2	4	14
−1	4	6
1	0	6

Ans.

Example 2

$$432_5 - 243_5$$

Step (i) Decompose the first number to yield

4	3^2	12
−2	4	3
		4

We can now perform the first step $12 - 3 = 4$. (We are now working in base five.)

Step (ii) Decompose the number further to yield the second step $12 - 4 = 3$

4^3	12	12
2	4	3
1	3	4

Ans.

Step (iii) Complete the subtraction.

On inspection the reader will see that the method is exactly the same as that for subtraction of denary numbers; all we have to remember is which base we are working in.

Exercise 7.4

1. Evaluate $14 - 5$ in bases 10, 9, 8, 7, 6.
2. Evaluate $13 - 7$ in bases 10, 9, 8.
3. Subtract 157_8 from 233_8.
4. Subtract 475_9 from 834_9.
5. Subtract 155_6 from 414_6.
6. Find the sum of 121_3, 221_3, 102_3.
7. Find the sum of 452_6, 501_6, 343_6.
8. Which is the larger number, 103_6 or 47_8?
9. Subtract 231_4 from 142_5 giving your answer as a denary number.
10. Subtract 34_8 from 31_9.

MULTIPLICATION

We could, of course, carry out multiplication as successive addition, but this would become tedious with the use of larger numbers. The only alternative is to learn new multiplication tables, such as those for base 4, say.

$1 \times 1 = 1$	$1 \times 2 = 2$	$1 \times 3 = 3$
$2 \times 1 = 2$	$2 \times 2 = 10$	$2 \times 3 = 12$
$3 \times 1 = 3$	$3 \times 2 = 12$	$3 \times 3 = 21$
$10 \times 1 = 10$	$10 \times 2 = 20$	$10 \times 3 = 30$
$11 \times 1 = 11$	$11 \times 2 = 22$	$11 \times 3 = 33$
$12 \times 1 = 12$	$12 \times 2 = 30$	$12 \times 3 = 102$
$13 \times 1 = 13$	$13 \times 2 = 32$	$13 \times 3 = 111$
$20 \times 1 = 20$	$20 \times 2 = 100$	$20 \times 3 = 120$

etc.

This really isn't a profitable exercise; it is a better idea to take each multiplication as it comes by working in denary and then translating into the required base.

Example 1

Multiply 352_6 by 4.

Step (i) $2 \times 4 = 8_{10} = 12_6$; register 2 and carry the 1 into the next column.

$$\begin{array}{r} 352 \\ \times \ \ 4 \\ \hline 2 \\ \hline 1 \end{array}$$

Step (ii) $5 \times 4 = 20_{10} = 32_6$; add the 1 carried over and we get 33; register 3 and carry 3 into the next column.

$$\begin{array}{r} 352 \\ \times \ \ 4 \\ \hline 32 \\ \hline 31 \end{array}$$

Step (iii) $3 \times 4 = 12_{10} = 20_6$; add the 3 carried over from step (ii) and we get 23.

$$\begin{array}{r} 352 \\ \times \ \ 4 \\ \hline 2332 \\ \hline 31 \end{array}$$

The complete result is $352_6 \times 4 = 2332_6$

Apart from the translation of each intermediate result, the method of multiplication is exactly the same as for denary numbers. Recall multiplication by 10 in denary numbers, i.e. $43_{10} \times 10_{10} = 430_{10}$, usually described as 'adding a nought on the end'. Suppose now that 43 was in base 5, 6, 7, 8, 9, would the answer have the same form?

Is $43_6 \times 10_6 = 430_6$? $43_7 \times 10_7 = 430_7$ and so on?
Well $3 \times 10 = 10 \times 3 = 30$ in any base greater than 3, therefore we have

Step (i) $10_6 \times 3 = 30_6$; register the 0 and carry 3 into the next column.
Step (ii) $10_6 \times 4 = 40_6$; add the 3 already carried over and we get 43_6.

The complete result is

$$43_6$$
$$10_6$$
$$\overline{430}$$
$$\overline{3}$$

or $$43_6 \times 10_6 = 430_6$$

More precisely for those who are a little more adept at algebra, if 43 is in base b then $43_b \times 10_b$ is equivalent to

$$[(b^1 \times 4) + (b^0 \times 3)] \times b \text{ [in denary numbers]}$$

Using the distributive laws this becomes

$$(b^2 \times 4) + (b^1 \times 3) \text{ which is } 430_b$$
$$43_b \times 10_b = 430_b$$

Furthermore, $$10_b \times 10_b = 100_b$$
and since $$43_b \times 100_b = 43_b \times 10_b \times 10_b$$
$$= (43_b \times 10_b) \times 10_b$$
$$= 430_b \times 10_b$$
$$= 4300_b$$

We see that multiplication by 100 in any base may be carried out in the same manner as for denary numbers, i.e. by attaching two noughts.

Example 2

Multiply 437_8 by 152_8.

		4	3	7
		1	5	2

Step (i) $437_8 \times 2 = 1076_8$
Step (ii) $437_8 \times 50_8 = (437_8 \times 5_8)10_8$
$$= 2633_8 \times 10_8$$
$$= 26330_8$$
Step (iii) $437_8 \times 100_8 = 43700_8$

		1	0	7	6
2	6	3	3	0	
4	3	7	0	0	

Total

7	3	3	2	6
1	1	1		

$$437_8 \times 152_8 = 73326_8 \quad \text{Ans.}$$

Of course binary numbers are the easiest to work with because 0 and 1 are the only figures involved. A look at the next example may convince the reader.

Example 3

Multiply 11011_2 by 11_2. We shall just set out the solution without the separate steps.

```
        11011
    ×      11
    ─────────
        11011    Multiplication by 1
       110110    Multiplication by 10
    ─────────
      1010001
```

$$\therefore \quad 11011_2 \times 11_2 = 1010001_2 \quad \text{Ans.}$$

Exercise 7.5

Carry out the following multiplications:

1. 152×100 in bases 10, 9, 8, 7, 6
2. $152_6 \times 103_6$
3. 11×11 in base 10, 9, 8, 7, 6, 5
4. 12×12 in base 10, 9, 8, 7, 6, 5
5. $111_2 \times 101_2$
6. $143_7 \times 105_7$
7. $1 \cdot 43_7 \times 10_7$
8. $1 \cdot 52_6 \times 10_6$
9. $1 \cdot 52_6 \times 10 \cdot 3_6$
10. $1 \cdot 43_7 \times 1 \cdot 1_7$

DIVISION

This is the most difficult operation in unfamiliar bases so we shall start by recalling simple division with denary numbers.

Example 1

Divide 265 by 3.

The usual method reads:
(i) $26 \div 3$ is approximately 8;
(ii) $8 \times 3 = 24$; subtract 24 from 26 then bring down the 5;
(iii) $25 \div 3$ is approximately 8;
(iv) $8 \times 3 = 24$; subtract 24 from 25.

$$\begin{array}{r} 88 \\ 3)\overline{265} \\ 24 \\ \hline 25 \\ 24 \\ \hline 1 \end{array}$$

The final result is 88 remainder 1

Now let us try the same example in base eight.

(i) $26_8 \div 3$ is approximately ?. This is where the trouble starts, because we haven't a very good repertoire of tables available. The best plan is to convert 26_8 to 22_{10} so that we are on more familiar ground. We can now say that $26_8 \div 3$ is approximately 7;
(ii) $7 \times 3 = 25_8$, subtract 25 from 26 then bring down the 5 as before.

$$\begin{array}{r} 74 \\ 3)\overline{265} \\ 25 \\ \hline 15 \\ 14 \\ \hline 1 \end{array}$$

(iii) $15_8 = 13_{10}$

$\therefore$ $15_8 \div 3$ is approximately 4

$4 \times 3 = 14_8$, subtract 14 from 15

The final result is 74_8 remainder 1.

Example 2

Divide 1432_7 by 24_7.
Before starting, let us make a multiplication table for 24.

Thus, $24_7 \times 1 = 24_7$
$24_7 \times 2 = 51_7$
$24_7 \times 3 = 105_7$ each of these results being obtained by a successive
$24_7 \times 4 = 132_7$ addition of 24_7
$24_7 \times 5 = 156_7$
$24_7 \times 6 = 213_7$

Step (i) $143_7 \div 24_7$ is approximately 4;

Step (ii) $24_7 \times 4 = 132_7$; subtract 132 from 143 and bring down 2;

Step (iii) $112_7 \div 24_7$ is approximately 3;

Step (iv) $24_7 \times 3 = 105_7$; subtract 105 from 112.

```
          43
24)1 432
    132
    ───
    112
    105
    ───
      4
```

The final result is 43_7 remainder 4.

Example 3

Divide 1335_6 by 35_6.
We start by making a useful multiplication table for 35_6

$35_6 \times 1 = 35_6$
$35_6 \times 2 = 114_6$ each of these results being obtained by a successive addition
$35_6 \times 3 = 153_6$ of 35_6.
$35_6 \times 4 = 232_6$
$35_6 \times 5 = 311_6$

Step (i) $133_6 \div 35_6$ is approximately 2;

Step (ii) $35_6 \times 2 = 114_6$; subtract 114 from 133 and bring down the 5;

Step (iii) $155_6 \div 35_6$ is approximately 3;

Step (iv) $35_6 \times 3 = 153_6$; subtract 153 from 155 and complete.

```
          23
35)1335
    114
    ───
    155
    153
    ───
      2
```

Final result is 23_6 remainder 2.

Exercise 7.6

Carry out the following divisions:

1. $366_7 \div 3$ 2. $448_9 \div 4$ 3. $120_5 \div 12_5$ 4. $240_6 \div 12_6$
5. $121_4 \div 11_4$ 6. $232_7 \div 15_7$ 7. $2\,430_6 \div 35_6$ 8. $3\,202_6 \div 35_6$
9. $6\,405_7 \div 24_7$ 10. $4\,612_7 \div 24_7$

11. If all the following numbers are to base 5, cancel the expressions down to the simplest form.

(a) $\frac{41}{12} \times \frac{10}{22} \times \frac{4}{30}$ (b) $\frac{31}{3} \times \frac{12}{4} \times \frac{14}{24}$

In the main our study of multibase arithmetic has served its purpose of making us reconcentrate on the basic methods we employ in our operations with denary numbers.

It would be reasonable to pursue the conversion of such expressions as $11_2{}^{11_2}$ or $\sqrt{31_5}$ but not really necessary to any increase in knowledge. Incidentally $11_2{}^{11_2}$ becomes 3^3 in denary numbers, i.e. 27, while $\sqrt{31_5}$ becomes $\sqrt{16} = 4$ in denary numbers.

Let us instead examine some of the more popular outlets for such work.

Examining the reference table for the binary numbers (base 2)

2^5	2^4	2^3	2^2	2^1	2^0
32	16	8	4	2	1

Fig. 7.5

We notice that with the use of 8, 4, 2, and 1 we can register binary numbers up to the value of fifteen using 0 and 1 only. The numbers are simply, 1, 10, 11, 100, 101, 110, 111, 1000, 1001, 1010, 1011, 1100, 1101, 1110, 1111.

It follows therefore that a shopkeeper could weigh up to 15 g using a set of four masses 8 g, 4 g, 2 g, 1 g, only (i.e. a whole number of grammes each time).

For example, 13 g = 8 g + 4 g + 1 g is symbolized by the binary number 1101.

Exercise 7.7

1. What would be the smallest number of masses which would enable the shopkeeper to weigh up to 31 g and 63 g?
2. Which masses would be used for weighing a 28 g parcel?
3. What mass is symbolized by the numbers 1011_2 and 1001_2?

We can make an interesting play with ternary numbers (base 3) in this way:

Suppose we try weighing articles of a whole number of grammes by using masses of 27 g, 9 g, 3 g, 1 g. We soon see that unlike the binary weighing above, some masses are unobtainable by the ordinary process of placing masses in one pan scale only. For example, 2, 5, 6, 7, etc. But suppose we permit (always possible) the masses to be placed in either scale, then 9 g on one side and 3 g + 1 g on the other will permit a mass of 5 g to be obtained.

Other examples are
$$35 = 27 + 9 - 1$$
$$22 = 27 + 3 + 1 - 9, \text{ etc.}$$

This is really a problem in ternary numbers where we are denied the opportunity of using the number 2 because we have only one mass of each size available. Thus, using Fig. 7.2, we see that

$$35_{10} = 1022_3$$
and
$$22_{10} = 211_3$$

This means that weighing by the use of one side only would involve one 27 g, two 3 g, two 1 g masses, but we are only allowed one of each.

But we can now see straight away that $1022_3 + 1 = 1100_3$

$$\therefore \quad 1022_3 = 1100_3 - 1$$

The right-hand side shows the mass of 35 g obtained by using both sides of the machine but only one each of the masses 27 g, 9 g, 1 g.

Similarly,
$$211_3 + 100_3 = 1011_3$$
$$\therefore \quad 211_3 = 1011_3 - 100_3$$

indicating that we may weigh 22 g by choosing 27 g + 3 g + 1 g − 9 g.

Let us now add an 81-g mass to the set and try to weigh a 59-g parcel

$$59_{10} = (27 \times 2) + (3 \times 1) + 2$$
$$= 2\ 0\ 1\ 2_3$$

As the reader is probably beginning to realize, the method is to get rid of $2s$ by substituting $10_3 - 1$

Thus
$$2012_3 = 10012_3 - 1000_3$$
$$= 10020_3 - 1001_3$$
$$= 10100_3 - 1011_3$$

So that a parcel of mass 59 g is weighed by $81 + 9$ on one side and $27 + 3 + 1$ on the other.

Exercise 7.7 (*continued*)

4. It is possible to weigh whole numbers of grammes from 1 g to 40 g by using one or more of the masses 27 g, 9 g, 3 g, 1 g, providing that where necessary one mass or more can be placed on each scale pan.

Show how to weigh parcels of the following masses: 8 g, 14 g, 29 g, 35 g.

5. If an 81-g mass is made available, express 45_{10}, 62_{10} to base 3 and show how parcels of mass 45 g and 62 g may be weighed.

6. Is it possible to carry out the same weighing scheme by using the set masses, 1 g, 4 g, 16 g, 64 g?

The game of Nim provides an interesting diversion for two players. Three separate sets of matchsticks containing any number of sticks are formed on a table as illustrated in Fig. 7.6.

Fig. 7.6

The players take turns in removing one or more sticks providing they are taken from any one single set. The player who removes the last matchstick is the winner.

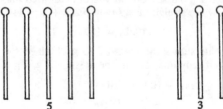

Fig. 7.7

With the arrangement of Fig. 7.6 suppose Player A removes the whole of the middle set. Player B is now confronted with the arrangement of Fig. 7.7 and subsequently removes 4 sticks to leave the position of Fig. 7.8, whereupon

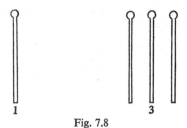

Fig. 7.8

Player A removes two sticks to leave the position of Fig. 7.9 and wins.

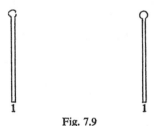

Fig. 7.9

Strangely enough, with the aid of binary numbers the first player of any game of Nim can be almost certain of winning.

We merely write down the binary number for the number of sticks in each separate set and total in the following way:

$$
\begin{array}{rcl}
5 & = & 1 \ | \ 0 \ | \ 1 \\
7 & = & 1 \ | \ 1 \ | \ 1 \\
3 & = & \quad \ | \ 1 \ | \ 1 \\
\hline
\text{Total} & & 2 \ | \ 2 \ | \ 3
\end{array}
$$

Now providing the first player ensures that each column yields an even total, he will win. He therefore removes one from the set of 3 so that the column totals are 2, 2, 2 and establishes a winning position.

Suppose Player B is now confronted by
$$
\begin{array}{rcl}
5 & = & 1 \ | \ 0 \ | \ 1 \\
7 & = & 1 \ | \ 1 \ | \ 1 \\
2 & = & \quad \ | \ 1 \ | \ 0 \\
\hline
\text{Totals} & & 2 \ | \ 2 \ | \ 2
\end{array}
$$

and whatever he does within the rule of only removing sticks from one set, he is bound to leave an odd total column. Of course, if the original set-up already gives even number totals you must encourage your opponent to go first. This is what we meant by saying 'almost certain of winning'.

Exercise 7.7 (*continued*)

7. In a game of Nim the starting set-up is 7, 4, 2. Is this a winning position?

8. After a few plays you are confronted with the position 0, 4, 1. What is your next move to ensure a win?

9. How would you ensure a win from the position: (*a*) 2, 2, 2? (*b*) *n*, *n*, *n*?

The binary system affords an interesting look at various fractions in **bicimal** form. For example, we have already seen the following relationships:

	2^{-1}	2^{-2}	2^{-3}	2^{-4}	2^{-5}	
	$\frac{1}{2}$	$\frac{1}{4}$	$\frac{1}{8}$	$\frac{1}{16}$	$\frac{1}{32}$	
$\frac{1}{2} = 0\cdot$	1					$= \frac{1}{2}$
$\frac{3}{4} = 0\cdot$	1	1				$= \frac{1}{2} + \frac{1}{4}$
$\frac{7}{8} = 0\cdot$	1	1	1			$= \frac{1}{2} + \frac{1}{4} + \frac{1}{8}$
$\frac{15}{16} = 0\cdot$	1	1	1	1		$= \frac{1}{2} + \frac{1}{4} + \frac{1}{8} + \frac{1}{16}$
$\frac{31}{32} = 0\cdot$	1	1	1	1	1	$= \frac{1}{2} + \frac{1}{4} + \frac{1}{8} + \frac{1}{16} + \frac{1}{32}$

Fig. 7.10

Other fractions may be written in the following easy manner:

$7_{10} = 111_2$ therefore $\frac{7}{8} = 0\cdot111_2$ and $\frac{7}{16} = 0\cdot0111_2$

$11_{10} = 1011_2$ „ $\frac{11}{16} = 0\cdot1011_2$ and $\frac{11}{32} = 0\cdot01011_2$

$19_{10} = 10011_2$ „ $\frac{19}{32} = 0\cdot10011_2$ and $\frac{19}{64} = 0\cdot010011_2$

Returning to Fig. 7.10, we notice that the fractions of the left-hand side gradually get closer to one. Perhaps this is easier to see if we examine the next few fractions in the scheme; they will be

$$\frac{63}{64}, \frac{127}{128}, \frac{255}{256}, \frac{511}{512}, \dots$$

with the bicimal for the last $\frac{511}{512}$ being $0\cdot111111111$ and we begin to see that the recurring bicimal $0\cdot\dot{1}$ is as close to $1\cdot0$ as we please.

Let us take the alternate fractions and sum them just to see what happens.

That is $\frac{1}{2} + \frac{1}{8} + \frac{1}{32} + \dots = 0\cdot10101\dots$

            $=$ the recurring bicimal $0\cdot\dot{1}\dot{0}$

And those fractions we have missed out give

$$\frac{1}{4} + \frac{1}{16} + \frac{1}{64} + \dots = 0\cdot010101\dots$$
$$= 0\cdot\dot{0}\dot{1}$$

Now $0\cdot\dot{1}\dot{0} \times 100 = 10\cdot\dot{1}\dot{0} = 10 + 0\cdot\dot{1}\dot{0}.$

This last equation is in binary notation. If we now change to denary numbers and call $0 \cdot 1\bar{0} = n$, then we have

$$n \times 4 = 2 + n$$

so that

$$n = \tfrac{2}{3}$$

therefore

$$\tfrac{2}{3} = 0 \cdot \bar{1}\bar{0}$$

and similarly by halving we have

$$0 \cdot \bar{0}\bar{1} = \tfrac{1}{3}$$

Let us try another one, say $0 \cdot \bar{0}\bar{0}\bar{0}\bar{1}\bar{1}$.

As before $0 \cdot \bar{0}\bar{0}\bar{0}\bar{1}\bar{1} \times 100000 = 11 \cdot \bar{0}\bar{0}\bar{0}\bar{1}\bar{1}$. (Remember, this is a binary equation.) Converting back to denary numbers and letting $0 \cdot \bar{0}\bar{0}\bar{0}\bar{1}\bar{1} = n$ we get

$$32n = 3 + n$$

whence

$$n = \tfrac{3}{31}$$

It follows therefore that any recurring bicimal represents a rational denary number.

Exercise 7.7 (*continued*)

Express the following recurring bicimals as rational denary numbers:

10. $0 \cdot \bar{0}\bar{0}\bar{1}$ 11. $0 \cdot \bar{0}\bar{0}\bar{1}\bar{1}$ 12. $0 \cdot \bar{1}\bar{0}\bar{0}$ 13. $0 \cdot \bar{1}\bar{1}\bar{0}\bar{1}$

CHAPTER EIGHT

SETS OF POINTS

THE LINES

Our environment contains many objects from which we develop mathematical ideas—especially our ideas about geometry. This has been true through the ages. The early Egyptians thought a geometrical fact had to be discovered by measurement. To them a taut rope was a straight line. The Greeks were more imaginative and felt that there were no objects that represented a straight line. To them a straight line could exist only in the imagination or in the make-believe world of geometry.

Mathematicians of today regard geometry much as the Greeks did in those days long ago. We use physical objects as models of geometric ideas. The best we can do is to draw pictures of the things we think about. But this is not strange; we never see or touch a number, yet we look at sets and invent ideas that we call numbers, we write names for them and we write phrases about operations on them. In much the same way, we look at physical objects and invent ideas that we then call points, lines, and planes.

The edge of a ruler or a tightly stretched wire serve as models of what we call a **line** (meaning straight line). (In this book line will mean *straight line*, and a line which is not straight will be called a **curve**.) A line is the property that these objects have in common. A line has no width, no thickness, but infinite length. A line is an idea.

The tip of a needle or a particle of dust in the air serve as models of what we call a **point**. A point is also an idea about an exact location. It has no thickness, no width, no length. It has no dimension whatsoever.

We usually draw dots to represent points. We label our drawings or dots with capital letters and refer to them as point A, point B, and so on.

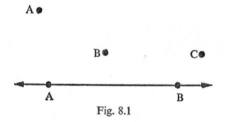

Fig. 8.1

To represent a line we make a drawing as shown above. We label two points on the line in order to refer specifically to that particular line. We refer to the line as line AB.

The arrows at either end of the drawing serve to remind us that a line extends indefinitely in both directions. We symbolize the name for line AB as $\overleftrightarrow{AB}$. We think of a line as an infinite set of points.

LINE SEGMENTS AND RAYS

We may not be concerned with the entire line, but only part of the line. In other words, we may not be concerned with the entire set of points on a line, only with a particular subset of these points.

We may be interested in only the points C and D and all points on $\overleftrightarrow{CD}$ that lie between the points C and D.

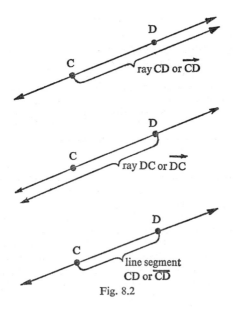

Fig. 8.2

A line segment has a definite length—the distance from C to D. Points C and D are called the **end points** of $\overline{CD}$. Line segment CD can also be called line segment DC or $\overline{DC}$. Similarly, $\overleftrightarrow{CD}$ and $\overleftrightarrow{DC}$ are names for the same line.

Another important subset of a line is called a **ray**. In the drawing above, ray CD (denoted by $\overrightarrow{CD}$) consists of point C and all the points on line CD that are on the same side of point C as point D.

Point C is called the **end** point of $\overrightarrow{CD}$, since it has only one end point. The arrow indicates that the ray extends indefinitely in this one particular direction.

Of course, we may also think of ray DC (denoted by $\overrightarrow{DC}$).

In this case point D is the end point of ray DC.

Exercise 8.1

K•

•B

R•   P•

M•   N•   Q•

1. Draw the line BK.
2. Draw line segment MN.
3. Draw ray RB.
4. Draw ray PQ.

Answer the following questions Yes or No:

5. Can you measure the length of a line?
6. Can you measure the length of a line segment?
7. Can you measure the length of a ray?
8. Does a line have end points?
9. In the drawing below, do $\overleftrightarrow{AB}$ and $\overleftrightarrow{AC}$ name the same line?

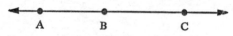

A   B   C

10. In the drawing for Question 9, do $\overrightarrow{AB}$ and $\overrightarrow{BC}$ name the same ray?

ASSUMPTIONS ABOUT POINTS AND LINES

By **space** we mean the set of all points—that is, the set of all exact locations you can possibly think of. Hence, a point or a line is a subset of space. In fact, the study of elementary geometry deals with the relationships that exist between the various subsets of space.

Think about some point A in space. How many lines are there that pass through point A? Use the following drawing as a model.

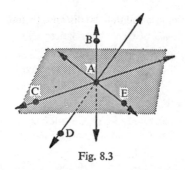

Fig. 8.3

Are there still more lines through point A? How many?

Assumption 1

There are an unlimited number of lines through any point. Now think of any two points in space. How many lines pass through both of these points?

What happens when you attempt to draw more than one line through both of the points?

Assumption 2

There is exactly one line through any two given points, or two points determine one, and only one, line.

Now consider the case where two lines intersect. Our geometrical concept of intersection is the same as that for sets, since a line is a set of points. The following is a drawing of two lines that intersect:

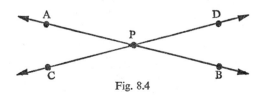

Fig. 8.4

How many points are common to both $\overleftrightarrow{AB}$ and $\overleftrightarrow{CD}$? Hence the intersection is the set whose only member is point P.

Assumption 3

Two lines can intersect in at most one point.

Let us use the following drawing to investigate the intersection of some of the subsets of a line.

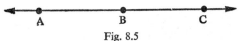

Fig. 8.5

What is the intersection of $\overrightarrow{AC}$ and $\overline{AC}$? In other words, what points do $\overrightarrow{AC}$ and $\overline{AC}$ have in common?

To answer this question, think about the drawing below.

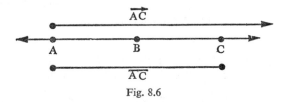

Fig. 8.6

We see that $\overline{AC}$ contains all the points common to both $\overrightarrow{AC}$ and $\overline{AC}$. Hence, $\overrightarrow{AC} \cap \overline{AC} = \overline{AC}$.

Exercise 8.2

Use the following drawing to complete the sentences below:

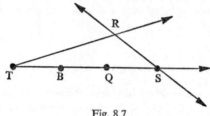

Fig. 8.7

1. $\overline{BQ} \cap \overline{QS} =$

2. $\overleftrightarrow{RS} \cap \overline{RS} =$

3. $\overline{TB} \cap \overleftrightarrow{BQ} =$

4. $\overline{TB} \cap \overline{QS} =$

5. $\overline{BQ} \cap \overline{TB} =$

6. $\overline{QS} \cap \overrightarrow{QS} =$

7. $\overleftrightarrow{RS} \cap \overrightarrow{TS} =$

8. $\overrightarrow{TS} \cap \overrightarrow{TR} =$

PLANES

The surface of a window-pane or the surface of a tabletop or any other flat surface serves as a model of what we call a **plane**. A plane has no thickness, but extends indefinitely without boundary. Obviously a plane consists of many exact locations (points) and is a subset of space. Notice that we have had to rely on intuition for the meaning of 'flat'. Since it is impossible to draw a picture of an entire plane, we draw only part of a plane. Below is such a drawing showing points A and B on the plane.

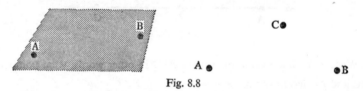

Fig. 8.8

Draw the line AB. Locate some other point Q on $\overleftrightarrow{AB}$. Is point Q a point on the plane? Is every point of $\overleftrightarrow{AB}$ also a point on the plane?

Assumption 4

If two points of a line are on a plane, then the line is on the plane.

Think of three points that are not on the same line, such as in Fig. 8.8.

How many planes may pass through point A? How many planes may pass through both points A and B? How many planes may pass through all three points?

There are infinitely many through A and infinitely many through A and B, but there is only one through A, B, and C.

Assumption 5

Three points that are not on the same line determine one and only one plane.

Think of the plane of one wall of a room and the plane of the ceiling. Do they intersect? What kind of geometric figure is their intersection?

Assumption 6

Two planes can intersect in at most one line.

PARALLEL LINES AND PLANES

The following drawing shows two lines on a plane which do not intersect, or their intersection is the empty set.

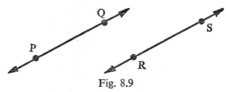

Fig. 8.9

We say that $\overleftrightarrow{PQ}$ is parallel to $\overleftrightarrow{RS}$ and denote this by $\overleftrightarrow{PQ} \| \overleftrightarrow{RS}$.

If two lines are on the same plane one, and only one, of the following situations must exist.

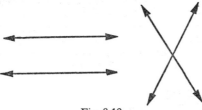

Fig. 8.10

Assumption 7

Two lines in the same plane are either parallel lines or intersecting lines.

Notice that the above discussions have the restriction that the two lines be on the same plane. In space, we might have two lines that do not intersect and that are not parallel. For example, the drawing below shows a rectangular box.

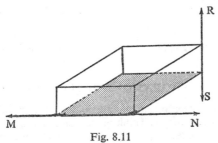

Fig. 8.11

Lines MN and RS would never intersect, but we would not call them parallel lines. They are called **skew lines**.

Think about the planes of the floor and the ceiling of a room. Do these planes intersect?

Definition 8.1

Two lines in a plane, two planes, or a line and a plane are **parallel** if their intersection is the empty set.

Exercise 8.3

Letters *m* and *n* refer to either of the two planes shown in the following figure. For each item 1 to 10 below, state the letter of the corresponding item in (*a*) to (*j*).

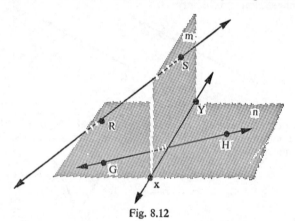

Fig. 8.12

1. Two skew lines.
2. Intersection of $\overline{RS}$ and $\overleftrightarrow{RS}$.
3. Intersection of planes *m* and *n*.
4. Intersection of $\overrightarrow{SR}$ and $\overrightarrow{XY}$.
5. Line segment on plane *n*.
6. Two intersecting lines.
7. Intersection of plane *n* and $\overleftrightarrow{RS}$.
8. A point not on plane *n*.
9. Intersection of $\overrightarrow{GH}$ and $\overrightarrow{HG}$.
10. A point common to planes *m* and *n*.

(*a*) $\overline{GH}$ (*b*) $\overline{XY}$ or $\overline{GH}$ (*c*) $\overleftrightarrow{XY}$ (*d*) $\overleftrightarrow{XY}$ and $\overleftrightarrow{GH}$

(*e*) ϕ (*f*) $\overleftrightarrow{RS}$ and $\overleftrightarrow{XY}$ (*g*) $\overline{RS}$ (*h*) point R

(*i*) point X   (*j*) point S

SEPARATION PROPERTIES

Another important relationship among points, lines, and planes is what we call **separation** or **betweenness**. Think of the plane represented by a wall of a room. This plane separates space into three distinct sets of points—those points on one side of the plane, those on the other side of the plane, and those on the plane. Those points on one side of the plane are usually referred to as a **half-space**. The plane itself is not in either half-space.

The following figure shows plane *m* separating space into three distinct

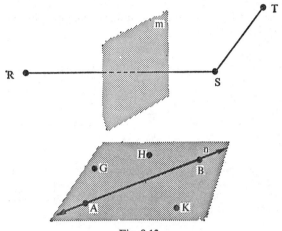

Fig. 8.13

sets of points: the half-space containing point R, the half-space containing points S and T, and the plane itself.

Points S and T are in the same half-space, since they can be joined by a line segment that does not intersect plane *m*. We say that points R and S are in different half-spaces.

In the same way a line separates a plane into three distinct sets of points— the line and the two half-planes. In the figure (8.13) line AB separates the plane *n* into the half-plane containing points G and H, the **half-plane** containing point K, and finally the line AB.

Since $\overline{GH} \cap \overleftrightarrow{AB} = \phi$, points G and H are in the same half-plane.

Since $\overline{GK}$ would intersect $\overleftrightarrow{AB}$, we say that points G and K are in different half-planes. A similar situation exists for a point and a line. In the following figure point P separates $\overleftrightarrow{AB}$ into two **half-lines** and the point P.

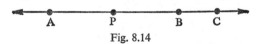

Fig. 8.14

Since $\overline{BC}$ does not contain P, we say that points B and C are on the same half-line. Since $\overline{AB}$ contains P, we say that points A and B are on different half-lines. Point P does not belong to either half-line.

SIMPLE CLOSED FIGURES

Every set of points in space constitutes a geometric figure. Let us think about a plane figure whose boundary never intersects itself. That is, we can draw such a figure by starting at some point in a plane, never lift the pencil tip from the paper, and trace any kind of curve we desire, provided it never intersects itself, and end at the starting-point. Examples of such figures are shown below (Fig. 8.15).

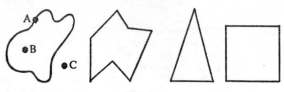

Fig. 8.15

These are called **simple closed** figures because they separate the plane into three distinct sets of points—the figure itself, the interior, and the exterior of the figure. In the first figure above point A is on the figure, point B is in the interior, and point C is in the exterior.

Definition 8.2

A polygon is a simple closed figure formed by the union of line segments that have common end points which are all distinct (i.e. different from one another). Each line segment is called a **side** of the polygon. The common end point of two sides is called a **vertex** of the polygon.

Study how this definition is applied to the following polygon, ABCD.

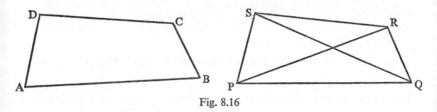

Fig. 8.16

This is called polygon ABCD (naming the vertices in order); $\overline{AB}$, $\overline{BC}$, $\overline{CD}$, and $\overline{DA}$ are sides of the polygon. Points A, B, C, and D are vertices (plural of vertex) of the polygon.

If two sides have a common end point they are called **adjacent sides**.

If two vertices are end points of the same side they are called **adjacent vertices**.

A line segment joining two non-adjacent vertices of a polygon is called a **diagonal** of the polygon. In the figure, PQRS, $\overline{SQ}$ and $\overline{PR}$ are diagonals.

We can now define the various kinds of polygons according to the number of sides each contains.

Definition 8.3

Type of polygon	Number of sides
Triangle	3
Quadrilateral	4
Pentagon	5
Hexagon	6
Heptagon	7
Octagon	8

Exercise 8.4

Answer the following:

1. A polygon has 6 sides. How many vertices does it have?
2. Is the numeral 8 a simple closed figure? Why or why not?
3. Which kind of polygon has no diagonals?
4. How many diagonals does a quadrilateral have?
5. How many diagonals does a pentagon have?

CIRCLES

Think of locating points in a plane that are a given distance from some fixed point in the plane, as shown in the figure at the left below.

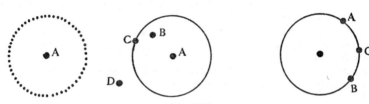

Fig. 8.17

We might continue to locate more and more points the given distance from A. If we think of locating all such points we would be thinking about a circle.

Definition 8.4

A **circle** is the set of all points in a plane at a given distance from some point in the plane.

In the figures on the left above, point A is called the centre of the circle; it is not a point of the circle. Point C is on the circle, point B is in the interior, and point D is in the exterior. Points in the interior are not considered to be part of the circle.

Definition 8.5

Any part of a circle containing more than one point is called an **arc** of the circle.

In the right-hand circle of Fig. 8.17 circle arc AB (denoted $\overset{\frown}{AB}$) is the set of all points on the circle between and including points A and B.

Of course, the above description does not tell which arc is meant. Therefore, let us agree to mean the shorter or minor of the two arcs. Or we might label another point C on the arc and refer to $\overset{\frown}{AB}$ as $\overset{\frown}{ACB}$.

ANGLES

Another important geometric figure or set of points is called an angle.

Definition 8.6

An **angle** is the union of two rays that have the same end point.
The following are pictures of angles:

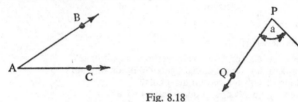

Fig. 8.18

The first figure is called angle BAC (denoted by ∠BAC) or angle CAB (denoted by ∠CAB). Notice that point A is the common end point of the two rays; it is called the **vertex** of the angle. Rays AB and AC are called the **sides** of the angle. In naming an angle we usually name a point on one side, the vertex, and then a point on the other side. If no confusion will arise, we may name an angle by giving only the letter of the vertex. That is ∠BAC might be called ∠A.

Another way of naming an angle is to draw a small curved arrow between the two sides and write a lower-case letter, as shown in ∠QPR above. This angle can then be called *a*.

An angle separates a plane into three distinct sets of points—the angle, its interior, and its exterior—as shown below.

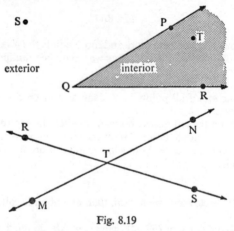

Fig. 8.19

In Fig. 8.19, $\overleftrightarrow{RS}$ and $\overleftrightarrow{MN}$ intersect in point T.

A pair of angles such as ∠RTM and ∠NTS are called **vertical angles**. What do these two angles have in common? Are angles RTN and MTS also a pair of vertical angles? If two angles have the same vertex and a common ray between them the angles are called **adjacent angles**. In the preceding figure ∠RTM and ∠MTS have the same vertex (point T) and a common ray (T̄M⃗) between them. Hence, ∠RTM and ∠MTS are adjacent angles.

Attention must be drawn to one difficulty which arises from this 'modern' definition of angle: the angles of a triangle.

Rays, which are infinite in length, are different from the sides of a triangle, which are line segments and finite in length. But the sides of the triangle do in fact determine three angles. We say that the angles are on the triangle and we refer to angles being determined by sides $\overline{AC}$ and $\overline{AB}$ or $\overline{BA}$ and $\overline{BC}$, etc.

Exercise 8.5

Use the following figure to answer the questions below:

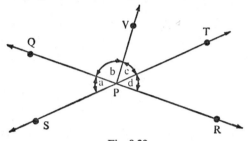

Fig. 8.20

1. Why are ∠QPT and ∠VPT not adjacent angles?
2. Name a point in the interior of ∠QPT.
3. Name a point in the exterior of ∠QPT.
4. ∠a and ∠ — are a pair of vertical angles.
5. ∠b and ∠ — or ∠ — are a pair of adjacent angles.
6. Is every point of P̄V⃗ in the interior of ∠QPT? Why, or why not?
7. What are the sides of ∠TPR?

MEASURING ANGLES

The size of an angle depends on the amount of opening between the sides (rays) of the angle. We need a suitable unit for measuring the size of an angle.

Think of separating a circle into 360 arcs of the same length. Suppose in the figure below that points E and F are end points of one of these arcs, and

Fig. 8.21

point Q is the centre of the circle. Then $\overrightarrow{QE}$ and $\overrightarrow{QF}$ form an angle whose measure is 1 degree (denoted by 1°).

A model showing such a separation for one half of a circle is called a **protractor**. A protractor is an instrument used to measure angles just as a ruler is an instrument to measure line segments.

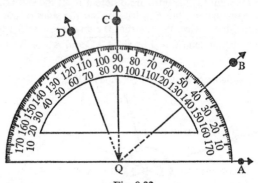

Fig. 8.22

Notice that the centre of the circle is marked on the protractor (point Q in the figure). Notice also how the protractor is placed to measure an angle—the centre of the circle is placed on the vertex of the angle and the straight edge of the protractor is placed on one side of the angle. We say that the degree measure of ∠AQB is 40. Let us agree to use the symbol m (∠AQB) to mean the degree measure of ∠AQB. Then we can say that m (∠AQB) = 40.

TYPES OF ANGLES

Angles are classified according to their degree measure. The angles referred to in the following definition are shown in the previous drawing of a protractor.

Definition 8.7

A **right angle** has a measure of 90. ∠AQC is a right angle.

An **acute angle** has a measure greater than 0 but less than 90. ∠AQB is an acute angle.

An **obtuse angle** has a measure greater than 90 but less than 180. ∠AQD is an obtuse angle.

A **straight angle** has a measure of 180.

A convenient way of indicating the degree measure of an angle is shown below.

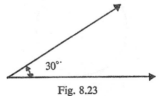

Fig. 8.23

Exercise 8.6

Use the following figure for the exercise below:

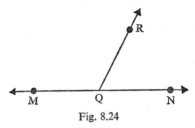

Fig. 8.24

1. What kind of an angle is ∠MQN? ∠MQR? ∠NQR?
2. What is the degree measure of ∠MQR? Of ∠NQR? What is the sum of their measures?
3. What is your conclusion about $\overrightarrow{QM}$ ∪ $\overrightarrow{QN}$?

PERPENDICULAR LINES AND PLANES

If two lines intersect so that all four angles have the same measure we say the lines are perpendicular.

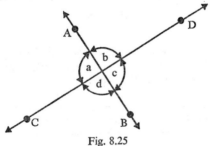

Fig. 8.25

If $m(\angle a) = m(\angle b) = m(\angle c) = m(\angle d)$, then $\overleftrightarrow{AB}$ is perpendicular to $\overleftrightarrow{CD}$ (denoted by $\overleftrightarrow{AB} \perp \overleftrightarrow{CD}$ and also $\overleftrightarrow{CD} \perp \overleftrightarrow{AB}$). On the other hand, if the two lines are perpendicular, then all four angles have the same measure. What kind of angle would each of them be?

Hence, we can say that perpendicular lines form right angles.

As a convenience in our drawings, let the symbol ⌐ drawn in an angle indicate that the rays are perpendicular or that the angle is a right angle.

∠ABC is a right angle
$\overrightarrow{BA} \perp \overrightarrow{BC}$

Think of drawing several lines on a plane through some given point P. Now place the tip of your pencil on point P and hold it so that it is perpendicular to each line you have drawn. Can the pencil be in more than one position so that this is true? If a line intersects a plane in only one point, let us call this point the **foot** of the line. In the following figure point P is the foot of $\overleftrightarrow{AB}$, and $\overleftrightarrow{AB}$ is perpendicular to lines $\overleftrightarrow{CD}$ and $\overleftrightarrow{EF}$ in plane *n*.

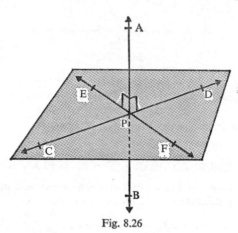

Fig. 8.26

Definition 8.8

If a line is perpendicular to two lines in a plane passing through its foot the line is perpendicular to every line in the plane. We simplify this by saying that the line is perpendicular to the plane.

Exercise 8.7

Answer the following:

1. How many lines can be drawn perpendicular to a line at a given point on the line?

2. How many lines can be drawn perpendicular to a plane at a given point on the plane?

3. Can you draw three lines through a point so that each line is perpendicular to the other two lines?

4. If $\overleftrightarrow{AB}$ is perpendicular to a vertical line, is $\overleftrightarrow{AB}$ a horizontal line?

5. If $\overleftrightarrow{CD}$ is perpendicular to a horizontal line, is $\overleftrightarrow{CD}$ a vertical line?

WHAT MEASUREMENT IS

Measure is a connecting link between the physical world and mathematics. We can count the number of buns we need for a picnic, but we measure the amount of milk we need.

To measure a line segment means to assign a number to it. We cannot count all of the points on the line segment, since there are infinitely many. So we select some line segments as an arbitrary unit and compare it to the line segment to be measured. Once this concept of measurement is developed, we find it is also useful in determining the measure of other things.

Suppose we want to find the measure of line segment AB shown below? We might choose some other line segment, such as $\overline{MN}$, as the arbitrary unit. By starting at A the unit $\overline{MN}$ can be laid off 4 times to reach point R, which is between A and B.

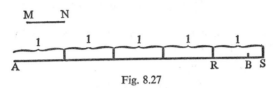

Fig. 8.27

If laid off 5 times we reach point S, which is beyond point B. Surely we can say that $\overline{AB}$ has a length greater than 4 units and less than 5 units. The best we can do is visually estimate that the length of $\overline{AB}$ is nearer to 5 units than to 4 units, and state this by saying that the length of $\overline{AB}$ is 5 units, to the nearest unit.

Exercise 8.8
Use 1 centimetre as a unit and find the length to the nearest centimetre of each line segment below.

Then use $\frac{1}{2}$ centimetre as the unit and find the length of each segment to the nearest $\frac{1}{2}$ centimetre.

1 _____

2 _____

3 _____

4 _____

APPROXIMATE NATURE OF MEASUREMENT

To help us determine whether the previously used $\overline{AB}$ has a length nearer to 4 units or 5 units, we can bisect the unit $\overline{MN}$ at T.

Now we see that the point B is more than $4\frac{1}{2}$ units from A, and hence the measurement is nearer to 5 units.

Suppose we use $\overline{MT}$ as the unit of measure? The length of $\overline{AB}$ is 9 units to the nearest unit.

We could continue to bisect each new unit (line segment) and would never find a unit that 'fits exactly' a whole number of times into $\overline{AB}$. Similarly, we are not surprised if someone tells us that it is 100 km to the city of Podunk, and then someone else says it is 98 km or 103 km.

Measurement is *only approximate, never exact.*

When we say that the 'length of a line segment is 5 units' we mean that *the measure of the line segment in terms of some unit is 5*. That is, a measure is a number. In a phrase such as '26 feet',

'26' names the measure,
'feet' names the unit, and
'26 feet' names the measurement.

Notice that a measurement contains two symbols—one for the measure and the other for the unit.

PRECISION

A ruler is simply a model showing different subdivisions of an inch. The figure below shows how we could find the measure of $\overline{AB}$.

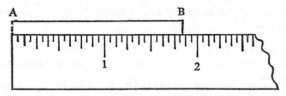

Fig. 8.28

Clearly, the length of $\overline{AB}$ is between 1 and 2 inches, and closer to 2 inches. A better measure would be that the length of $\overline{AB}$ is between $1\frac{1}{2}$ and 2 inches, but closer to 2 inches. A still better measure would be that the length of $\overline{AB}$ is between $1\frac{3}{4}$ and 2 inches, but closer to $1\frac{3}{4}$ inches.

If we let x represent the measurement in inches of $\overline{AB}$ we can indicate the value of x as follows:

$$1 < x < 2$$
$$1\tfrac{1}{2} < x < 2\tfrac{0}{2}$$
$$1\tfrac{3}{4} < x < 2\tfrac{0}{4}$$
$$1\tfrac{6}{8} < x < 1\tfrac{7}{8}$$
$$1\tfrac{13}{16} < x < 1\tfrac{14}{16}$$

The units on the ruler could be subdivided into still smaller units, since between any two points on a line there is a third point.

We say that a measurement of $1\frac{14}{16}$ inches is more precise or has greater precision than a measure of $1\frac{7}{8}$ inches because a smaller unit of measure is used. Similarly, a measurement of $1\frac{3}{4}$ inches is more precise than a measurement of $1\frac{1}{2}$ inches. *The smaller the unit of measure, the greater is the precision.* When a measurement is made to the nearest inch we say the precision of measurement is 1 inch. When a measurement is made to the nearest $\frac{1}{2}$ inch we say the precision of the measurement is $\frac{1}{2}$ inch.

What does it mean to say that $\overline{AB}$ is 2 inches long, to the nearest inch?

It means: if the 0-point of the ruler is placed at point A of AB, then point B might fall anywhere between the $1\frac{1}{2}$-inch mark and the $2\frac{1}{2}$-inch mark. In other words, the inch measure of $\overline{AB}$ is a number between $1\frac{1}{2}$ and $2\frac{1}{2}$.

Therefore, we say that when a line segment is measured to the nearest inch *the greatest possible error is $\frac{1}{2}$ inch.* This does not mean that we have made a mistake in finding the measure; it simply means that the measurement is approximate within these limits.

We can indicate the greatest possible error of such a measurement by writing $(2 \pm \frac{1}{2})$ inches.

The symbol $\pm$ is read 'plus or minus'.

The greatest possible error of a measurement is equal to one-half of the unit of measure.

If a measurement is made to the nearest $\frac{2}{4}$ inch such as $7\frac{2}{4}$ inches, do not reduce the fraction.

The fraction $\frac{2}{4}$ indicates that the unit of measure is $\frac{1}{4}$ inch and the greatest possible error is $\frac{1}{8}$ inch. But the fraction $\frac{1}{2}$ indicates that the unit of measure is $\frac{1}{2}$ inch and the greatest possible error is $\frac{1}{4}$ inch. Furthermore, a measurement such as $23\frac{0}{4}$ inches means that the unit of measure is $\frac{1}{4}$ inch while 23 inches means that the unit of measure is 1 inch.

Exercise 8.9

State the unit of measure, the greatest possible error, and how you would write the measurement to show the greatest possible error for each of the following:

1. $14\frac{1}{2}$ in.   2. $3\frac{1}{4}$ ft.   3. $5\frac{7}{8}$ in.   4. 9 in.   5. $57\frac{9}{6}$ yd.   6. $15\frac{9}{16}$ in.

CONGRUENCE AND SIMILARITY
Congruent Line Segments

In the physical world we never expect two objects to have exactly the same length, but in our imaginary world of geometry every line segment is assumed to have many exact copies.

Given $\overline{AB}$ below, we could use a pair of compasses to construct $\overline{CD}$ so that the measures of the two line segments are the same.

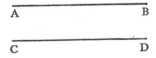

We do not say $\overline{AB} = \overline{CD}$ because they are obviously two different line segments, and $=$ means that two symbols name precisely the same thing. Instead, we say that $\overline{AB}$ is congruent to $\overline{CD}$, or $\overline{AB} \cong \overline{CD}$.

Definition 8.9

Two line segments are congruent if, *and only if,* they have the same measure. The symbol $\cong$ means 'is congruent to'.

Exercise 8.10

Answer these questions:

1. Can a horizontal line segment be congruent to a vertical line segment?
2. If $\overline{RS} \cong \overline{UV}$ and the measure of $\overline{RS}$ is $5\frac{1}{2}$, what is the measure of $\overline{UV}$?
3. Are any two sides of a square congruent line segments?
4. If point K is the midpoint of $\overline{AB}$ is $\overline{AK} \cong \overline{KB}$?

BISECTING LINE SEGMENTS AND ANGLES

The word bisect means to separate into two parts of equal measure. To bisect $\overline{AB}$ means to locate some point P on $\overline{AB}$ such that $\overline{AP} \cong \overline{PB}$.

To bisect $\overline{AB}$, open a pair of compasses so that the distance between the point of the compasses and the tip of the pencil is greater than half the length of $\overline{AB}$. Place the point on A and then B to draw equal arcs which intersect as shown below.

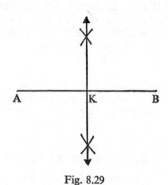

Fig. 8.29

Draw the line determined by the points of intersection of the arcs. This line bisects $\overline{AB}$. Point K is called the **midpoint** of $\overline{AB}$, and $\overline{AK} \cong \overline{KB}$.

To bisect $\angle RST$ means to locate some point P in the interior of the angle so that $m(\angle RSP) = m(\angle PST)$. ($m$ stands for degree measure. See page 162.)

To locate such a point, place the point of the compasses at the vertex of the angle (point S in this case) and draw an arc to intersect the sides of the angle, as shown below at A and B.

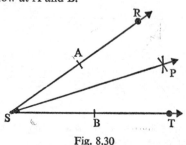

Fig. 8.30

With A and B centres, draw two equal intersecting arcs in the interior of the angle. Call that point P. Draw $\overrightarrow{SP}$. Then $m(\angle RSP) = m(\angle PST)$ and we say that $\angle RSP \cong \angle PST$.

Definition 8.10

Two angles are congruent if, *and only if*, they have the same measure.

CONSTRUCTING CONGRUENT ANGLES

Given $\angle ABC$ below, how can we construct $\angle DEF$ so that the two angles are congruent?

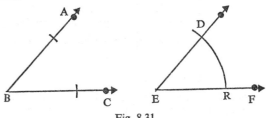

Fig. 8.31

Draw ray EF. Place the point of the compasses on vertex B and draw an arc that intersects the sides of $\angle ABC$. Without changing the opening of the compass, use point E as a centre and draw an arc as shown so that it intersects $\overrightarrow{EF}$ at some point R.

Now place the point of the compasses at the point of intersection of one side of $\angle ABC$ and the arc. Open the compass so that the pencil tip is on the point of intersection of the other side of $\angle ABC$ and the arc. Without changing the opening of the compasses use point R as centre and draw an arc that intersects the previously drawn arc at point D. Draw $\overrightarrow{ED}$. This construction makes $\angle ABC \cong \angle DEF$.

Exercise 8.11

Construct any angle congruent to each of the angles shown below:

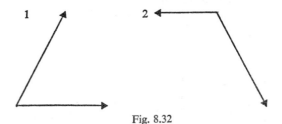

Fig. 8.32

PERPENDICULAR LINES

Suppose we are given $\overleftrightarrow{AB}$ below and asked to construct some line RQ so that $\angle AQR \cong \angle BQR$. We might consider $\overleftrightarrow{AB}$ as a straight angle whose vertex is point Q. Then we can proceed to bisect $\angle AQB$.

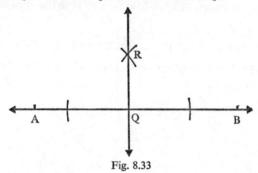

Fig. 8.33

Since $m(\angle AQB)$ is 180°, and since $\angle AQR \cong \angle BQR$, then $\angle AQR$ is a right angle. Each of these newly formed angles has a degree measure of 90.

In this case we say that $\overleftrightarrow{AB}$ is **perpendicular** to $\overleftrightarrow{RQ}$. This is denoted by $\overleftrightarrow{AB} \perp \overleftrightarrow{RQ}$ (see page 163).

Definition 8.11

If two lines intersect so that right angles are formed at their point of intersection, then the lines are called **perpendicular lines**.

Suppose we are given $\overleftrightarrow{CD}$ and some point P not on $\overleftrightarrow{CD}$, and we are asked to construct a line through P that is perpendicular to $\overleftrightarrow{CD}$. Hence, we must locate some point M on $\overleftrightarrow{CD}$ so that $\overleftrightarrow{PM} \perp \overleftrightarrow{CD}$. This can be done by doing the steps in the previous construction but in the reverse order. Place the point of the compasses on point P and draw arcs that intersect $\overleftrightarrow{CD}$ at some points called G and H, as shown below. Then bisect $\overline{GH}$; this is done by drawing the two arcs that intersect at J.

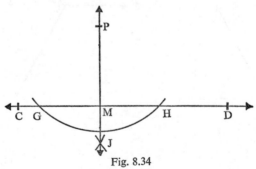

Fig. 8.34

Draw, $\overleftrightarrow{PJ}$ and label its intersection with $\overleftrightarrow{CD}$ as M. Since the angles formed by these two lines are all right angles, $\overleftrightarrow{CD} \perp \overleftrightarrow{PM}$.

Exercise 8.12

Use compasses to construct the following:

1. A line perpendicular to $\overleftrightarrow{UV}$ at point P.

2. A line through point S perpendicular to $\overleftrightarrow{ST}$.

CONGRUENT TRIANGLES

We have learned that a general idea of congruence is that two figures are congruent if they have the same size and the same shape.

By drawing the necessary diagonals in any polygon we can form a figure composed entirely of triangles. Since triangles are so fundamental to much of mathematics and science, we need to know the conditions under which two triangles are congruent.

The two triangles below have the same size and the same shape. In other words, they are congruent triangles. Intuitively, this means that the picture of either triangle can be moved so that it fits exactly on the picture of the other, or that the two triangles can be made to coincide, always assuming that the shape of the triangle has not been changed because of the movement. In such a process each side and each vertex of one triangle is matched with a side or a vertex of the other triangle.

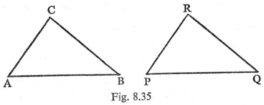

Fig. 8.35

To be more specific about what we mean by saying that triangle ABC is congruent to triangle PQR, we have to tell which parts of the two triangles are matched or paired. In this way we tell which points of one triangle correspond to which points of the other triangle.

If we should move the picture of △PQR on to the picture of △ABC as follows the two triangles will coincide.

Point P on point A

Point Q on point B

Point R on point C

We indicate this pairing in our notation of congruent triangles by arranging the letters of the vertices of each triangle so that the first letters correspond, the second letters correspond, and the third letters correspond.

This statement is read: triangle ABC is congruent to triangle **PQR**.

From this arrangement of the letters we also see that $\overline{AB}$ corresponds to $\overline{PQ}$, $\overline{BC}$ corresponds to $\overline{QR}$, and $\overline{CA}$ corresponds to $\overline{RP}$.

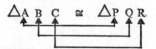

If two triangles are congruent, then for each side or angle there is a congruent side or angle in the other triangle. Each such pair of sides or angles are called corresponding parts of the two triangles.

Exercise 8.13

Assume that the two triangles in each exercise are congruent. For each pair of triangles, state the congruence between the triangles so that the corresponding parts are indicated by the arrangements of the letters of the vertices.

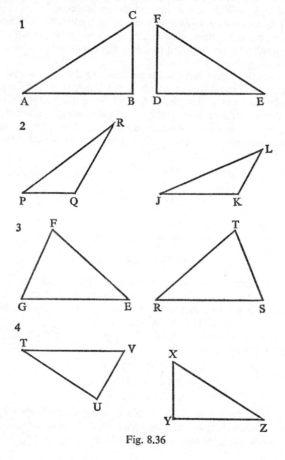

Fig. 8.36

CONDITIONS FOR CONGRUENT TRIANGLES

Suppose we were given △ABC below and asked to construct DEF so that △ABC ≅ △DEF.

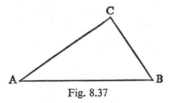

Fig. 8.37

We could start by constructing $\overline{DE} \cong \overline{AB}$. This would locate two vertices of △DEF.

We have three ways of locating the third vertex.

Case 1

Construct ∠DER ≅ ∠ABC

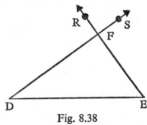

Fig. 8.38

Then construct ∠EDS ≅ ∠BAC. The point of intersection of $\overrightarrow{ER}$ and $\overrightarrow{DS}$ is the third vertex. In this case $\overline{DE}$ is part of both ∠DEF and ∠EDF, so let us refer to this case as two angles and the included side. Let us agree to accept the following condition for two triangles to be congruent.

Two triangles are congruent if two angles and the included side of one triangle are congruent respectively to two angles and the included side of the other triangle. Further, let us abbreviate this condition by A.S.A. (angle, side, angle).

Case 2

Construct ∠DET ≅ ∠ABC

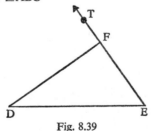

Fig. 8.39

Then construct $\overline{EF} \cong \overline{BC}$. Then draw $\overline{DF}$.

This could also be done by constructing $\angle EDS \cong \angle BAC$ and $\overline{DF} \cong \overline{AC}$. In this case $\angle E$ is formed by sides $\overline{DE}$ and $\overline{EF}$.

We refer to this as two sides and the included angle. Let us agree to accept the following condition for two triangles to be congruent.

Two triangles are congruent if two sides and the included angle of one triangle are congruent respectively to two sides and the included angle of the other triangle (S.A.S.—side, angle, side).

Case 3

Open a pair of compasses to the length of $\overline{BC}$. Use the point E as centre and draw an arc as shown below.

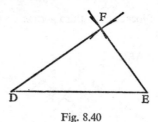

Fig. 8.40

Now open the compasses to the length of $\overline{AC}$. Use point D as centre and draw an arc that intersects the previously drawn arc. This point of intersection is the third vertex F of the triangle, so draw lines DF and EF.

Let us agree to accept the following conditions for two triangles to be congruent:

Two triangles are congruent if the three sides of one triangle are congruent respectively to the three sides of the other triangle (S.S.S—side, side, side).

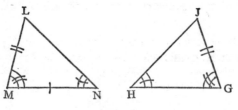

Fig. 8.41

The two marks on each of $\overline{ML}$ and $\overline{GJ}$ indicate that they are congruent. We use the same idea for angles of two triangles. The marks above indicate that $\angle M \cong \angle G$ and that $\angle N \cong \angle H$. Which side is congruent. to $\overline{MN}$?

Exercise 8.14

Decide whether the pair of triangles in each exercise are congruent or not. If they are congruent, use A.S.A., S.A.S., or S.S.S. as a reason for your conclusion.

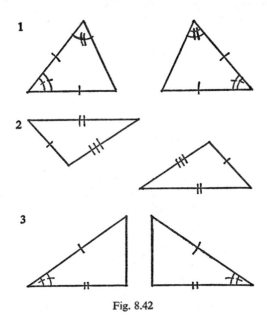

Fig. 8.42

IDENTITY CONGRUENCE

Is every geometric figure congruent to itself?
Consider $\overline{AB}$ below.

Certainly $\overline{AB}$ coincides with itself. That is, $\overline{AB} \cong \overline{AB}$. We could also match the end points so that $\overline{AB} \cong \overline{BA}$. A congruence between a geometric figure and itself is called the identity congruence.

Now consider $\angle PQR$ below.

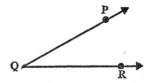

If we match each ray with itself we have $\angle PQR \cong \angle PQR$. Another way to establish the identity congruence is to match each ray with the other ray. Then $\angle PQR \cong \angle RQP$.

Obviously every triangle is congruent to itself only if we match every vertex with itself. This produces the identity congruence for triangles.

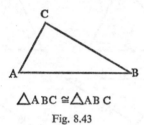

$$\triangle ABC \cong \triangle ABC$$

Fig. 8.43

In order to establish a congruence between the two triangles shown below, we make use of the identity congruence.

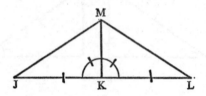

We are given the following:

$$\angle JKM \cong \angle LKM$$
$$\overline{JK} \cong \overline{LK}$$

By the identity congruence we know

$$\overline{MK} \cong \overline{MK}$$

Therefore, $\triangle JKM \cong \triangle LKM$, because two sides and the included angle of one triangle are congruent to two sides and the included angle of the other triangle (S.A.S.).

The whole purpose of this procedure is to gain information other than that which is given. In this case we may go on to say that $\overline{MJ} \cong \overline{ML}$ and $\angle MJK \cong \angle MLK$ and $\angle KMJ \cong \angle KML$.

Exercise 8.15

In each figure below, congruent sides or angles are indicated by the small marks. Give four statements for each figure—three statements for congruence between corresponding parts and the fourth stating the congruence of two triangles and the condition (A.S.A., S.A.S., or S.S.S.) for this congruence.

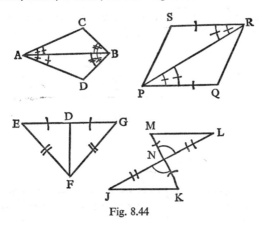

Fig. 8.44

VERTICAL ANGLES

In the following figure $\overleftrightarrow{AB}$ intersects $\overleftrightarrow{CD}$ at point T.

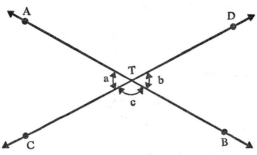

Fig. 8.45

Recall that $\angle a$ and $\angle b$ are a pair of vertical angles.

$$m(\angle a) + m(\angle c) = 180$$
$$m(\angle b) + m(\angle c) = 180$$

Therefore, $\qquad m(\angle a) + m(\angle c) = m(\angle b) + m(\angle c)$

Since $m(\angle c)$ names a number, we can use the addition property of equations to obtain the following:

$$m(\angle a) = m(\angle b)$$
or $\qquad\qquad \angle a \cong \angle b$

A similar argument holds for every pair of vertical angles. We state this as: **vertical angles are congruent.**

PARALLEL LINES AND TRANSVERSALS

Parallel lines $\overleftrightarrow{AB}$ and $\overleftrightarrow{CD}$ below are intersected by line $\overleftrightarrow{RS}$. Line $\overleftrightarrow{RS}$ is called a **transversal** of the parallel lines.

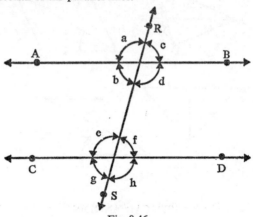

Fig. 8.46

Since points A and D are on different lines and on opposite sides of the transversal, we can think of alternating from one side of the transversal to the other.

Furthermore, in the figure above we refer to pairs of angles as follows:

$\angle b$ and $\angle f$ are alternate interior angles

$\angle d$ and $\angle e$ are alternate interior angles

$\angle a$ and $\angle h$ are alternate exterior angles

$\angle c$ and $\angle g$ are alternate exterior angles

Since $\angle c$ and $\angle f$ are in corresponding positions with regard to the two points of intersection, we can think of each pair of angles given below as a pair of corresponding angles:

$\angle c$ and $\angle f$ $\angle b$ and $\angle g$

$\angle a$ and $\angle e$ $\angle d$ and $\angle h$

Use a protractor to find the measure of $\angle b$ and the measure of $\angle f$. How do their measures compare? Do the same for $\angle d$ and $\angle e$. Let us accept the following conditions about parallel lines and alternate interior angles:

If two parallel lines are cut by a transversal, then alternate interior angles are congruent.

If two lines are cut by a transversal so that alternate interior angles are congruent, then the lines are parallel.

Since $\overleftrightarrow{AB} \| \overleftrightarrow{CD}$ in the previous figure, then $\angle b \cong \angle f$. Furthermore, $\angle b \cong \angle c$ and $\angle f \cong \angle g$ because they are pairs of vertical angles. These statements of congruence can be translated into equalities of measures

$$m(\angle b) = m(\angle f) \qquad m(\angle b) = m(\angle c) \qquad m(\angle f) = m(\angle g)$$

Therefore, by the transitive property of equality (see Chapter Nine), $m(\angle c) = m(\angle g)$, or $\angle c \cong \angle g$. Since $\angle c$ and $\angle g$ are alternate exterior angles, let us accept the following conditions about parallel lines and alternate exterior angles:

If two parallel lines are cut by a transversal, then the alternate exterior angles are congruent.

If two lines are cut by a transversal so that alternate exterior angles are congruent, then the lines are parallel.

Refer once again to the previous figure of parallel lines cut by a transversal. Since $\overleftrightarrow{AB} \| \overleftrightarrow{CD}$, $\angle b \cong \angle f$ because they are alternate interior angles. Also, $\angle b \cong \angle c$ because they are vertically opposite angles. Let us translate these statements into equalities of measures.

$$m(\angle b) = m(\angle f) \qquad m(\angle b) = m(\angle c)$$

Therefore, $m(\angle f) = m(\angle c)$ or $\angle f \cong \angle c$. What kind of angles are $\angle f$ and $\angle c$?

Let us accept the following conditions about parallel lines and corresponding angles:

If two parallel lines are cut by a transversal, then corresponding angles are congruent.

If two lines are cut by a transversal so that corresponding angles are congruent, then the lines are parallel.

Exercise 8.16

Use the figure below to complete the sentences that follow. Assume $\overleftrightarrow{JK} \| \overleftrightarrow{MN}$.

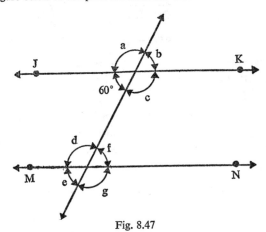

Fig. 8.47

1. $\angle a$ and $\angle$ — are alternate exterior angles.
2. $\angle c$ and $\angle$ — are alternate interior angles.
3. $\angle g$ and $\angle$ — are corresponding angles.
4. $m(\angle b) =$ —   5. $m(\angle e) =$ —   6. $m(\angle c) =$ —
7. $m(\angle g) =$ —   8. $m(\angle d) =$ —   9. $m(\angle f) =$ —

PROVING TWO TRIANGLES CONGRUENT

To prove that two triangles are congruent we must establish one of the conditions A.S.A., S.A.S., or S.S.S. To make our proof easier to follow, let us arrange the congruence statements and a reason for each as shown in the following example:

Given:

$$\overline{AB} \cong \overline{AC}$$
$$\overline{AD} \text{ bisects } \overline{BC}$$

Prove:

$$\triangle ADC \cong \triangle ADB$$

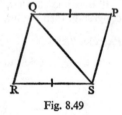

Fig. 8.48

Statements	Reasons
1. $\overline{AB} \cong \overline{AC}$	Given
2. $\overline{AD}$ bisects $\overline{BC}$	Given
3. $\overline{BD} \cong \overline{DC}$	Definition of bisection
4. $\overline{AD} \cong \overline{AD}$	Identity congruence
5. $\triangle ADC \cong \triangle ADB$	S.S.S.

We may now say that

$$m(\angle ADB) = m(\angle ADC)$$

Given:

$$\overline{QP} \cong \overline{RS}$$
$$\overleftrightarrow{QP} \parallel \overleftrightarrow{RS}$$

Prove:

$$\triangle RSQ = \triangle PQS$$

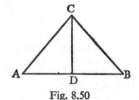

Fig. 8.49

Statements	Reasons
1. $\overline{QP} \cong \overline{RS}$	Given
2. $\overleftrightarrow{QP} \parallel \overleftrightarrow{RS}$	Given
3. $\angle PQS \cong \angle RSQ$	Alternate interior angles
4. $\overline{QS} \cong \overline{QS}$	Identity congruence
5. $\triangle RSQ \cong \triangle PQS$	S.A.S.

Exercise 8.17

Write the necessary statements and reasons to prove each of the following:

1. *Given:*

$\overrightarrow{CD}$ bisects $\angle ACB$
$\overline{AC} = \overline{BC}$

Prove:

$\triangle ADC \cong \triangle BDC$

Fig. 8.50

2. *Given:*

$\overleftrightarrow{MJ} \parallel \overleftrightarrow{KL}$
$\overleftrightarrow{ML} \parallel \overleftrightarrow{JK}$

Prove:

$\triangle JKL \cong \triangle LMJ$

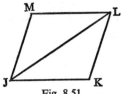

Fig. 8.51

3. *Given:*

$\overline{SR}$ bisects $\overline{PQ}$
$\overline{PQ}$ bisects $\overline{RS}$

Prove:

$\triangle PTS \cong \triangle QTR$

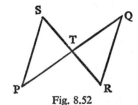

Fig. 8.52

SIMILAR TRIANGLES

Generally speaking, any two geometric figures are similar if they have the same shape but not necessarily the same size, e.g. enlarging photos yields similar figures. For example, we say that any two circles are similar, any two squares are similar, or any two line segments are similar.

Study the two triangles below. Do they appear to have the same shape? Are they the same size? Are they congruent? The triangle DEF has been obtained from triangle ABC simply by doubling the length of the sides.

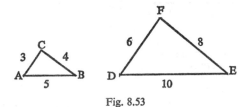

Fig. 8.53

In order that these two triangles should have the same shape, the corresponding angles must be congruent. Just as for congruent triangles, let us arrange the letters of the vertices of two similar triangles so that they indicate the matching of corresponding vertices. For the above triangles we can say

$$\triangle ABC \sim \triangle DEF$$

and read this statement as *triangle ABC is similar to triangle DEF*. Now let us state the ratios of the corresponding sides of these two triangles. Let us use the symbol $\dfrac{\overline{AC}}{\overline{DF}}$ to denote the ratio of the measure of $\overline{AC}$ to the measure of $\overline{DF}$

$$\frac{\overline{AC}}{\overline{DF}} = \frac{3}{6} = \frac{1}{2} \qquad \frac{\overline{AB}}{\overline{DE}} = \frac{5}{10} = \frac{1}{2} \qquad \frac{\overline{BC}}{\overline{EF}} = \frac{4}{8} = \frac{1}{2}$$

We notice that the same ratio exists for each pair of corresponding sides. We refer to this as *corresponding sides are proportional*. Hence, we can conclude the following condition for two triangles to be similar:

Two triangles are similar if corresponding angles are congruent and corresponding sides are proportional.

Furthermore, *if two triangles are similar, then corresponding angles are congruent and corresponding sides are proportional.*

If you construct two triangles so that corresponding angles are congruent you will also find that corresponding sides are proportional. Furthermore, if you construct two triangles such that corresponding sides are proportional you will find that corresponding angles are congruent.

Exercise 8.18

Assume that the two triangles below are similar. Complete the sentences that follow.

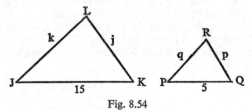

Fig. 8.54

1. If $m(\angle J) = 40$, then $m(\angle P) = ?$
2. If $m(\angle Q) = 70$, then $m(\angle K) = ?$
3. If $k = 12$, then $q = ?$
4. If $p = 2$, then $j = ?$

5. In the figure below $\overline{DE} \| \overline{AB}$. Give the necessary statements and reasons to show that $\triangle ABC \sim \triangle DEC$.

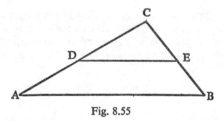

Fig. 8.55

ANGLES OF A TRIANGLE

In the figure below $\overleftrightarrow{AB}$ is drawn parallel to $\overleftrightarrow{PQ}$.

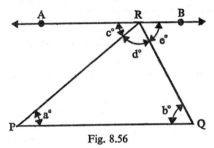

Fig. 8.56

Since $\overleftrightarrow{AB}$ is a straight line, we know that $c + d + e = 180$.

Since $\overleftrightarrow{AB} \| \overleftrightarrow{PQ}$, we know that $a = c$ and $b = e$ because alternate interior angles are congruent when two parallel lines are cut by a transversal.

Since a and c name the same number, we can replace c by a in the sentence $c + d + e = 180$. Similarly, we can replace e by b, since they name the same number. The resulting sentence is as follows:

$$a + b + d = 180$$

Notice that a, b, and d are the measures of the angles of the triangle. Since this argument holds for any triangle, we make the following conclusion:

The sum of the degree measures of the angles of a triangle is 180.

Exercise 8.19

Use $\triangle PQR$ above to answer the following questions:

1. If $a = 30$ and $b = 60$, then $d = $?
2. If $d = 106$ and $a = 24$, then $b = $?
3. If $a = b = d$, then the measure of each angle is ?

MORE ABOUT SIMILAR TRIANGLES

Assume in the two triangles below that $\angle A \cong \angle D$ and $\angle B \cong \angle E$.

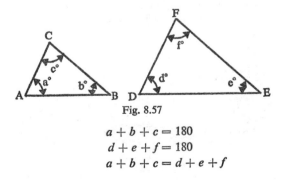

Fig. 8.57

$$a + b + c = 180$$
$$d + e + f = 180$$
$$a + b + c = d + e + f$$

Since $\angle A \cong \angle D$ and $\angle B \cong \angle E$, we know that $a = d$ and $b = e$. Hence, we can use the addition property of equations to deduce from $a + b + c = d + e + f$ that $c = f$ or $\angle C \cong \angle F$.

Therefore, if two angles of one triangle are congruent to two angles of another triangle the third angles are congruent.

USING SIMILAR TRIANGLES

To find the distance between points J and K on opposite banks of a river, Roger used a transit to measure the angles in forming the following figure:

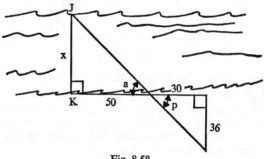

Fig. 8.58

What kind of angles are a and p? Is $\angle a$ congruent to $\angle p$? Since two angles of one triangle are congruent to two angles of the other, the two triangles are similar. Therefore the following proportion holds:

$$\frac{x}{36} = \frac{50}{30}$$

$$x = \frac{50 \times 36}{30}$$

$$x = 60$$

Therefore the measurement of $\overline{JK}$ is 60 units.

Exercise 8.20

Use proportional sides of similar triangles to solve these problems:

1. A pole 8 metres tall casts a shadow 12 metres long at the same time that a tree casts a shadow 42 metres long. How tall is the tree?

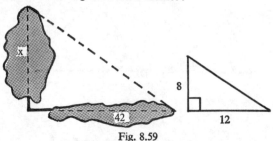

Fig. 8.59

2. To find the distance between points P and Q at opposite ends of a pond a transit was used to measure the angles to form this figure. What is the length of $\overline{PQ}$?

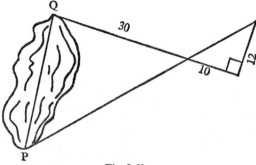

Fig. 8.60

3. The structure of a roof truss formed the triangles shown below. Find the length of $\overline{AB}$.

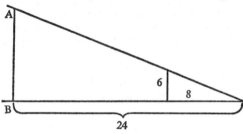

Fig. 8.61

CHAPTER NINE

RELATIONS

Up to now we have discussed operations with numbers or quantities which have enabled us to discover simple relationships between the numbers or quantities concerned. But what precisely is a relation in the mathematical sense? Since the relations are so fundamental in mathematics, we must attempt to acquire some deeper notions about them.

The most fundamental idea in mathematics is that of a set. This embodies the collecting together of quantities or elements to be examined in the very first place. Thus the set is useful in bringing together elements which have common properties. For example, the set of even numbers bring together numbers which are divisible by two. When we begin to do something with the elements which involves some form of comparison, then we are beginning to obtain a relation between the elements. For example, $14 < 16$ means that we are thinking of the relation 'is less than'. The idea of relation is a universal one and is not restricted to mere numbers. Thus the following are all **relations**:

> 'is the wife of'
> 'is the husband of'
> 'is colder than'
> 'is heavier than'
> 'is the owner of'
> 'is slower than'
> 'is a member of'
> 'is younger than'   etc.

Relations which involve the comparison of two objects (as above) are known as **binary relations**, and these will be our main concern. It is, of course, possible to have **ternary relations**, involving three quantities or elements at a time, for example, the relation 'is the sum of two integers' is a ternary relation, and 'is in between' is another ternary relation.

We employ a shorthand notation for this work by letting R represent the relation so that if two elements a and b have the relation R between them or 'are in the relation R' we write a R b.

Suppose R represents 'is older than', then a R b means that

$$a \text{ is older than } b \qquad \text{(i)}$$

It also shows that b R a would be untrue and draws attention to our need to observe the order of the pair of elements a and b. Furthermore, we need some idea of the sets A and B to which a and b belong. The sentence (i) is an example of an open sentence (Chapter Four, page 71) connecting the two

186

variables *a* and *b* so that any open sentence which connects two variables is an example of a binary relation.

Definition 9.1 (*a*)

An open sentence in two variables *a* and *b* where $a \in A$ and $b \in B$ is called a **binary** relation from A to B. If the relation is represented by R, then the binary relation is written *a* R *b*. For the moment we will discuss relations between numbers of the same set. In these cases A = B and we speak of the 'relation *on* the set A'. We can use diagrams to represent binary relations between a small number of elements in the following manner. (These are called **Papygrams**.)

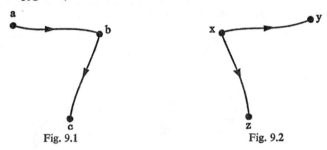

Fig. 9.1   Fig. 9.2

In these diagrams if the relation exists between two elements they will be joined by a line (straight or otherwise) and the direction of the relation indicated by an arrow on the line. Suppose Fig. 9.1 describes the relation 'is older than' on set $\{a, b, c\}$, then we see that

a is older than *b*,
b is older than *c*,

and we could complete the diagram by joining *a* to *c* with an arrow pointing from *a* to *c* because we also know that *a* is older than *c*. The reader must not take it for granted that we can always close up the diagram like this, for consider the same relation on the set $\{x, y, z\}$ of Fig. 9.2.

Here we are informed that

x is older than *y*,
x is older than *z*,

which is insufficient information to say whether or not *y* is older than *z* (e.g. ages of 42, 27, 22 or 42, 22, 27 satisfy the diagram).

Alternatively, the relation concerned might not exist between *y* and *z* at all. For example, consider the relation 'is the father of', then Fig. 9.2 informs us that

x is the father of *y*,
x is the father of *z*

in which case this relation cannot possibly exist between *y* and *z*.

We can also tabulate the relations in the following manner (Figs. 9.3, 9.4). If *a* R *b* exists we enter 1 in the table, if not, then we enter 0. Thus in Fig. 9.1 we see that *a* R *b* exists but *b* R *a* does not. Thus the entry with *a* taken first and *b* second is 1, while the entry with *b* taken first and *a* second is 0. The table for the relation of Fig. 9.1 should now be examined. The actual

second element

		a	*b*	*c*
first	*a*	0	1	1
element	*b*	0	0	1
	c	0	0	0

Fig. 9.3

second element

		x	*y*	*z*
first	*x*	0	1	1
element	*y*	0	0	0
	z	0	0	0

Fig. 9.4

table of the relation which is contained by the brackets is called a **matrix**. The matrix of Fig. 9.4 carries the information for the relation 'is the father of' on the set {*x*, *y*, *z*}, which we have already discussed in Fig. 9.2.

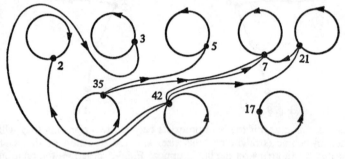

Fig. 9.5

Let us examine another relation diagram in Fig. 9.5. The relation concerned is 'is a multiple of' on the set {2, 3, 5, 7, 21, 35, 42, 17}. The first thing the reader probably notices is the circular path which joins each element to itself. Take, for example, 17, this being 17 × 1. We are saying that each number is a multiple of itself, and consequently each element of the set is in

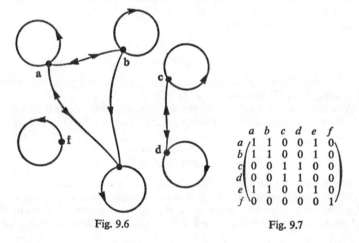

	a	*b*	*c*	*d*	*e*	*f*
a	1	1	0	0	1	0
b	1	1	0	0	1	0
c	0	0	1	1	0	0
d	0	0	1	1	0	0
e	1	1	0	0	1	0
f	0	0	0	0	0	1

Fig. 9.6   Fig. 9.7

relation to itself. The writing of the matrix for this relation is left as an exercise in Exercise 9.1. One join has been omitted from the figure, which one?

So far all the relations we have discussed have been 'one way', that is, there has been only one arrow on the joins of the diagrams. An example of a 'two-way' relation is given in Fig. 9.6, which gives the relation 'drives the same type of car as' on the set of drivers $\{a, b, c, d, e, f\}$. Now clearly each driver 'drives the same type of car as himself', so we have the self-joining circular paths as for Fig. 9.5, and also if 'a drives the same type of car as b', then 'b drives the same type of car as a'. Two entries on the figure need to be completed by the reader.

If we now look at the matrix for this relation in Fig. 9.7 we notice the same sort of symmetry as we observed in Chapter Six, page 119. But, notice particularly the presence of 1s instead of 0s in the leading diagonal of the matrix.

Exercise 9.1

1. Draw a relation diagram for the matrix

$$
\begin{array}{c}
\begin{array}{ccc} a & b & c \end{array} \\
\begin{array}{c} a \\ b \\ c \end{array}
\begin{pmatrix} 1 & 0 & 1 \\ 0 & 1 & 0 \\ 1 & 0 & 1 \end{pmatrix}
\end{array}
$$

2. Suggest a relation which satisfies the matrix of Question 1.
3. a is 70 years of age and has a son c and a daughter d; b is 65 years of age and is the brother of a; c is 40 years of age and d is 38 years of age and also has a son e of age 10.

Draw diagrams for the relations:

   (a) 'is older than'
   (b) 'is the brother of'
   (c) 'is the parent of'
   (d) 'is the grandfather of'

4. The steamer service between a set of Greek islands $\{\alpha, \beta, \gamma, \delta\}$ is given by the network diagram of Fig. 9.8.

Fig. 9.8

   (a) Construct the matrix of this network.
   (b) Obtain the relation within the set.

5. A family consists of three brothers b_1, b_2, b_3 and two sisters s_1, s_2. Draw relation diagrams for

   (a) 'is the sister of'
   (b) 'is the brother of'

6. Construct the matrix for the relation diagram of Fig. 9.5.

CLASSIFICATION OF RELATIONS

Since certain types of relations are more common in mathematics, we give them special names.

Reflexive Relations

If every element under consideration is in relation with itself, then the relation is said to be reflexive. For example, the relation 'is a multiple of' on the set of numbers {2, 3, 5, 7, 21, 35, 42, 17} was a reflexive relation. Such a relation leads to 'self-join' paths on the diagram and 1s in the leading diagonal of the matrix of the diagram. 'Drives the same type of car as' (see Fig. 9.6) was another reflexive relation. (*e* needed to be inserted and also an arrow-head pointing from *e* to *b*.)

Symmetric Relations

We referred to this relation earlier on as being two way. This meant that the relation was such that the two elements in relation could in fact be interchanged. The relation 'drives the same type of car as' is a symmetric relation because

from *a* R *b* it follows that *b* R *a*.

In this case one statement implies the other.

Another example is R 'is equal to' because if *a* = *b*, then *b* = *a*.
Notice that the relation of Fig. 9.6 is both reflexive and symmetric.
The relation of Fig. 9.5 is reflexive but not symmetric.

Anti-symmetric Relations

These are the obvious ones to speak about next. Here, of course, the elements cannot be interchanged, thus if *a* R *b*, then *b* is not in relation to *a*, i.e. *b* ℟ *a*. The relation 'is older than' is an anti-symmetric relation (see Fig. 9.1) because clearly if '*a* is older than *b*', then '*b* can never be older than *a*'. The relation 'is the father of', which was illustrated in Fig. 9.2, was also an anti-symmetric relation. Note that some relations are neither symmetric nor anti-symmetric, e.g. 'is the sister of' on a family of boys and girls.

Transitive Relation

This is a very important relation, as its name implies it has a crossing over or transfer property. Let us look first at this in symbolic form.

If   *a* R *b*   and   *b* R *c*,
then   *a* R *c*

The very first example we discussed (in Fig. 9.1) was the relation 'is older than'. You remember that we had

a is older than *b*
b is older than *c*

and from this we concluded that a is older than c; in other words, we were able to cross over b and go straight from a to c with the relation. Another example is the relation 'is greater than', thus if $a > b$ and $b > c$, then $a > c$, so we say that 'is greater than' is a transitive relation.

Notice, by the way, that the relation 'is equal to' is reflexive, symmetric, and transitive.

The relation 'is greater than' is anti-symmetric and transitive.

It is clear that some relations have more to commend them than others, that is, they belong to more of the above classifications than others. We therefore pick out a further classification or type and these are called:

Equivalence Relations

An equivalence relation is a relation which is reflexive, symmetric, and transitive. As already pointed out 'is equal to' is an equivalence relation, but 'is greater than' is not an equivalence relation.

During the course of reading the above the reader must have noticed the frequency with which the phrasing 'if . . . then', or 'implies that' occurred. Since this is so common, we have a shorthand form for this expression. It is written $\Rightarrow$

For example:

    (i) R is symmetric when $a\ R\ b \Rightarrow b\ R\ a$
    (ii) R is transitive when $(a\ R\ b$ and $b\ R\ c) \Rightarrow a\ R\ c$

Order Relations

Certain relations in a set allow the elements to be arranged in order. Any such relation is called an order relation. For example, the relations 'is less than', 'is greater than', 'is older than', 'is younger than', 'is heavier than', etc., are all order relations. An order relation, however, must have certain properties. The relation must be:

    (i) anti-symmetric;
    (ii) transitive.

These ideas are not new to us, but the following is:

    (iii) the relation must hold for any two members of the set.

The last point is substantiated simply by considering what to do with the elements of the set which are not in the relation—they are therefore impossible to order. This third property (iii) is called **completeness**.

We express the completeness relation more formally by saying that if a and b are any two distinct elements of the set S, i.e. $a \in$ S, $b \in$ S, then the relation is complete if $a\ R\ b$ or $b\ R\ a$. Thus any relation which is anti-symmetric, transitive, and complete, is an order relation.

NOTE: All the above are the properties of the relations and are quite independent of the sets in which the relations hold.

Exercise 9.2

(i) is a factor of	on (set of all natural numbers)
(ii) is half of	on (set of all natural numbers)
(iii) is equal to	on (set of all natural numbers)
(iv) is the father of	on (set of all people)
(v) is the child of	on (set of all people)
(vi) is a member of the same team as	on (set of all people)
(vii) is shorter than	on (set of all people)
(viii) is in love with	on (set of all people)
(ix) is a brother of (boy to girl or boy to boy)	on the set of all people.

Examine the relations above, numbers (i)–(ix), and then answer the following questions:

1. Which of the above relations are reflexive?
2. Which of the above relations are symmetric?
3. Which of the above relations are antisymmetric?
4. Which of the above relations are transitive?
5. Which of the above relations are equivalence relations?
6. Which of the above relations are order relations?
7. Suggest an order relation in the set $\{2, 4, 8, 16\}$.
8. Why isn't the relation 'is a factor of' an order relation on the set $\{2, 3, 6\}$.
9. If a R b is '$a + b$ is divisible by 5', where a and b belong to the set of natural numbers, classify R.
10. If a R b is '$a - b$ is divisible by 5', where a and b are integers, classify R.

We have thus seen that a pair of elements is associated with each binary relation defined on a set. Furthermore, we have seen that each pair is ordered, i.e. a R b or b R a.

Consider the relation 'is greater than' on the set $\{1, 3, 7, 15\}$. Here we have

$$15 > 7, 15 > 3, 15 > 1$$
$$7 > 3, 7 > 1$$
$$3 > 1$$

We can dispense with $>$ and write these statements as the following ordered pairs:

$$(15, 7); (15, 3); (15, 1); (7, 3); (7, 1); (3, 1)$$

(ordered in the sense that the first number is the greater member of the pair).

The relation 'is factor of' on the same set may be represented by the ordered pairs.

$$(1, 1); (1, 3); (1, 7); (1, 15)$$
$$(3, 3); (3, 15)$$
$$(7, 7)$$
$$(15, 15)$$

We now discuss relations which compare or associate the elements of two different sets, but first of all we have the following definition:

Definition 9.1 (*b*)

An open sentence in two variables a and b where $a \in$ A and $b \in$ B is called a binary relation R from set A to set B, and written a R b.

To get a clearer idea of this definition consider the two sets

$$A = \{1, 3, 4, 8\} \qquad\qquad B = \{2, 3, 5, 8, 9\}$$

We are able to form the following twenty ordered pairs:

<u>(1, 2)</u>	(3, 2)	(4, 2)	(8, 2)
[(1, 3)]	(3, 3)	(4, 3)	(8, 3)
<u>(1, 5)</u>	[(3, 5)]	<u>(4, 5)</u>	(8, 5)
<u>(1, 8)</u>	<u>(3, 8)</u>	<u>(4, 8)</u>	(8, 8)
<u>(1, 9)</u>	<u>(3, 9)</u>	<u>(4, 9)</u>	<u>(8, 9)</u>

Fig. 9.9

This set of ordered pairs is called the cartesian product and written $A \times B$.

The elements in each ordered pair are called the **first co-ordinate** and **second co-ordinate** respectively and Fig. 9.9 is called the **co-ordinate diagram of** $A \times B$.

Since we have written down all the possible ordered pairs, it follows that if a relation exists it is completely specified by a subset of $A \times B$.

Definition 9.2

A binary relation R from set A to set B is any subset of $A \times B$.

We have now arrived at a much wider definition of a relation which employs the essential property that it is represented by a set of ordered pairs.

The term 'wider' definition is used because we can now select the set of ordered pairs without looking for the name of the relation such as 'is greater than' and so on, because one way of naming a set is to list its elements.

In Fig. 9.9 the underlined ordered pairs specify the relation 'is less than', while those in square brackets specify the relation 'is two less than'.

The set of first co-ordinates in the relation R is called the **domain** of R. The set of second co-ordinates in the relation R is called the **range** of R. Note that the domain is a subset of A, the range is a subset of B, and R is a subset of the cartesian product $A \times B$.

THE INVERSE RELATION

Definition 9.3

The inverse relation of R is the relation obtained by **interchanging** the domain and range, and will be written R^{-1}.

Thus if R is given by the set of ordered pairs

$$(a_1, b_1); \quad (a_2, b_2); \quad (a_3, b_3); \quad (a_4, b_4)$$

then the inverse relation is given by the set of ordered pairs

$$(b_1, a_1); \quad (b_2, a_2); \quad (b_3, a_3); \quad (b_4, a_4)$$

For example, in the relation between sets A and B illustrated by Fig. 9.9 relation R 'is less than' was underlined.

The inverse relation is the subset of ordered pairs given by Fig. 9.10.

(2, 1)			
(3, 1)			
(5, 1)	(5, 3)	(5, 4)	
(8, 1)	(8, 3)	(8, 4)	
(9, 1)	(9, 3)	(9, 4)	(9, 8)

Fig. 9.10

This is clearly the relation 'is greater than' in the inverse relation. The original range is now the domain, and the original domain is now the range for the inverse relation.

THE GRAPH

The graph of a relation is simply the indication of which ordered pairs are in the relation within the diagram of the cartesian product A × B. Thus in Fig. 9.9 the graph of the relation 'is less than' is given by underlining the ordered pairs, while the graph of the relation 'is two less than' is given by placing the pairs inside square brackets. Any such method is said to be **graphing the relation**. The ordered pairs in the relation are also referred to as **the set of the binary relation**, a phrase with which we are already familiar (see page 192).

If R_1 'is less than' and R_2 'is two less than' (both graphs in Fig. 9.9), then we can see straight away that $R_1 \cap R_2$ has the graph (1, 3), (3, 5) in Fig. 9.9, while $R_2 \subset R_1$.

Exercise 9.3

1. If A = {1, 3, 10}; B = {2, 3, 11} draw the diagram for the cartesian product A × B.

2. Using your answer to Question 1, graph the relation R_1 'is greater than'.

3. The relation 'is mother of' has A = {women} as domain and B = {children} as the range. Classify this relation and state its inverse.

4. Using your answer to Question 1, graph the relation R_2 'is one more than' and give the elements of the sets (*a*) $R_1 \cap R_2$; (*b*) $R_1 \cup R_2$; (*c*) if R_3 'is one less than' what is $R_2 \cup R_3$? and $R_2 \cap R_3$?

5. If A = {1, 2, 3, 4}; B = {1, 2, 3, 4}, draw the cartesian product for A × B and graph the relation $a = b + 1$ where $a \in$ A and $b \in$ B.

CO-ORDINATE AXES

When the domain and range of a relation are each A = {the set of integers} the diagram of the cartesian product is infinite in size, but we can get a good idea of its pattern or form by examining a small section illustrated in Fig. 9.12, in which the ordered pairs of the product A × A have been deliberately arranged.

We may now graph the relation by underlining the elements of the relation subset of the above diagram. A moment's thought suggests that we ought to be able to streamline the writing of the cartesian product by the following method:

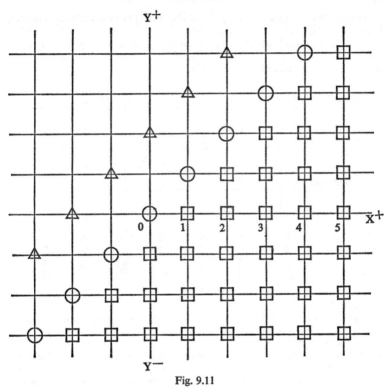

Fig. 9.11

We observe that everywhere along the line $Y^- Y^+$ the first co-ordinate is zero, while everywhere along the line $X^- X^+$ the second co-ordinate is zero. Consequently, these two lines could be taken as reference axes or starting lines for the values of the co-ordinates. These two number lines are called the **co-ordinate axes.**

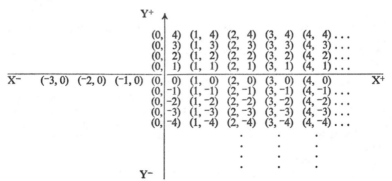

Fig. 9.12

We now construct a complete grid for the diagram of the cartesian product, as given in Fig. 9.13, until we can identify any ordered pair by the intersections on the grid. Furthermore, we can draw a graph of any relation in the product A × A simply by using dots to pick out the relevant ordered pairs. We give one or two examples of this in Fig. 9.11.

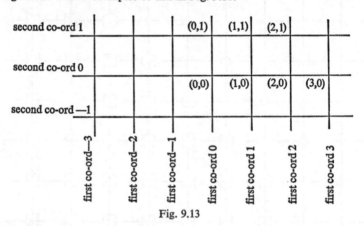

Fig. 9.13

Example 1

Consider the relation 'is equal to' with A = {the set of integers} for a product A × A. The relation is the set of ordered pairs (−4, −4), (−3, −3), (−2, −2), (−1, −1), (0, 0), (1, 1), (2, 2), (3, 3). . . . .

A graph of this relation is given by the set of round dots in Fig. 9.11. The reader will probably notice that these dots all lie on a straight line passing through the origin. (Clearly we can only give some of the dots because the set has an infinite number of elements.)

Example 2

Consider now the relation 'is greater than' in the same product A × A. This will be the set of all ordered pairs in which the first co-ordinate is greater than the second co-ordinate. The graph of this relation is given by the set of square dots ■, and as before we can only draw some of them.

Example 3

Consider the relation 'is 2 less than' in the same cartesian product A × A. A few ordered pairs of this set are (1, 3), (2, 4), (−1, 1), and the graph of the relation is given by the triangular dots ▲, and we notice incidentally that these lie on a straight line (see Example 1).

In the last three examples we chose A as the set of integers. In Examples 1 and 3 the graph of the relation was a set of isolated dots on a straight line. It would be incorrect to say that the graph of the relation *is* the straight line, because this implies that all the possible points or dots on the straight line belonged to the graph of the relation. For example, in the relation 'is equal to' on the set of integers the ordered pair $(2\frac{1}{2}, 2\frac{1}{2})$ would lie on the straight line, but is not an element of the relation.

Suppose we were to take A = {set of rational numbers} this would fill the spaces in, but we would still be unable to say that the graph of the relation *was* the straight line because it would not contain ordered pairs such as $(\sqrt{2}, \sqrt{2})$, for $\sqrt{2}$ is irrational.

In fact, in order to be able to say that the graph of the relation is a straight line, we must take A as the set of real numbers. As already pointed out, the definition of real numbers is beyond the scope of the book. But you can probably see how helpful it is to define a straight line as being a set of points. Let us have another look at our graphs of relations using the set of real numbers as the domain and range.

To refer to the elements of each ordered pair in the cartesian product A × A as being first co-ordinate and second co-ordinate is a little tedious, also we shall only be able to write down a small number of elements of the relation set. It would be very convenient if we could speak of the general element of the relation set, and just as we found it convenient to let the letter *n* stand for a general number, so we refer to the ordered pair (x, y) as a general element in the cartesian product A × A.

Most readers who already have some knowledge of Analytic Geometry will recognize this as **the cartesian co-ordinate system**. The relation 'is equal to' will therefore be adequately stated as

$$R_1: \{(x, y) \mid x = y\}$$

This reads: the relation R_1 is the set of all order pairs (x, y) such that $x = y$.

As we have already discussed, the graph of this relation is a straight line— meaning the set of all points in the straight line.

In Example 2, which is given above, was the relation 'is greater than'. In shorthand this may be stated as

$$R_2: \{(x, y) \mid x > y\}$$

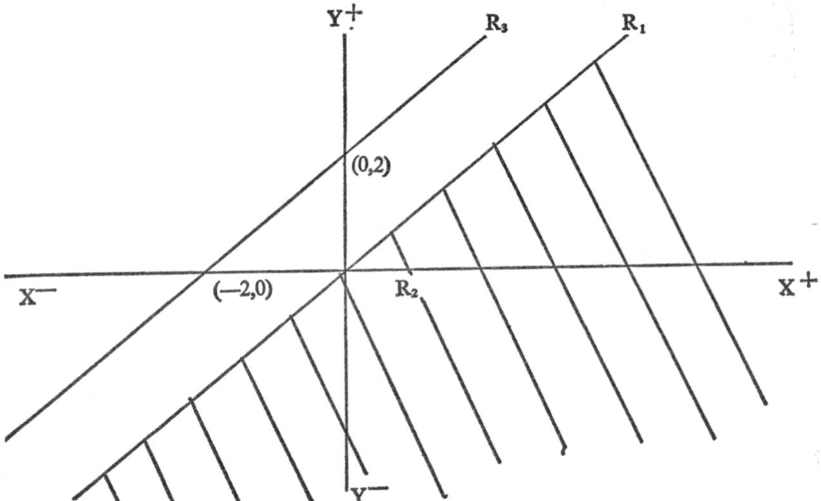

Fig. 9.14

This reads: the relation R_2 is the set of all ordered pairs (x, y) such that $x > y$.

Example 3 without more ado may be written as

$$R_3: \{(x, y) \mid x = y - 2\}.$$

Fig. 9.14 shows the relations R_1, R_2, and R_3.

The really important relation (see page 196) is R_2. *Anywhere* in the shaded region of Fig. 9.14 there is an ordered pair which belongs to the relation R_2 but not on the boundary, which, of course, is the graph of R_1.

Note: The whole of the shaded area is the graph of R_2.

A selection of graphs of other relations now follows in Figs. 9.15–9.20.

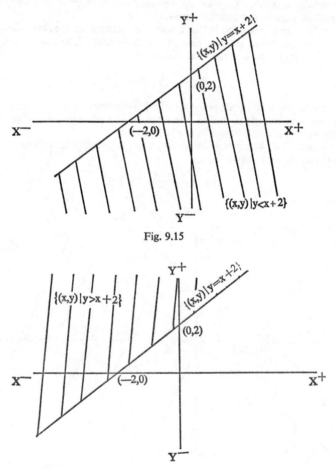

Fig. 9.15

Fig. 9.16

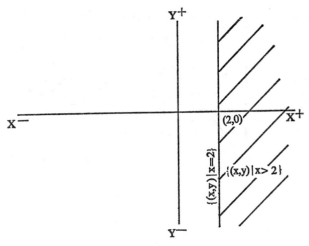

Fig. 9.17

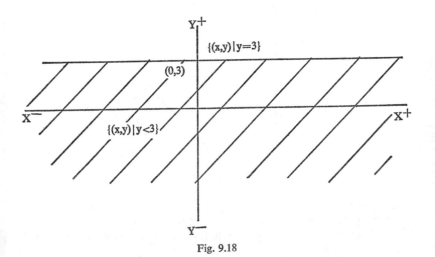

Fig. 9.18

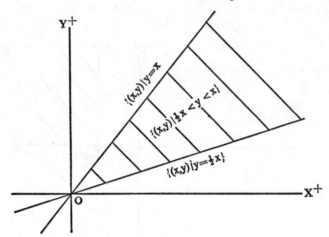

Fig. 9.19

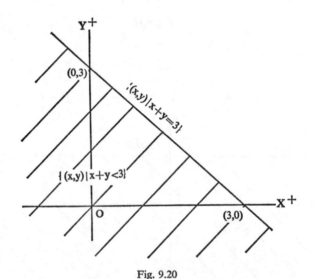

Fig. 9.20

Exercise 9.4

Each of the following relations is on the set of real numbers.

1. In Fig. 9.15 shade the region which is the graph of the relation $\{(x, y) \mid x < y - 2\}$.

2. In Fig. 9.16 shade the region which is the graph of the relation $\{(x, y) \mid x > y - 2)\}$.

3. In Fig. 9.17 shade the region which is the graph of the relation $\{(x, y) \mid x < 2\}$.

4. In Fig. 9.18 shade the region which is the graph of the relation $\{(x, y) \mid y > 3\}$.

5. In Fig. 9.19 shade the region which is the graph of the relation $\{(x, y) \mid \frac{1}{2}x > y\}$.

6. In Fig. 9.20 shade the region which is the graph of the relation $\{(x, y) \mid x + y > 3\}$.

7. In Fig. 9.14 shade the region which is the graph of the relation $\{(x, y)x > y - 2\} \cap \{(x, y) \mid x < y\}$.

8. Draw the graph of the relation R: $\{(x, y) \mid x > 1, y > 2\}$.

FUNCTIONS

In some of the relations we have discussed the reader may have noticed that the relation does not always provide one and only one partner to each element of the set A in the products $A \times B$ or $A \times A$.

For example, the relation 'is equal to' R_1: $\{(x, y) \mid x = y\}$ on the set of real numbers is such that to each member or element of the domain there corresponds one, and only one, member or element of the range. This is an example of a one-to-one correspondence.

On the other hand, the relation R_2: $\{(x, y) \mid x > y\}$ on the set of real numbers is such that to each element x of the domain there corresponds any number of elements y of the range. For example, suppose we take $x = 7$ as the first co-ordinate, then $y = 6\frac{1}{2}, 6, 5\frac{1}{2}, 5\frac{1}{4}, 5, 4\frac{1}{2}, 4, \ldots$ may be taken as the corresponding second co-ordinate. This is an example of a one-to-many correspondence. In the relation R: $\{(x, y) \mid y = 3\}$, the graph of which is given in Fig. 9.18, we see that to each element x of the domain there corresponds only one element y of the range. This is an example of a many-to-one correspondence.

It is clear that there is something special about the one-to-one correspondence, so we choose to give this sort of relation the special name of **function**. (We also call many-to-one correspondences, functions.)

Definition 9.4

A function f from set A into set B is a subset of $A \times B$ such that for each and every element of A there corresponds one, and only one, element of B. Note that the domain of f *is* the set A and the range of f is either set B *or* a subset of B, as indicated in the diagrams.

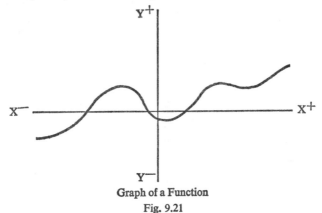

Graph of a Function
Fig. 9.21

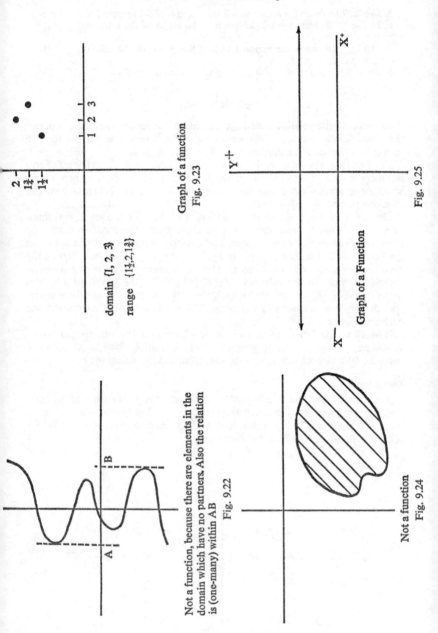

domain {1, 2, 3}

range {1½, 2, 1¾}

Graph of a function
Fig. 9.23

Not a function, because there are elements in the domain which have no partners. Also the relation is (one-many) within AB

Fig. 9.22

Not a function
Fig. 9.24

Graph of a Function

Fig. 9.25

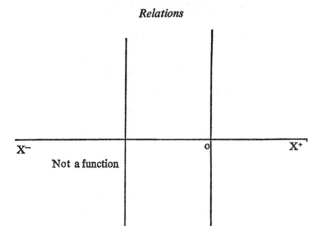

Fig. 9.26

Reminder—when the diagram uses the number lines XX, YY it is to be understood that the domain and the range are in the set of real numbers.

Another type of diagram to underline the idea of a function is given in the following; the joins of the elements give the correspondence.

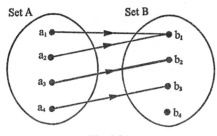

Fig. 9.27

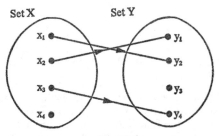

Fig. 9.28

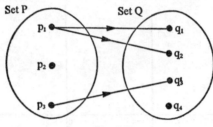

Fig. 9.29

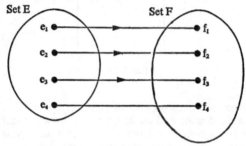

Fig. 9.30

Details

Fig. 9.27. To each *a* corresponds, one and only one, *b*. Therefore the relation is a function.

Fig. 9.28. There is no corresponding element to x_4. Therefore the relation is not a function.

Fig. 9.29. There are two corresponding elements to p_1. Therefore the relation is not a function.

Fig. 9.30. To each and every *e* there corresponds one, and only one, element *f*. Therefore the relation is a function. An arrow should be inserted on the last join.

Summary

For a relation with domain A and range in B to be a function, each element of A must correspond with or be matched with one, and only one, element of B.

INVERSE RELATIONS AND FUNCTIONS

The diagrams of Figs. 9.27, 9.28, 9.29, 9.30 will provide us with the inverse relation if we simply reverse the direction of the arrows. We notice at once the following details:

Fig. 9.27. Relation was a function; inverse relation not a function.
Fig. 9.28. Relation not a function; inverse relation not a function.
Fig. 9.29. Relation not a function; inverse relation not a function.
Fig. 9.30. Relation a function; inverse relation a function.

Exercise 9.5

1. Given A = {1, 2, 3, 4} and B = {2, 3, 4, 5, 9}, which of the following relations are functions in A × B?

    (*a*) 'is one less than'
    (*b*) 'is less than'
    (*c*) 'is equal to'

2. Draw diagrams similar to Figs. 9.27–9.30 for 1 (*a*), (*b*), (*c*).
3. Which of the following are graphs of a function?

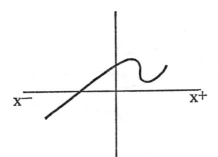

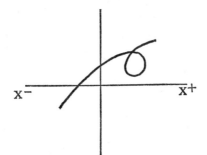

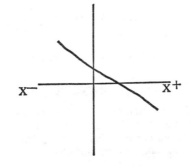

4. Which of the following diagrams illustrates a function?

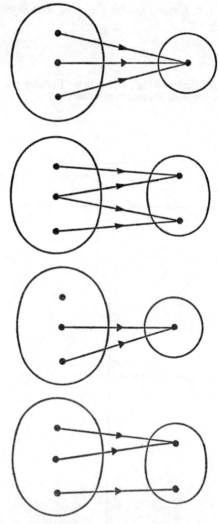

5. Which inverse relations in Question 4 are functions?

FUNCTION NOTATION

In a relation which is a function, since the elements of the domain correspond to one, and only one, element in the range, we also call the correspondence a many-to-one correspondence or **mapping**. Consequently, we use a mapping sign → in our notation for functions.

Thus we say that the function $\{(x, y) \mid y = 2x\}$ on the set of real numbers **maps** each x to twice its value $2x$, or, using the mapping notation, $x \to 2x$.

For y we sometimes write $f(x)$; read 'f of x' or 'f at x' (sometimes called the value of the function at the point x). We use the notation $f: A \to B$ to indicate that the relation between elements of A and B is a function. If the range is a proper subset of B then the mapping is said to be 'into'. If the range is the whole set B then the mapping is said to be 'onto'.

Example 1

$f_1 : \{(x, y) \mid y = x^2\}$ in mapping notation is written $x \to x^2$.
Typical ordered pairs of the function are:

(1, 1), (2, 4), (3, 9), (4, 16), (5, 25), etc.
(−1, 1), (−2, 4), (−3, 9), (−4, 16), (−5, 25) . . .

A graph of the function is:

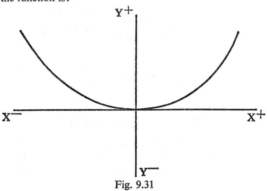

Fig. 9.31

Example 2

$f_2 : \{(x, y) \mid y = x^3\}$ in mapping notation is written $x \to x^3$.
Typical ordered pairs of the function are:

(1, 1), (2, 8), (3, 27), (4, 64), (5, 125) . . .
(−1, −1), (−2, −8), (−3, −27), (−4, −64), (−5, −125) . .

A graph of the function is:

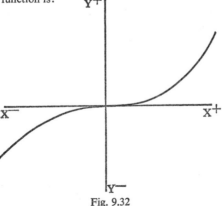

Fig. 9.32

Example 3

$f_3 : \{(x, y) \mid y = x + 3\}$ in mapping notation written as $x \to x + 3$.

A graph of the function is:

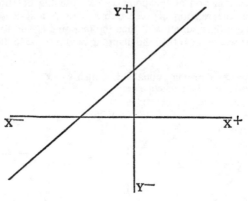

Fig. 9.33

COMPOSITION OF FUNCTIONS

If we have two functions f and g, then the **composite function** $f \cdot g$ is to mean that we shall obtain the results of mapping g first and then carry out the mapping f on these results second. (The reader is warned that some authors take the reverse order.)

Example 1

g is the mapping $x \to x + 2$

f is the mapping $x \to x^2$

Then $f \cdot g$ is the mapping $x \to (x + 2)^2$ but $g \cdot f$ is the mapping $x \to x^2 + 2$. Thus $f \cdot g$ is not equal to $g \cdot f$ or $f \cdot g$ is not the same relation as $g \cdot f$.

One or two numerical values will probably help in understanding the above.

Suppose $x = 7$ then $g : 7 \to 9$; $f : 7 \to 49$

and $f : 9 \to 81$; $g : 49 \to 51$

$f \cdot g$: $7 \to 81$ and $g \cdot f : 7 \to 51$

Again suppose $x = -2$, then $g : -2 \to 0$; $f : -2 \to 4$

f: $0 \to 0$; g: $4 \to 6$

$f \cdot g : -2 \to 0$; $g \cdot f : -2 \to 6$

Example 2

Given that $g : x \to 2x$

$f : x \to 3x$ show that $f \cdot g = g \cdot f$

We have $f \cdot g : x \to 3(2x)$; $g \cdot f : x \to 2(3x)$; $f \cdot g = g \cdot f$

For a numerical value take $x = 11$, then

$g : 11 \to 22$; $f : 22 \to 66$

$f : 11 \to 33$; $g : 33 \to 66$

Example 3

We shall now look at a composite function in diagram form:

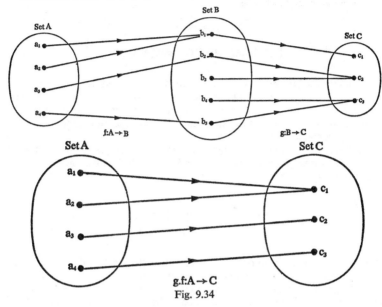

g.f:A → C
Fig. 9.34

Is the composition of the two functions also a function? Yes. It can thus be seen that we are often able to construct new functions by the composition of other functions. This is common mathematical practice, and is particularly useful in the case of functions which have the same domain and range, for under these circumstances the composite functions always exist.

Exercise 9.6

1. In the example 3 above, is the inverse of $g \cdot f : A \to C$ a function?

2. If $f : x \to 7x$ and $g : x \to x^2$ where the domain and range is the set of real numbers, state in mapping notation the functions.

    (a) $f \cdot g$ (b) $g \cdot f$ (c) $f \cdot f$ (d) $g \cdot g$

and map the value $x = 2$ in each case.

3. If $f : \{(x, y) \mid y = x^2\}$ what is the inverse relation?
Is this inverse relation a function?

4. Examine the graph of Figs. 9.31, 9.33 and then draw the graphs of the following relations:

    (a) $f_1 \cup f_3$ (b) $f_1 \cap f_3$

5. Draw the graph of the relations:

$$g_1 : \{(x, y) \mid y = x\}$$
$$g_2 : \{(x, y) \mid y = -x\}$$
$$g_3 : \{(x, y) \mid x = 2\}$$

and indicate the graphs of

    (a) $g_1 \cap g_2$ (b) $g_2 \cap g_3$ (c) $g_3 \cap g_1$
    (d) $g_4 : \{(x, y) \mid -x < y < x; x < 2\}$

Which of the relations g_1, g_2, g_3, g_4 are not functions? Try to give a reason.

CHAPTER TEN

LINEAR PROGRAMMING

The method of linear programming depends almost entirely on being able to draw the graphs of equality and inequality relations. We have given attention to these relations in the previous chapter, but just in case some readers have found difficulty with their study of Chapter Nine, the authors offer the following introduction.

It will be assumed that the reader is familiar with the equation to the straight line in the three given forms. (See *Additional Mathematics Made Simple*, Chapter Six.)

$$1.\ y = mx + c$$
$$2.\ Ax + By = C$$
$$3.\ \frac{x}{a} + \frac{y}{b} = 1$$

Clearly, each of these equations will be satisfied by a set of ordered pairs (x, y) which the reader evaluates of his own choice.

For example, suppose we are given a straight line $y = 2x + 3$. Then it is up to the reader to decide which pairs he wishes to quote. It is usual to start with any suggestion for the first co-ordinate and then use the given equation to work out the second, but one must be sensible about it. For instance, we could suggest $x = 1\ 907\ 845{\cdot}6$, but think of the scale and size of paper in order to get this value on the cartesian diagram. So, we choose 'sensible' values such as $x = 1$ or 2 or 3 or $-1, -2, -3$, and so on.

Let us go ahead with the suggestion $x = 1$, then the corresponding value of y is $y = (2 \times 1) + 3 = 5$.

Therefore $(1, 5)$ is an ordered pair which belongs to the graph of the relation $\{(x, y) \mid y = 2x + 3\}$.

Another suggestion is $x = 2$, in which case the corresponding value of y is

$$y = (2 \times 2) + 3 = 7$$

so that $(2, 7)$ is another ordered pair which belongs to this relation.

Pausing here for a moment, we see that we now know: (*a*) the graph will be a straight line; (*b*) two points on the line. Consequently, we are in a position to draw the graph, but suppose we have made an error? We had better work out another ordered pair to make certain. This will be the procedure for the future, i.e. we recognize the equation as being that of a straight line, and then we evaluate three ordered pairs for the line, thus giving us two pairs for actually drawing the line and one for a check.

For the line $y = 2x + 3$ to get the third pair (for a check), let us take $x = -1$, then the corresponding value of y is

$$y = -2 + 3 = 1$$

A tidy procedure is to enter these pairs in an array as follows:

x	-1	1	2
y	1	5	7

We have drawn the line in Fig. 10.1 and labelled it *l*.

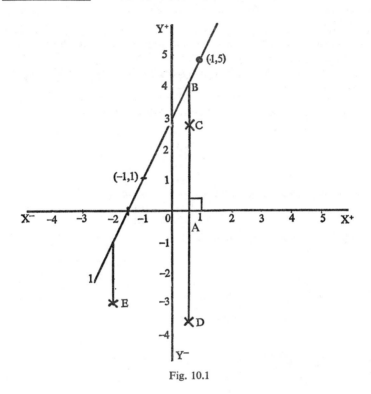

Fig. 10.1

Now this line divides the cartesian plane into three sets.

1. The set of points for which $y = 2x + 3$, i.e. the line itself.
2. The set of points for which $y > 2x + 3$, which is called an open half-plane. (If we put $y \geqslant 2x + 3$ it would be a closed half-plane.)
3. The set of points for which $y < 2x + 3$. The reader should now label the intersection of BD and X⁻X⁺ as the point A.

We demonstrate 3 with the aid of the lines AB and AC. C is any point in the lower half-plane. Through C draw a line parallel to the y-axis meeting the line at B and the x-axis at A. Then B and C are two points in the plane which have the same x co-ordinate. But the corresponding y co-ordinate is measured by the length of AB for B, and the length of AC for C.

Since AC < AB for any point C in the lower half-plane, then y for C is less than y for B. But y for B is $2x + 3$

$$\therefore \quad y < 2x + 3 \text{ in the lower half-plane.}$$

To complete the demonstration consider the point D below the x-axis. D also has the same x co-ordinate as B, but the corresponding second co-ordinate is measured by AD, a negative quantity, so, as before, D is a member of the relation $y < 2x + 3$.

The equation in form $\dfrac{x}{a} + \dfrac{y}{b} = 1$ is very interesting because it enables the line to be drawn very quickly.

Once again we choose our own ordered pairs (x, y), but by this time we realize that $x = 0$ is the easiest one to try.

Consider the line $\dfrac{x}{2} + \dfrac{y}{3} = 1$
If $x = 0$, then $y = 3$
If $y = 0$, then $x = 2$
If $x = 1$, then $y = 1\frac{1}{2}$

The array of results is given by

x	0	2	1
y	3	0	$1\frac{1}{2}$

and the line is drawn in Fig. 10.2.

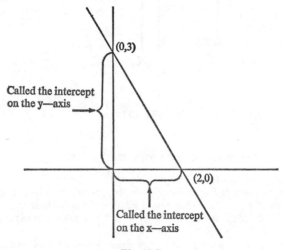

(0,3)

Called the intercept
on the y—axis

(2,0)

Called the intercept
on the x—axis

Fig. 10.2

From the labelling in this figure we see why this equation is called the **intercept form**.

Examine the graphs in Fig. 10.3 and write down their equation in intercept form immediately.

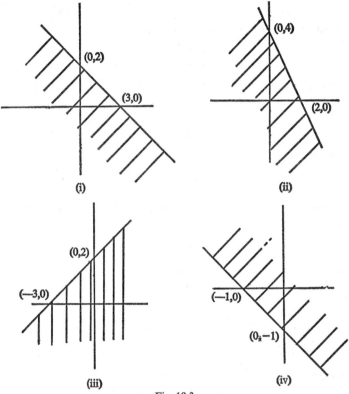

Fig. 10.3

The equations were:

(i) $\dfrac{x}{3} + \dfrac{y}{2} = 1$ (ii) $\dfrac{x}{2} + \dfrac{y}{4} = 1$

(iii) $\dfrac{x}{-3} + \dfrac{y}{2} = 1$ (iv) $\dfrac{x}{-1} + \dfrac{y}{-1} = 1$

Or we can modify these to the form

(i) $y = 2 - \dfrac{2x}{3}$ (ii) $y = 4 - 2x$ (iii) $y = 2 + \dfrac{2x}{3}$ (iv) $y = -x - 1$

You can see how easy it is to obtain the equation if we know the intercepts The shaded half-planes are regions for which the following inequalities hold.

(i) $y < 2 - \dfrac{2x}{3}$ (ii) $y < 4 - 2x$ (iii) $y < 2 + \dfrac{2x}{3}$ (iv) $y > -x - 1$

or the shaded half-planes are graphs of the relations (i), (ii), (iii), and (iv).

Exercise 10.1

1. Draw the graph of the relation $R_1 : \{(x, y) \mid x + y = 1\}$ and identify the two half-planes.

2. Repeat Question 1 for $R_2 : \{(x, y) \mid x + y = -1\}$.

3. Repeat Question 1 for $R_3 : \{(x, y) \mid y = 2x + 3\}$.

Write down the three relations whose graphs are represented by the following:

4.

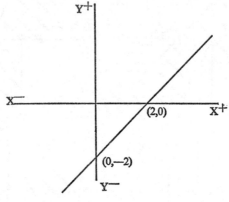

5.

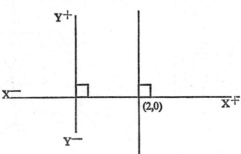

6.

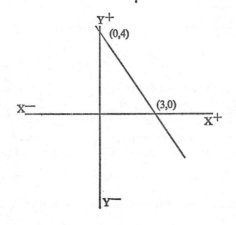

7.

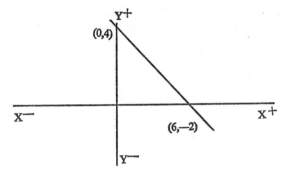

8.

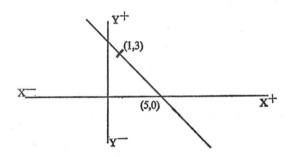

So far we are only finding the graphs of one inequality relation at a time. The object of the next exercise is to determine the graph of ordered pairs which belong to two, three, or more inequality relations simultaneously.

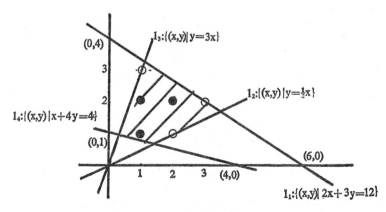

Fig. 10.4

The shaded region in Fig. 10.4 is the graph of all the ordered pairs which satisfy the following four inequalities simultaneously:

$$R_1: \{(x, y) \mid 2x + 3y < 12\}$$
$$R_2: \{(x, y) \mid y > \tfrac{1}{2}x\}$$
$$R_3: \{(x, y) \mid y < 3x\} \qquad\qquad \text{(A)}$$
$$R_4: \{(x, y) \mid x + 4y > 4\}$$

We could put these together in two pairs thus:

$$\tfrac{1}{2}x < y < 3x \qquad\qquad \text{(i)}$$

$$1 - \frac{x}{4} < y < 4 - \frac{2x}{3} \qquad\qquad \text{(ii)}$$

And even these forms could be modified further by multiplying:

(i) by 2 to get $x < 2y < 6x$;
(ii) by 12 to get $12 - 3x < 12y < 48 - 8x$.

Suppose now that the replacement set for the inequalities in (A) above is restricted to the set of whole numbers? What is the solution set?

An examination of Fig. 10.4 reveals the solution set to be the set of ordered pairs (1, 1); (1, 2); (2, 2).

Note especially that the circled dots are ordered pairs which belong to the equalities relation but do not satisfy the inequalities of (A) above.

Knowing that the solution set was restricted in this manner means that we could have arrived at the solution set by trial and error methods, i.e. try all the pairs

$$(0, 1), (0, 2), (0, 3) \ldots \text{etc.}$$
$$(1, 1), (1, 2), (1, 3) \ldots \text{etc.}$$
$$(2, 1), (2, 2), (2, 3) \ldots \text{etc.}$$

(This, of course, is the cartesian product which was discussed in Chapter Nine.)

This is perfectly all right and works very well here, but remember that this is a very much simplified situation, and what we are really interested in is the general method of handling this type of problem.

The second form for the straight line is $Ax + By = C$, and will be encountered frequently in our current work. The important point to realize is that the different straight lines obtained by changing the value of C only, are all parallel to one another.

Consider the line $3x + 2y = C$ and the set of lines obtained from this equation by choosing $C = 6, 12, 18$. You are probably wondering 'why choose 6, 12, 18?' Well suppose we rewrite the equation as

$$\frac{3x}{C} + \frac{2y}{C} = 1$$

It is now in the intercept form we have already discussed, and clearly if we choose values of C which contain the factors 3 and 2 we shall be able to simplify the equation further. The first suggestion is C = 6, so that the line becomes

$$\frac{x}{2} + \frac{y}{3} = 1$$

The second suggestion is C = 12, so that the line becomes

$$\frac{x}{4} + \frac{y}{6} = 1 \quad \text{and so on}$$

As another example consider $4x + 7y = C$. The set of C values which are most convenient here are 28, 56, 84, . . .

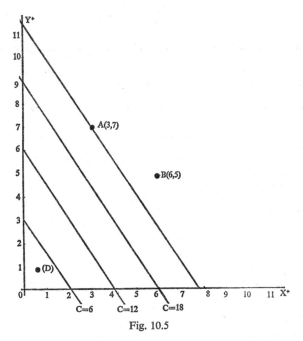

Fig. 10.5

However, returning to $3x + 2y = C$ with C = 6, 12, 18, we obtain the three parallel lines of Fig. 10.5, and do notice that as we increase the value of C the line progresses from left to right across the cartesian plane, i.e. the line drifts away from the origin as C increases. Suppose we fix two points A(3, 7) and B(6, 5) in the cartesian plane and suggest the following problem:

What is the maximum value of $3x + 2y$ obtained without passing over the points A or B?

All we have to do is to draw a set of parallel lines $3x + 2y = C$ as we have in Fig. 10.5 which get closer and closer to one or other of the points; in this case A, until the line is parallel to $3x + 2y = C$ passing through A. In fact, we need only draw a first line (for C = 12, say) and a second parallel line to pass through A(3, 7).

The value of C for this second line is $3\times$ (the x intercept), or $2\times$ (the y intercept). In this case the x intercept is 7·7 approximately.

$$\therefore \quad C = 23\cdot1 \text{ approximately.}$$

This process is called **maximizing** the function $3x + 2y$.

There is, of course, a similar process of **minimizing** a function; this could be brought about by not allowing the line $3x + 2y = C$ to pass to the left of the point D.

One final remark. Sometimes we do not know the co-ordinates of the end points like A or D, so that we have to **construct** this line through these points and actually measure the intercepts to calculate C, the maximum or minimum of the function concerned in satisfying the restriction of the fixed points. In this particular example the point A was known, and since A belongs to the line when the function is maximized, then

$$3 \times 3 + 2 \times 7 = 23 \text{ is the calculated value of C.}$$

We shall now consider some examples of the programming method:

Example 1

Plans are being drawn up for a new motel. On the basis of experience it has been discovered that there is no advantage in having less than 2 double bedrooms to 1 single bedroom. Neither is there any advantage in having less than 1 single bedroom to 6 double bedrooms. The type of service being provided is such that the charges for a double room and single room will be fixed by a national scale at £5 and £3 per night respectively. The builders have only permission to build not more than 12 bedrooms. During the open season it is hoped to make more than £30 a night. What is the best allocation of double and single rooms for the maximum possible nightly charge, and what is the smallest number of rooms that need to be booked in order to clear £30 a night?

Solution

Clearly, trial and error will reveal the necessary solutions, this is left for the reader as a check on the following general method.

Our first task is to obtain the inequality relations. Let $d =$ the number of double rooms and $s =$ the number of single rooms. The relations are:

(i) 'no advantage in having less than 2 double to 1 single'
$$R_1 : \{(s, d) \mid d \geqslant 2s\}$$
(ii) 'no advantage in having less than 1 single to 6 double'
$$R_2 : \{(s, d) \mid d \leqslant 6s\}$$
(iii) 'not more than 12 bedrooms' $R_3 : \{(s, d) \mid s + d \leqslant 12\}$
(iv) 'more than £30 a night' $R_4 : \{(s, d) \mid 5d + 3s > 30\}$
(v) 'maximum possible nightly charge' $5d + 3s$ maximized.

The graph of these relations is given in Fig. 10.6.

Obviously the solution set is from {whole numbers}. The rings on the diagram lie in the region of ordered pairs which satisfy the inequality relations. We see at once that seven rooms is the smallest number which needs to be booked to clear the £30. The seven rooms may be such that he takes £33 for 1 single and 6 doubles or £31 for 2 singles and 5 doubles. By examining Fig. 10.7 insert the missing 3 rings in Fig. 10.6.

We now maximize the function $3s + 5d$.

For example, in Fig. 10.6 the line has been drawn in the position $3s + 5d = 45$, whereupon we notice that there are still six arrangements of rooms which, if booked up, will realize more than £45 a night. We continue to move the 'takings' line parallel

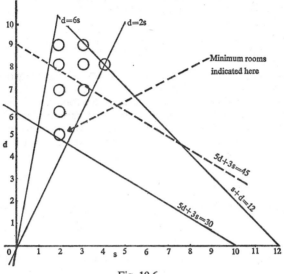

Fig. 10.6

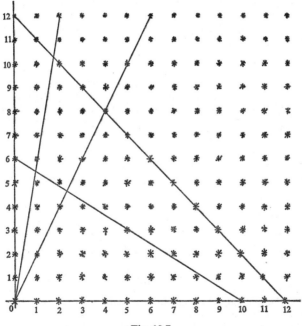

Fig. 10.7

to itself until there is only the one ordered pair left for the solution set (inserted by the reader). We have thus shown that 10 doubles and 2 singles yield the maximum night's takings of £56.

Obtaining the best possible solutions by this method is known as **linear programming**.

Tackling the problem in the manner of Chapter Nine, we write down the cartesian product A × A, where A = {0, 1, 2, 3, 4, 5, 6, 7, 8, 9, 10, 11, 12}, since we know there cannot be more than twelve rooms.

This product is given in Fig. 10.7 using the cartesian axes as given in Chapter Nine, page 195. The 169 ordered pairs of the product are represented by the asterisk dots. Now follow through the next steps, marking Fig. 10.7 with a coloured pencil.

(i) Mark the ordered pairs which are members of the relation R_1: $\{(s, d) \mid d \geqslant 2s\}$ (mark 1).

(ii) Mark the ordered pairs which are members of the relation R_2: $\{(s, d) \mid d \leqslant 6s\}$ (mark —).

(iii) Mark the ordered pairs which are members of the relation R_3: $\{(s, d) \mid s + d \leqslant 12\}$ (mark /).

(iv) Mark the ordered pairs which are the members of the relation R_4: $\{(s, d) \mid 3s + 5d \geqslant 30\}$ (mark \).

The dots with four marks is the solution set. Clearly the first method is much easier and certainly more efficient. But this diagram does enable us to demonstrate the solution set as being given quite clearly by $R_1 \cap R_2 \cap R_3 \cap R_4$. Comparison of Figs. 10.6 and 10.7 should give a good insight into the general method required.

Example 2

A hire-van service is ordering a fleet of small and medium vans. A small van needs 50 m² of parking space, a medium van 80 m² of parking space. But the owner only has 4000 m² of parking space available. The average weekly profit for the small and medium vans is £20 and £40 respectively. The cost of the small and medium vans is £400 and £900 respectively. What is the maximum profit from a maximum outlay of £36 000?

Solution

Let s and m be the number of small and medium vans in the fleet. Our inequalities will arise from:

(i) 'only 4000 m² of parking space' $R_1 : \{(s, m) \mid 50s + 80m \leqslant 4000\}$
                                 or $R_1 : \{(s, m) \mid 5s + 8m \leqslant 400\}$

(ii) 'maximum outlay of £36 000' $R_2 : \{(s, m) \mid 400s + 900m \leqslant 36\,000\}$
                                 or $R_2 : \{(s, m) \mid 4s + 9m \leqslant 360\}$

(iii) profit line is parallel to $20s + 40m = C$. Putting $C = 800$ we get

$$\frac{s}{40} + \frac{m}{20} = 1$$

Draw this line and move it parallel to itself in the usual way (Fig. 10.8) until the function $20s + 40m$ is maximized.

The graph shows that the maximum profit will be obtained from 55 small and 15 medium vans [i.e. (55, 15) is the last point to be reached by the profit line as it is

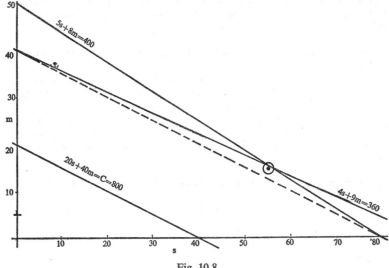

Fig. 10.8

moved parallel to itself]. Since $(20 \times 55) + (40 \times 15) = 1100 + 600$, then the maximum profit is £1700 per week.

Graphical work is only as accurate as our eyes allow, so that on occasions like this it is always a good idea to run a test on some possible results close to the one we have obtained.

Let us try 54 small + 16 medium, and 56 small + 14 medium as possible results.

$50s + 80m = 2700 + 1280$	$50s + 80m = 2800 + 1120$
$= 3980$	$= 3920$
The ordered pair (54, 16) belongs to R_1	The ordered pair (56, 14) belongs to R_1
$4s + 9m = 216 + 144$	$4s + 9m = 224 + 126$
$= 360$	$= 350$
The ordered pair (54, 16) belongs to R_2	The ordered pair (56, 14) belongs to R_2
The profit gained is	The profit gained is
$20s + 40m = 1080 + 640$	$20s + 40m = 1120 + 560$
$= £1720$	$= £1680$

A check on the result we did offer reveals the following:

$$50s + 80m = 2750 + 1200$$
$$= 3950 \qquad (55, 15) \in R_1$$
$$4s + 9m = 220 + 135$$
$$= 355 \qquad (55, 15) \in R_2$$

Profit £1700.

We now see by calculation that the optimum arrangement is 54 small vans and 16 medium. But do note the linear programme did, even to those with poor eyesight, reduce a possible $60 \times 60 = 3600$ ordered pairs down to three!

The arrangement and examination we have just carried out in this example is typical of this type of work.

Example 3

A naval force consisting of submarines and destroyers finds the following limitations placed upon its effectiveness when setting out upon an exercise:

(i) Shortage of port facilities, which means that there cannot be more than 5 destroyers to every 4 submarines.

(ii) Naval policy that there must be at least 1 destroyer to every 4 submarines.

(iii) United Nations agreement that the number of vessels must be not greater than 50.

(iv) On the basis of a strike capability of 60 units per submarine and 25 units per destroyer, the total strike capability of force must exceed 1 500 units.

(v) If the cost of one destroyer for the exercise is twice the cost of a submarine, What is the cheapest arrangement for the exercise?

Solution

Let s be the number of submarines and d the number of destroyers in the exercise. The inequality relations are:

(i)　$R_1 : \left\{ (s, d) \mid \dfrac{d}{5} \leqslant \dfrac{s}{4} \right\}$ modification $R_1 : \left\{ (s, d) \mid d \leqslant \dfrac{5s}{4} \right\}$

(ii)　$R_2 : \left\{ (s, d) \mid d \geqslant \dfrac{s}{4} \right\}$ This combines with (i) as $R_1 \cap R_2 : \left\{ (s, d) \mid \dfrac{s}{4} \leqslant d \leqslant \dfrac{5s}{4} \right\}$

(iii)　$R_3 : \{ (s, d) \mid s + d \leqslant 50 \}$

(iv)　$R_4 : \{ (s, d) \mid 60s + 25d > 1500 \}$

(v)　$s + 2d$ to be minimized.

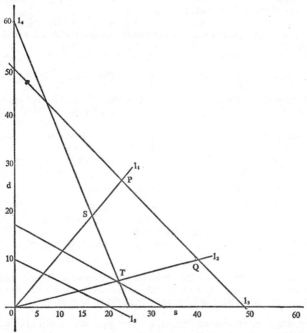

Fig. 10.9

We obtain the solution set of Fig. 10.9 by drawing the lines:

$$l_1 : 5s = 4d$$
$$l_2 : s = 4d$$
$$l_3 : s + d = 50$$
$$l_4 : 60s + 25d = 1500$$

or

$$\frac{s}{25} + \frac{d}{60} = 1$$
$$l_5 : s + 2d = C$$

The replacement set is the set of whole numbers and the relation $R_1 \cap R_2 \cap R_3 \cap R_4$ is the region TSPQ excluding the side of the boundary ST. For the cost line choose $C = 20$ and minimize the cost function $s + 2d$ by the line through T; the intersection of l_2 and l_4. We know that T is not a member of the solution set, and examination of the graph reveals that the ordered pair (23, 6) is our solution.

The strike capability of (23, 6) is $(23 \times 60) + (6 \times 25) = 1530$ units, and the cost is $(23 \times 1) + (6 \times 2) = 35$ units.

Example 4

A shopkeeper has 60 m^3 available for the storage of supplies of two brands of plastic ceiling tiles X and Y. The volume of a crate of X is 3 m^3, while the volume of a crate of Y is 2 m^3. A crate of X costs the shopkeeper £1·5, but a crate of Y costs £3. His profit is £0·5 per crate of each type of tile, but he has only £45 to spend on an order.

Obtain the order which will yield the highest profit.

Solution

Let x be the number of crates of brand X and y the number of crates of brand Y. The necessary relations are:

(i) $R_1 : \{(x, y) \mid 3x + 2y \leqslant 60\}$ Draw the line $l_1 : \frac{x}{20} + \frac{y}{30} = 1$

(ii) $R_2 : \{(x, y) \mid 15x + 30y \leqslant 450\}$ Draw the line $l_2 : \frac{x}{30} + \frac{y}{15} = 1$

(iii) $R_3 : \{(x, y) \mid 5x + 5y = C\}$ the function $5x + 5y$ to be maximized

To maximize the function $5x + 5y = C$, choose $C = 100$ and then proceed in the usual manner, as indicated in Fig. 10.10.

It will be found that there are three different orders which yield a profit of £11 namely (16, 6); (15, 7); (14, 8).

Example 5

A works plant produces two types of smokeless coke fuel. One type is for industrial use the other is for household use. The plant produces fuel in the ratio of 2 tonnes of industrial to 1 tonne of household, but the maximum production of household coke is only 10 tonnes a day. The plant has two standing orders per day, they are a delivery to depot D, 5 km away, of at least 5 tonnes of household coke and a delivery to depot E, 7 km away, for at least 8 tonnes of industrial coke.

Unfortunately the transport resources are only 175 tonne-km per day (e.g. 25 tonnes carried through 7 km or 35 tonnes carried 5 km etc.).

If all the household coke can be sold at D at a profit of £5 per tonne and all the industrial coke can be sold at E at a profit of £3 per tonne, obtain the tonnage of each coke per day which yields the greatest profit.

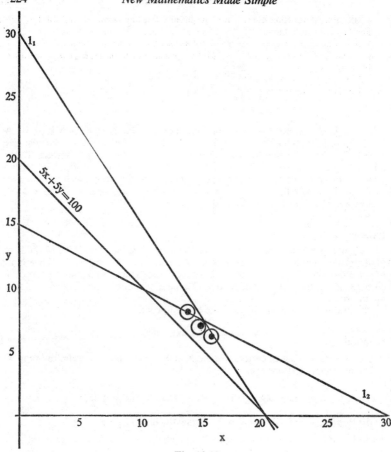

Fig. 10.10

Solution

Let i = number of tonnes of industrial coke to be produced and h = number of tonnes of household coke to be produced.

The inequality relations are:

(i) $R_1 : \{(i, h) \mid 8 \leqslant i \leqslant 20\}$

(ii) $R_2 : \{(i, h) \mid 5 \leqslant h \leqslant 10\}$

(iii) $R_3 : \{(i, h) \mid 7i + 5h \leqslant 175$ modified $\left\{(i, h) \mid \dfrac{i}{25} + \dfrac{h}{35} \leqslant 1\right\}$

(iv) $R_4 : 3i + 5h$ to be maximized.

The required lines are drawn in Fig. 10.11 and the replacement set is in the region ABCDE, including all the boundary. Before proceeding further the reader should try to identify each line in Fig. 10.11, observing that there is an extra line in the figure arising out of Question 6, Exercise 10.2. To maximize the function $3i + 5h = C$, choose $C = 60$ and draw $\dfrac{i}{20} + \dfrac{h}{12} = 1$ first, then maximize the function in the usual manner, by moving $3i + 5h = C$ across ABCDE as far as possible.

We obtain our maximum profit by passing the line through the point A. From the diagram the intercept on the h-axis is 21, the maximum profit is therefore £105 approximately. If we are restricted to the set of whole numbers for replacement set, then examining Fig. 10.11 we see two possible ordered pairs (i, h); they are (17, 10) and (18, 9).

For (17, 10) the profit is £101.

For (18, 9) the profit is £99.

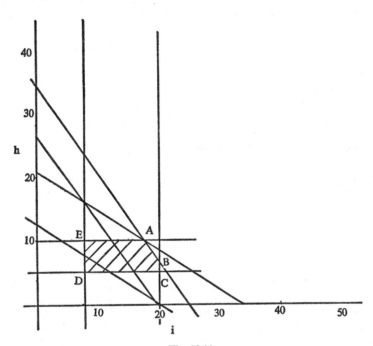

Fig. 10.11

Exercise 10.2

1. Obtain the solution set of whole numbers for

$$x < 13; y < 5, x + y > 14.$$

2. Obtain the solution set of whole numbers for

$$x < y < 2x; x + y < 7.$$

3. Obtain the solution set of whole numbers for $x + y < 10$; $20x + 5y < 100$; $2x + y > 10$.

4. Which of the ordered pairs (2, 7); (3, 5); (3, 6) maximize the function $3x + 5y$? What is the maximum under these conditions?

5. In Example 3, suppose the strike capability had to exceed 1 800 units. What would have been the cheapest arrangement?

6. In Example 5 as a result of servicing requirements the effective transport resources are reduced to 140 tonne-km day. What is the most profitable production of each type of coke now?

7. An exporter wishes to use the method of container transport in order to export two different types of machines which have been packed in separate crates. He can pack 1 machine of type X in a container, but he can only pack 1 machine of type Y in 2 containers. His outlay is £450 for a type X machine and £250 for a type Y machine. The shipping authority offer him 50 crates as a maximum number available at the moment. He decides to take the offer. What is the most profitable consignment for a maximum outlay of £11 250 if he is prepared to take £50 profit on each type of machine?

8. A farmer has 30 hectares of land to prepare for two different crops, X and Y. The cost per hectare for sowing crop X is £15 and for crop Y is £9. Crop X takes 4 labour units per hectare, while crop Y takes 5 labour units per hectare. A labour unit being one man for 4 hours or 2 men for 2 hours, etc. In view of a long-range weather forecast he wishes to get the job done as soon as possible, but only has 100 labour units available for this purpose. If he hopes for a profit of £40 per hectare for crop X and £48 per hectare for crop Y, what is his most profitable arrangement for a maximum outlay of £270 for sowing?

9. A company has room for 50 machines in a new factory. The machines are of two types, X and Y. The average weekly cost of running these machines (including depreciation) is £70 and £42 for X and Y respectively. The profit from the machine production is £100 and £200 for X and Y respectively. What is the most profitable number of machines to install, set against running costs of less than £2940 per week?

NON-LINEAR PROGRAMMING

The method of linear programming arises from the linear relations involved. That is relations such as R: $\{(x, y) \mid x + y > 4\}$ requires the construction of the boundary line $x + y = 4$ in order to obtain the graph of the relation R. But the graph of any function yields a curve (possibly a straight line) which divides the cartesian plane into different sets so the basic method should still apply for finding solution sets to non-linear inequality relations such as R: $\{(x, y) \mid x^2 > y\}$, etc.

(Reminder—we are still working with the domain and range in the set of real numbers.)

The drawing of these curves will require a little more patience, but the procedure is the same. We suggest values (sensible ones) of x (domain) and work out the corresponding values of y (range).

Consider $y = x^2$. Suggest $x = 0, 1, -1, 2, -2, 3, -3, \ldots$ (we suggested negative values because of x^2). The corresponding values of y are then calculated, and the set of ordered pairs thus obtained are neatly assembled in the following array:

x	0	1	-1	2	-2	3	-3
y	0	1	1	4	4	9	9

We plot the points in the usual manner on the cartesian diagram (Fig. 10.12) and join up the points so as to yield the most natural curve through the points.

If the points (pairs) are too far apart, then $x = \pm\frac{1}{2}, \pm1\frac{1}{2}$ will have to be suggested and more points inserted for extra guidance.

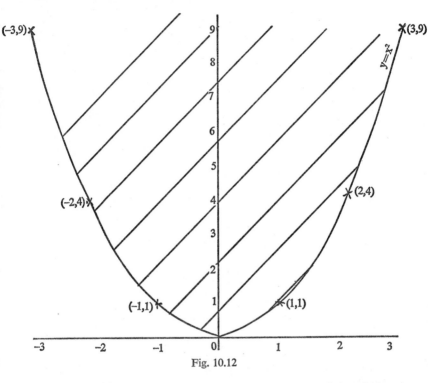

Fig. 10.12

Now it is easy to see that the shaded region is the graph of the relation

$$R_1: \{(x, y) \mid y > x^2\}$$

The curve itself is the set of points which form the graph of the relation

$$R_2: \{(x, y) \mid y = x^2\}$$

Finally, the unshaded region is the graph of

$$R_3: \{(x, y) \mid y < x^2\}$$

Suppose we now add the relation $R_4: \{(x, y) \mid y > 2 + x\}$.

We are practised enough to recognize $y = 2 + x$ to be a straight line. We now add this to Fig. 10.12 as Fig. 10.13.

The double-shaded region is clearly the graph of $R_1 \cap R_4$, the members

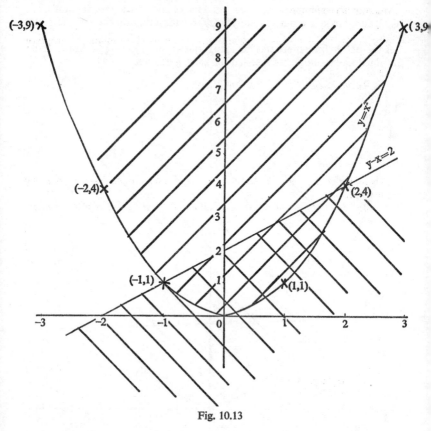

Fig. 10.13

of which may now be picked out in the usual manner of the earlier problems in this chapter.

We have thus solved the inequalities

$$y < 2 + x$$

and $$x^2 < y \quad \text{simultaneously.}$$

Putting these together, we see that we have solved the inequality

$$x^2 < 2 + x$$

or $$x^2 - x - 2 < 0.$$

For those who are familiar with the necessary algebra, this last expression is

$$(x - 2)(x + 1) < 0$$

putting $$(x - 2)(x + 1) = 0$$

will yield $$x = 2, \quad x = -1$$

the intersection of $y = 2 + x$ with $y = x^2$.

One further example:

If R_5: $\left\{(x, y) \mid y < \dfrac{1}{x}; \; x > 0\right\}$ and R_6: $\{(x, y) \mid y < 4\}$ obtain the graph of

$R_1 \cap R_5 \cap R_6$.

For R_5 we draw up the following table:

x	1	2	3	4	$\frac{1}{2}$	$\frac{1}{3}$	$\frac{1}{4}$	$\frac{1}{5}$
y	1	$\frac{1}{2}$	$\frac{1}{3}$	$\frac{1}{4}$	2	3	4	5

The relations are now plotted in Fig. 10.14, and the shaded region is the required graph.

We are now able to solve the inequalities (*a*), (*b*), and (*c*) simultaneously.

$$y < 1/x \qquad\qquad (a)$$
$$y < 4 \qquad\qquad (b)$$
$$x^2 < y \qquad\qquad (c)$$

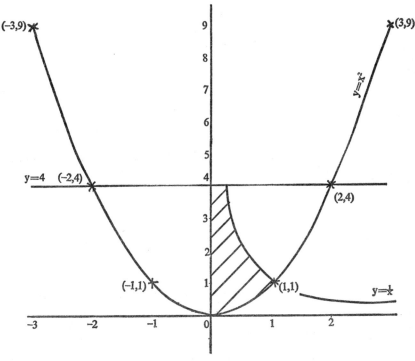

Fig. 10.14

Exercise 10.3

1. Obtain the region for solutions to the inequalities

$$x^2 < y, y < 1/x, y < 2x. (x \neq 0)$$

2. Obtain the region for solutions to the inequalities

$$y < \sqrt{x}, 0 \leqslant x \leqslant 9, y < 1/x.$$

3. What is the greatest value of y which satisfies Question 2?

The graphical methods of this chapter apply only to problems which involve two variables. It would be a mistake to assume that all problems of profitable economics and factory management lend themselves to such simple expressions as those considered here. Considerable algebraic manipulation is needed to solve comparatively easy problems such as the one expressed by the following inequations:

$$39 \leqslant w + x + y + z \leqslant 58,$$
$$0 < 5w < y,$$
$$x + y < 3z + w,$$
$$0 < 3z < 2x.$$

A solution such as $x = 20$, $y = 21$, $z = 13$, $w = 4$ could be obtained by a computer with the greatest of ease, but that is another story.

CHAPTER ELEVEN

MATRICES

TRANSFORMATIONS

We have now had much experience in handling the equations and graphs of straight lines. We have plotted the graphs in the cartesian plane using the two number lines, $X^- X^+$, $Y^- Y^+$, as reference axes, and we have learned a little more about the different regions of the plane which supply solution sets for our inequality relations. We now wish to study changes of position and shape, which can be effected by rotations, shrinking or expanding, and reflections, accepting for the moment intuitive notions about these operations. Note we call them operations just as we called $+$, $-$, $\times$, $\div$ operations in Chapter Two because they are carried out on pairs of elements with unique results. These elements are in fact ordered pairs. For example, consider rotating a line. Since a line is a set of points, we must rotate every point of the line, and as each point is designated by an ordered pair, it follows therefore that an examination of the effects of our transformation on every relevant ordered pair will tell us all we need to know about the effects of the transformation on the complete plane. Visualizing the process is fairly straightforward; what does not come so easily is the manner of communicating what we have in mind. In other words, to find the algebraic statement which enables others to repeat our transformations exactly.

REFLECTIONS

As a first instance let us consider the effects of reflection, since most people find this to be the simplest transformation to understand. Obviously we need something to reflect in. In this work we shall only consider reflections in straight lines. Now in any reflection the image is the same distance behind the mirror line as the object is in front. So suppose we have the point $(2, 3)$ to reflect in the x-axis (i.e. the x-axis is the mirror line). To obtain its image we draw a line through the object point $(2, 3)$ perpendicular to the mirror line, and then extend the perpendicular an equal distance on the other side to obtain the required image. Thus the mirror image of $(2, 3)$ in the line $X^- X^+$ is the point $(2, -3)$. Similarly, the reflection of the point $(1, 1\frac{1}{2})$ in $X^- X^+$ is $(1, -1\frac{1}{2})$, and incidentally we note that if we reflect $(2, -3)$ in the line $X^- X^+$, we get back to where we started so that the operation of reflection appears to be reversible.

In general, a point P_0 (x_0, y_0) is mapped on to the point P_1 (x_1, y_1), and as we choose P_0 from everywhere in the plane so P_1 takes every different position in the plane as well.

In a case like this we say that this reflection has mapped the plane on to itself.

Furthermore, if we obtain the reflection of P_1 we arrive back at P_0. We therefore call this a one–one mapping, and we describe these mappings as transformations or operations.

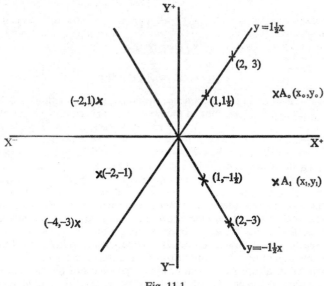

Fig. 11.1

Now consider reflecting $(-4, -6)$ in $X^- X^+$. As before, draw a perpendicular from $(-4, -6)$ to the axis $X^- X^+$ and then extend an equal distance on the other side to obtain the image $(-4, 6)$, try this for $(-4, -3)$ in Fig. 11.1.

From these particular points we should be able to see that instead of drawing perpendiculars each time and measuring the required distances, all we need to do is multiply the y co-ordinate by -1. If we now consider any point A (x, y), then its image in the x-axis is $A_1 (x, -y)$. It is general practice to refer to the original point as (x_0, y_0) and the point obtained after one operation as (x_1, y_1).

We have seen that the operation of reflection maps one set of points on to another set of points. For example, so far we have

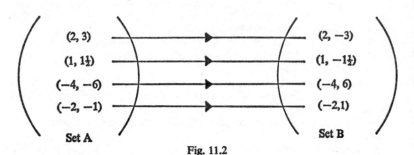

Fig. 11.2

The pattern of these results leads us to deduce that

$$(x, y) \rightarrow (x, -y)$$

If there was something special about set A, would this 'something' be retained after the operation? Well there is something special about the elements of set A: three lie on the straight line.

$$y = \tfrac{3}{2}x$$

Clearly, after the reflection in $X^- X^+$ the points (except one) which belong to this line will no longer lie on a line with this equation, but they will lie on a line because, if

$$(x, y) \rightarrow (x, -y)$$

then
$$(x, 1\tfrac{1}{2}x) \rightarrow (x, -1\tfrac{1}{2}x)$$

and then the new line is

$$-y = 1\tfrac{1}{2}x.$$

In this case the one point which does not change is (0, 0). This lies on the original line, $y = 1\tfrac{1}{2}x$, and the image line $y = -1\tfrac{1}{2}x$. We call this point an **invariant** for this particular transformation.

Suppose we had reflected this line in the y-axis $Y^- Y^+$? Would the result still be a straight line? Just to get an idea about this, we shall use set A as before. By drawing the perpendicular to Y^-Y^+, and marking off the same distance behind the new mirror line as before, we see that this reflection gives us the mapping.

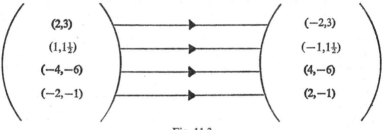

Fig. 11.3

The pattern of these results, together with the construction of the image of the general point (x, y), gives us the mapping.

$$(x, y) \rightarrow (-x, y)$$

The original line $y = 1\tfrac{1}{2}x$ becomes $y = -1\tfrac{1}{2}x$, the same result as before.

It would seem at first glance that it makes no difference which axis is used for the reflection, but this is a conclusion which is dangerously drawn from insufficient evidence.

After all, both of these lines passed through the origin (0, 0), making them rather special. But nevertheless it has given us the basic idea about reflection.

Consider the line $y = x + 2$. On reflection, in the x-axis this line becomes

$-y = x + 2$ or $y = -x - 2$, while on reflection in the y-axis it becomes $y = -x + 2$, and these certainly are not the same line (so the conclusion was a dangerous one!). The final results are given in Fig. 11.4.

We notice at once from the equations that the two images are parallel to one another, but examination of the order of the points A, B, C shows that the lines have 'turned' in different directions. (Construct the images as follows. The original line cuts the y-axis at Q_0; find the image of Q_0 at Q_1 then join P_0Q_1 for the image line. Similarly, find the image of P_0 at P_2 and join Q_0P_2.)

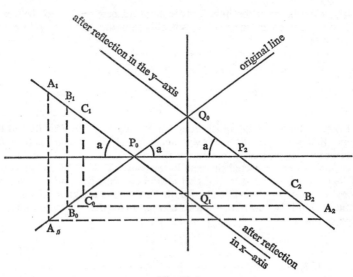

Fig. 11.4

To obtain the same final position as for the reflection in the x-axis we could have rotated the original line clockwise about P_0, while to obtain the same final position as for reflection in the y-axis we could have rotated anti-clockwise about Q_0.

ROTATIONS

Let us now see the results of rotating all of the points of the plane anti-clockwise through 90° about the origin O. This means that any point $A(x_0, y_0)$ is mapped on to a point $B(x_1, y_1)$ such that $m(\angle AOB) = 90°$.

The reader should insert O in Fig. 11.5 and Fig. 11.6.

From the geometry of Fig. 11.5 we see that $\triangle OBC \cong \triangle OAD$, consequently

$$\overline{CO} \cong \overline{DA} \quad \therefore \quad x_1 = -y_0$$
$$\overline{CB} \cong \overline{OD} \quad \therefore \quad y_1 = x_0$$

We notice here that not only has one of the co-ordinates been multiplied by -1 but they have also been interchanged. Suppose we had rotated the points of the plane clockwise through 90° about O? Once again Fig. 11.5 gives the geometry of the rotation with A(x_0, y_0) now mapped on to E(x_2, y_2). We see, as before, that $\triangle$OEF $\cong$ $\triangle$OAD, and consequently

$$\overline{FE} \cong \overline{DA} \quad \therefore \quad x_2 = y_0$$
$$\overline{OF} \cong \overline{OD} \quad \therefore \quad y_2 = -x_0$$

As for the anti-clockwise rotation, there has been an interchange of co-ordinates, but a different one has been multiplied by -1. (In mathematics

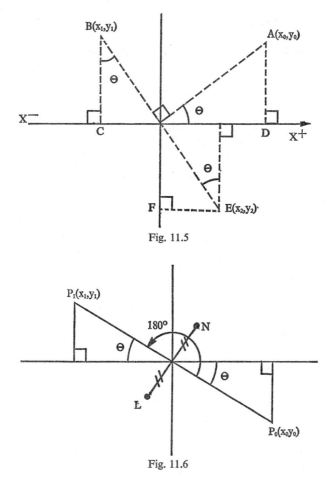

Fig. 11.5

Fig. 11.6

clockwise rotation is usually called a negative rotation, while a positive rotation is anti-clockwise.)

It is obvious that we investigate next the effects of rotating our points through 180° about the origin. It makes no difference whether we rotate clockwise or anti-clockwise, since the final position will be the same in either case. The original point $P_0(x_0, y_0)$ is mapped on to $P_1(x_1, y_1)$. From the geometry of Fig. 11.6, we see that

$$x_1 = -x_0$$
$$y_1 = -y_0$$

So that both the original co-ordinates are multiplied by -1. A second glance at Fig. 11.6 suggests that the point P_1 is the reflection of P_0 in the origin O. So far we have only spoken of reflections in lines, but the idea of reflection in a point is quite simple. Using Fig. 11.6, all we need to do is join P_0O and extend through to the other side as far as P_1 such that $\overline{P_0O} \cong \overline{OP_1}$, then P_1 is said to be the image of P_0 in O. For example, in Fig. 11.6 the image of L in O is the point N because: (i) LON is a straight line; (ii) $\overline{LO} \cong \overline{ON}$.

RADIAL EXPANSION

If we join a point to the origin by a straight line and then move the point along this line we are said to be moving the point radially through O. If we do this for every point in the plane, then we are said to be giving the plane a radial expansion or contraction.

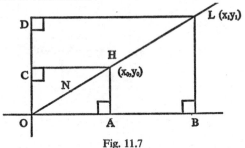

Fig. 11.7

For example, in Fig. 11.7 we have joined the point H to the origin O. If we double OH we arrive at L, the image of H. If this is done for all points of the plane, then we have given the plane a radial expansion which has doubled the distance of each point from the origin. The same applies to A and C, which are mapped on to B and D respectively. Putting these results together, we see that $H(x_0, y_0)$ is mapped on to $L(x_1, y_1)$ such that

$$x_1 = 2x_0$$
$$y_1 = 2y_0$$

If we had halved OH to map H on to N, then we would have obtained

$$x_1 = \tfrac{1}{2}x_0$$
$$y_1 = \tfrac{1}{2}y_0$$

We can quickly generalize these results to suggest that

$$x_1 = kx_0$$
$$y_1 = ky_0$$

represents any radial expansion or contraction of the plane, and we notice that taking $k = 1$ leaves the plane as it was. We pause now to list the following transformations and their matrices. (The matrices are for reference to the next chapter.)

Matrix Notation.

1. Reflection in x-axis $M_{y=0}$: $x_1 = x_0$ $y_1 = -y_0$ $\begin{pmatrix} x_1 \\ y_1 \end{pmatrix} = \begin{pmatrix} 1 & 0 \\ 0 & -1 \end{pmatrix}\begin{pmatrix} x_0 \\ y_0 \end{pmatrix}$

2. Reflection in y-axis $M_{x=0}$: $x_1 = -x_0$ $y_1 = y_0$ $\begin{pmatrix} x_1 \\ y_1 \end{pmatrix} = \begin{pmatrix} -1 & 0 \\ 0 & 1 \end{pmatrix}\begin{pmatrix} x_0 \\ y_0 \end{pmatrix}$

3. Reflection in $y = x$ $M_{y=x}$: $x_1 = y_0$ $y_1 = x_0$ $\begin{pmatrix} x_1 \\ y_1 \end{pmatrix} = \begin{pmatrix} 0 & 1 \\ 1 & 0 \end{pmatrix}\begin{pmatrix} x_0 \\ y_0 \end{pmatrix}$

4. Reflection in $y = -x$ $M_{y=-x}$: $x_1 = -y_0$ $y_1 = -x_0$ $\begin{pmatrix} x_1 \\ y_1 \end{pmatrix} = \begin{pmatrix} 0 & -1 \\ -1 & 0 \end{pmatrix}\begin{pmatrix} x_0 \\ y_0 \end{pmatrix}$

5. Rotation about O through 90° anti-clockwise $R(0, +90°)$: $x_1 = -y_0$ $y_1 = x_0$ $\begin{pmatrix} x_1 \\ y_1 \end{pmatrix} = \begin{pmatrix} 0 & -1 \\ 1 & 0 \end{pmatrix}\begin{pmatrix} x_0 \\ y_0 \end{pmatrix}$

6. Rotation about O through 90° clockwise $R(0, -90°)$: $x_1 = y_0$ $y_1 = -x_0$ $\begin{pmatrix} x_1 \\ y_1 \end{pmatrix} = \begin{pmatrix} 0 & 1 \\ -1 & 0 \end{pmatrix}\begin{pmatrix} x_0 \\ y_0 \end{pmatrix}$

7. Rotation about O through 180° H_0: $x_1 = -x_0$ $y_1 = -y_0$ $\begin{pmatrix} x_1 \\ y_1 \end{pmatrix} = \begin{pmatrix} -1 & 0 \\ 0 & -1 \end{pmatrix}\begin{pmatrix} x_0 \\ y_0 \end{pmatrix}$

8. Leave everything unchanged I: $x_1 = x_0$ $y_1 = y_0$ $\begin{pmatrix} x_1 \\ y_1 \end{pmatrix} = \begin{pmatrix} 1 & 0 \\ 0 & 1 \end{pmatrix}\begin{pmatrix} x_0 \\ y_0 \end{pmatrix}$

9. Radial expansion $E(0, k)$: $x_1 = kx_0$ $y_1 = ky_0$ $\begin{pmatrix} x_1 \\ y_1 \end{pmatrix} = \begin{pmatrix} k & 0 \\ 0 & k \end{pmatrix}\begin{pmatrix} x_0 \\ y_0 \end{pmatrix}$

Numbers 1, 2, 5, 6, 7, 8, 9 have already been derived, while numbers 3, 4 will arise later on.

An inspection of these results shows that each operation yields a new point whose co-ordinates are a combination of the original co-ordinates. So far the combination has been a very simple one consisting mainly of an interchange of single co-ordinates coupled with changes in sign. A linear combination of both co-ordinates x and y would result in an expression such as $x + y$, $3x - 4y$, $-17x + 2y$, etc., and clearly a general combination would be represented by $ax + by$.

Returning to the nine results above, we could gather them all together under the heading

$$x_1 = ax_0 + by_0$$
$$y_1 = cx_0 + dy_0 \qquad \text{(A)}$$

because we may easily give values to a, b, c, d which pick out whichever transformation we require of the above nine, e.g.

Taking $\left.\begin{matrix} a = 1, & b = 0 \\ c = 0, & d = -1 \end{matrix}\right\}$ gives $M_{y=0}$

$\left.\begin{matrix} a = 0, & b = 1 \\ c = -1, & d = 0 \end{matrix}\right\}$ gives $R(0, -90°)$

But the equations of (A), by allowing us to choose any constant values for a, b, c, and d, do in fact enable us to examine many more transformations than we have discussed here. Since the equations (A) are linear equations, we call the (A) a set of linear transformations of the plane onto itself.

Since we would wish to pick out particular members of the set by allocating the relevant values to a, b, c, and d, it follows that it would be convenient to detach these from the body of the equations for separate study. But the need for clear identification dictates that we set them out in a special order, thus if written

$$\begin{pmatrix} 4 & 7 \\ 11 & 9 \end{pmatrix} \text{ we agree that this means } \begin{array}{ll} a = 4, & b = 7 \\ c = 11, & d = 9 \end{array}$$

and we employ the larger round brackets to keep the arrangement separate from other numbers that may be around. The printer's type is moulded or cast in what is called a matrix. The way in which the numbers a, b, c, d are cast in the brackets leads us to call such an arrangement a matrix, the plural of which is matrices. Some printers use square brackets instead of round brackets. A matrix acts as a store of information; we have already encountered this use of a matrix when studying the relation networks of Chapter Nine; it specifies completely the transformation concerned.

The reader will notice that each of the transformations has been written in matrix form on page 237, and a little practice in recognizing the meaning of the notation now follows.

Example 1

Find the result of the operation R(0, $-90°$) on the point $(-6, 7)$.

Solution

This operation is given by the equations

$$\begin{aligned} x_1 &= \quad y_0 \\ y_1 &= -x_0 \end{aligned}$$

and here we have been given $x_0 = -6$, $y_0 = 7$

$$\begin{aligned} \therefore \quad x_1 &= 7 \\ y_1 &= 6 \end{aligned}$$

Therefore the effect of the operation R(0, $-90°$) on $(-6, 7)$ is to map $(-6, 7)$ on to $(7, 6)$.

Example 2

Find the image of the point $(-2, -3)$ under the operation R(0, $+90°$).

Solution

This is the same type of question as Example 1 but using different wording.

R(0, $+90°$) is given by

$$\begin{aligned} x_1 &= \quad -y_0 \\ y_1 &= x_0 \end{aligned}$$

and we have been given $x_0 = -2$, $y_0 = -3$

$$\begin{aligned} \therefore \quad x_1 &= 3 \\ y_1 &= -2 \end{aligned}$$

therefore the results of R(0, $+90°$) on $(-2, -3)$ is to map $(-2, -3)$ on to $(3, -2)$.

Example 3

A new operation is given by the equations

$$x_1 = 8x_0 + 10y_0$$
$$y_1 = 7x_0 - 4y_0$$

Express this in matrix form and determine the point which corresponds to $(-4, 5)$ in this transformation.

Solution

Here we have $x_0 = -4$ and $y_0 = 5$. Substitution in the given equations yields

$$x_1 = -32 + 50 = 18$$
$$y_1 = -28 - 20 = -48$$

Therefore the corresponding point is $(18, -48)$.

In matrix form the transformation is given by $\begin{pmatrix} x_1 \\ y_1 \end{pmatrix} = \begin{pmatrix} 8 & 10 \\ 7 & -4 \end{pmatrix} \begin{pmatrix} x_0 \\ y_0 \end{pmatrix}$

Exercise 11.1

1. What values are given to a, b, c, and d in order to pick out the transformations $R(0, +90°)$ and $E(0, k)$?
2. Find the result of each of the transformations 1, 2, 5, 6, 7, 9 on the points $(3, 2)$ and $(-1, 2)$.
3. Write the following pairs of equations in matrix form

$$x_1 = 4x_0 + 2y_0 \qquad x_1 = -x_0 + y_0 \qquad x_1 = 2x_0 + 2y_0$$
$$y_1 = x_0 - 3y_0 \qquad y_1 = x_0 - y_0 \qquad y_1 = -3x_0 + y_0$$

4. Write the following in separate equation form:

$$\begin{pmatrix} x_1 \\ y_1 \end{pmatrix} = \begin{pmatrix} 1 & 2 \\ 1 & 1 \end{pmatrix}\begin{pmatrix} x_0 \\ y_0 \end{pmatrix}; \begin{pmatrix} x_1 \\ y_1 \end{pmatrix} = \begin{pmatrix} 2 & 3 \\ 0 & 1 \end{pmatrix}\begin{pmatrix} x_0 \\ y_0 \end{pmatrix}; \begin{pmatrix} x_1 \\ y_1 \end{pmatrix} = \begin{pmatrix} -4 & 0 \\ 0 & 4 \end{pmatrix}\begin{pmatrix} x_0 \\ y_0 \end{pmatrix}$$

Having experimented with one or two individual points we shall now try to see what happens to a whole set all in one step. Just to get the basic idea, consider a linear transformation of the line $y = x$ so that all our original points are such that $y_0 = x_0$.

Substituting in the equations

$$x_1 = ax_0 + by_0$$
$$y_1 = cx_0 + dy_0$$

we get

$$x_1 = ax_0 + bx_0 = (a + b)x_0$$
$$y_1 = cx_0 + dx_0 = (c + d)x_0$$

If $a + b = 0$, $c + d \neq 0$, then $x_1 = 0$ is our final result. (the y-axis)

If $c + d = 0$, $a + b \neq 0$, then $y_1 = 0$ is our final result. (the x-axis)

If $c + d = 0$ and $a + b = 0$, then $x_1 = 0$, $y_1 = 0$, which is too awkward to consider, so we usually state that $a + b = 0$ and $c + d = 0$ may not occur together, since it would mean that every point became $(0, 0)$.

If $a + b \neq 0$ and $c + d \neq 0$, then $(a + b)y_1 = (c + d)x_1$ is the final result.

In the above case we have seen that the linear transformation of a straight line which passes through the origin is another straight line through the origin.

We now try to generalize the situation by considering a linear transformation of the line $Ax + By = C$ (a form of equation which we met quite often in Chapter Ten). The trouble is that the algebra though straightforward is a little heavier than usual, but as long as the reader appreciates the final result, the intermediate work may be taken for granted.

For our original points we have $By_0 = C - Ax_0$

and therefore $y_0 = \dfrac{C - Ax_0}{B}$

Substituting in the equations for the general transformation we have

$$x_1 = ax_0 + by_0 = ax_0 + b\frac{(C - Ax_0)}{B} = \frac{(Ba - bA)x_0 + bC}{B} \quad \ldots (p)$$

Similarly, $y_1 = cx_0 + dy_0 = \dfrac{(Bc - dA)x_0 + dC}{B}$ $\ldots (q)$

$$Bx_1 - bC = (Ba - bA)x_0 \quad \text{from } (p)$$

and $$By_1 - dC = (Bc - dA)x_0 \quad \text{from } (q)$$

$$\therefore \quad \frac{Bx_1 - bC}{Ba - bA} = \frac{By_1 - dC}{Bc - dA}$$

which is another straight line given by

$$(Bc - dA)Bx_1 - By_1(Ba - bA) = bC(Bc - dA) - dC(Ba - bA) \quad \text{(i)}$$

We have now proved that the linear transformation of any straight line is another straight line.

Just as important is the following:

Two parallel lines are transformed into parallel lines.

A line parallel to $Ax + By = C$ is $Ax + By = E$, say, and substituting in the previous result (i) (i.e. substituting E for C) above its image is

$$(Bc - dA)Bx_1 - By_1(Ba - bA) = bE(Bc - dA) - dE(Ba - bA)$$

which is parallel to (i).

A numerical example will possibly make this a little clearer.

Example 4

Find the image of the line $2y_0 + 3x_0 = 5$ in the transformation

$$x_1 = 2x_0 + 6y_0$$
$$y_1 = 7x_0 - 8y_0$$

Solution

Here we are given $2y_0 + 3x_0 = 5$

from which $y_0 = \dfrac{5 - 3x_0}{2}$

$$\therefore \quad x_1 = 2x_0 + \frac{6(5 - 3x_0)}{2} = 2x_0 + 3(5 - 3x_0) = -7x_0 + 15 \quad \text{(i)}$$

and $y_1 = 7x_0 - \dfrac{8(5 - 3x_0)}{2} = 7x_0 - 4(5 - 3x_0) = 19x_0 - 20$ (ii)

we now have $19x_1 = -133x_0 + 285$ by (i) $\times$ 19
$$7y_1 = 133x_0 - 140 \text{ by (ii)} \times 7$$

Adding these last two equations we have

$$19x_1 + 7y_1 = 145$$

The transformation has thus mapped points on the line $2y + 3x = 5$ onto the line $19x_1 + 7y_1 = 145$.

One point for a check is (1, 1). This will yield

$$x_1 = 2 + 6 = 8$$
$$y_1 = 7 - 8 = -1$$

The image point $(8, -1)$ lies on the image line $19x + 7y = 145$, as expected of course.

To get an idea of what happens to the plane under the general transformation $\begin{pmatrix} a & b \\ c & d \end{pmatrix}$ consider the result of this transformation on the cross-shaped figure in the plane of Fig. 11.8 made up of $x = 2$ and $y = 1$.

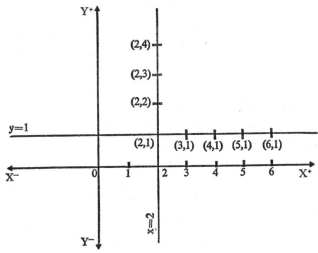

Fig. 11.8

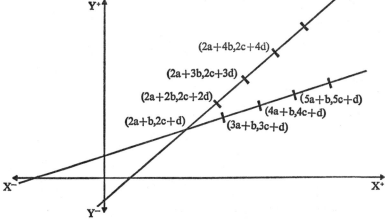

Fig. 11.9

To carry out the transformation of the line $x = 2$ we have

$$x_0 = 2, \quad y_0 = y_0$$
$$\therefore \quad x_1 = 2a + by_0 \qquad \qquad \text{(i)}$$
$$y_1 = 2c + dy_0 \qquad \qquad \text{(ii)}$$

In particular,

the point $(2, 1)$ is mapped on to $(2a + b, 2c + d)$
$(2, 2)$ is mapped on to $(2a + 2b, 2c + 2d)$
$(2, 3)$ is mapped on to $(2a + 3b, 2a + 3d)$
$(2, k)$ is mapped on to $(2a + kb, 2a + kd)$

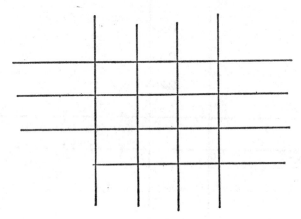

Fig. 11.10

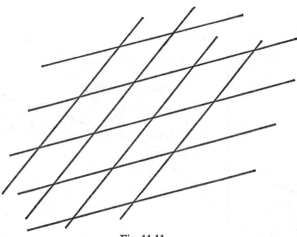

Fig. 11.11

The line $y = 1$ has an image given by

$$x_1 = ax_0 + b$$
$$y_1 = cx_0 + d$$

In particular,

the point $(2, 1)$ is mapped on to $(2a + b, 2c + d)$
$(3, 1)$ is mapped on to $(3a + b, 3c + d)$
$(4, 1)$ is mapped on to $(4a + b, 4c + d)$
$(h, 1)$ is mapped on to $(ha + b, hc + d)$

Putting the results together and using the fact that parallel lines are still parallel after the transformation, we see that Fig. 11.8 is transformed into Fig. 11.9 and any figure such as Fig. 11.10 is transformed into Fig. 11.11.

We have now seen the result of the transformation $\begin{pmatrix} a & b \\ c & d \end{pmatrix}$.

Exercise 11.2

1. Find the image of the points $(0, 1)$, $(0, 2)$, $(0, 3)$, $(0, 4)$ under the transformation $\begin{pmatrix} a & b \\ c & d \end{pmatrix}$. Now deduce the image of the line $x = 0$.

2. Find the image of the point $(1, 0)$, $(2, 0)$, $(2, 0)$, $(4, 0)$ under the transformation $\begin{pmatrix} a & b \\ c & d \end{pmatrix}$. Now deduce the image of the line $y = 0$.

3. Find the image of the line $y = 2x$ under the transformation $\begin{pmatrix} 1 & 2 \\ 3 & 11 \end{pmatrix}$.

4. Find the image of the point (h, k) under the transformation $\begin{pmatrix} 1 & 0 \\ 0 & 1 \end{pmatrix}$.

5. If the transformation $\begin{pmatrix} a & b \\ c & d \end{pmatrix}$ carries out the following mappings, determine a, b, c, d:

$$(x_0, y_0) \longrightarrow (x_1, y_1)$$
$$(1, 0) \longrightarrow (2, 3)$$
$$(0, 1) \longrightarrow (1, 0)$$

MATRIX MULTIPLICATION

We shall now carry out two transformations in the cartesian plane, one directly after the other. Let us try a reflection in the x-axis followed by a rotation about 0 through 90° anti-clockwise. Returning to page 237, we see that this means the operation $M_{y=0}$ followed by $R(0, +90°)$.

For $M_{y=0}$

$$x_1 = x_0$$
$$y_1 = \quad -y_0$$

Since we now follow this with $R(0, +90°)$, it means that x_1 above becomes x_0 and y_1 becomes y_0 for $R(0, +90°)$. It would be confusing to have two different meanings for (x_0, y_0), so we decide to use the following obvious form to describe the operation $R(0, +90°)$.

$$x_2 = \quad -y_1$$
$$y_2 = x_1$$

and using the results of M_{y-0}: this becomes

$$x_2 = \qquad +y_0 = \begin{pmatrix} 0 & 1 \\ 1 & 0 \end{pmatrix}\begin{pmatrix} x_0 \\ y_0 \end{pmatrix} \qquad \text{(i)}$$
$$y_2 = x_0$$

In matrix form we see that

$$\begin{pmatrix} x_1 \\ y_1 \end{pmatrix} = \begin{pmatrix} 1 & 0 \\ 0 & -1 \end{pmatrix}\begin{pmatrix} x_0 \\ y_0 \end{pmatrix}; \quad \begin{pmatrix} x_2 \\ y_2 \end{pmatrix} = \begin{pmatrix} 0 & -1 \\ 1 & 0 \end{pmatrix}\begin{pmatrix} x_1 \\ y_1 \end{pmatrix}$$

Putting these two together by substituting for $\begin{pmatrix} x_1 \\ y_1 \end{pmatrix}$ we have

$$\begin{pmatrix} x_2 \\ y_2 \end{pmatrix} = \begin{pmatrix} 0 & -1 \\ 1 & 0 \end{pmatrix}\begin{pmatrix} 1 & 0 \\ 0 & -1 \end{pmatrix}\begin{pmatrix} x_0 \\ y_0 \end{pmatrix}$$

But the final result was

$$\begin{pmatrix} x_2 \\ y_2 \end{pmatrix} = \begin{pmatrix} 0 & 1 \\ 1 & 0 \end{pmatrix}\begin{pmatrix} x_0 \\ y_0 \end{pmatrix} \quad \text{from (i)}$$

Comparison of these last two results informs us that we would like this to mean that:

$$\begin{pmatrix} 0 & -1 \\ 1 & 0 \end{pmatrix}\begin{pmatrix} 1 & 0 \\ 0 & -1 \end{pmatrix} = \begin{pmatrix} 0 & 1 \\ 1 & 0 \end{pmatrix}$$

or in the symbolic form of page 237

$$R(0, +90°). M_{y-0} = \begin{pmatrix} 0 & 1 \\ 1 & 0 \end{pmatrix}$$

As already suggested on page 237, this last result represents a reflection in the line $y = x$, as the following diagram shows. Here we have $\triangle OAP \cong \triangle OBQ$, where O is the origin.

Consequently $\qquad \overline{BQ} \cong \overline{AP}$ i.e. $x_1 = \qquad y_0$
$\qquad\qquad\qquad \overline{OB} \cong \overline{OA} \qquad\quad y_1 = x_0$

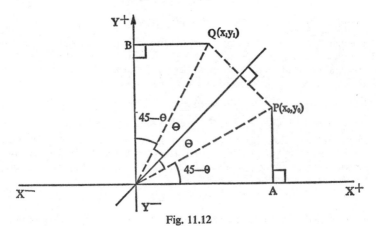

Fig. 11.12

In which case $\begin{pmatrix} 0 & 1 \\ 1 & 0 \end{pmatrix}$ is the operation of reflection in the line $y = x$.

Again suppose we consider $M_{x=0}$ followed by H_0, then we have

$$M_{x=0}: x_1 = -x_0 \qquad H_0: x_2 = -x_1 \qquad \therefore \quad x_2 = x_0$$
$$y_1 = y_0 \qquad y_2 = -y_1 \qquad y_2 = -y_0$$

In matrix form we see that

$$\begin{pmatrix} x_1 \\ y_1 \end{pmatrix} = \begin{pmatrix} -1 & 0 \\ 0 & 1 \end{pmatrix}\begin{pmatrix} x_0 \\ y_0 \end{pmatrix}; \quad \begin{pmatrix} x_2 \\ y_2 \end{pmatrix} = \begin{pmatrix} -1 & 0 \\ 0 & -1 \end{pmatrix}\begin{pmatrix} x_1 \\ y_1 \end{pmatrix}$$

Putting these two together we have

$$\begin{pmatrix} x_2 \\ y_2 \end{pmatrix} = \begin{pmatrix} -1 & 0 \\ 0 & -1 \end{pmatrix}\begin{pmatrix} -1 & 0 \\ 0 & 1 \end{pmatrix}\begin{pmatrix} x_0 \\ y_0 \end{pmatrix}$$

But the final result was

$$\begin{pmatrix} x_2 \\ y_2 \end{pmatrix} = \begin{pmatrix} 1 & 0 \\ 0 & -1 \end{pmatrix}\begin{pmatrix} x_0 \\ y_0 \end{pmatrix}$$

$$\therefore \quad \begin{pmatrix} -1 & 0 \\ 0 & -1 \end{pmatrix}\begin{pmatrix} -1 & 0 \\ 0 & 1 \end{pmatrix} = \begin{pmatrix} 1 & 0 \\ 0 & -1 \end{pmatrix}$$

is an assumption we would like to make, or in the symbolic form of page 237,

$$H_0 M_{x=0} = M_{y=0}$$

By direct substitution we proved that:

(i) a reflection in the x-axis followed by a rotation about 0 through 90° anti-clockwise is equivalent to a reflection in the line $y = x$;

(ii) a reflection in the y-axis followed by a rotation about 0 through 180° is equivalent to a reflection in the x-axis.

We are now to be more interested in the matrices. Since the matrices are written side by side, we say we are 'multiplying' the matrices together. In the first case considered

$$\begin{pmatrix} 0 & -1 \\ 1 & 0 \end{pmatrix}\begin{pmatrix} 1 & 0 \\ 0 & -1 \end{pmatrix} = \begin{pmatrix} 0 & 1 \\ 1 & 0 \end{pmatrix},$$

while in the second case

$$\begin{pmatrix} -1 & 0 \\ 0 & -1 \end{pmatrix}\begin{pmatrix} -1 & 0 \\ 0 & 1 \end{pmatrix} = \begin{pmatrix} 1 & 0 \\ 0 & -1 \end{pmatrix}$$

What we wish to discover is some rule for multiplying two matrices together. Examination of these two results is not very promising, but this, of course, is a typical mathematical situation. We fail to spot the rule from these results, so we put them to one side so as to use them later on as a check—when we do discover the rule.

We shall try again by returning to the general matrix and consider the two following transformations:

$$x_2 = ex_1 + fy_1 \qquad x_1 = ax_0 + by_0$$
$$y_2 = gx_1 + hy_1 \qquad y_1 = cx_0 + dy_0$$

$$\begin{pmatrix} x_2 \\ y_2 \end{pmatrix} = \begin{pmatrix} e & f \\ g & h \end{pmatrix}\begin{pmatrix} x_1 \\ y_1 \end{pmatrix}; \quad \begin{pmatrix} x_1 \\ y_1 \end{pmatrix} = \begin{pmatrix} a & b \\ c & d \end{pmatrix}\begin{pmatrix} x_0 \\ y_0 \end{pmatrix}$$

Therefore by ordinary substitution

$$x_2 = e(ax_0 + by_0) + f(cx_0 + dy_0)$$
$$y_2 = g(ax_0 + by_0) + h(cx_0 + dy_0)$$

Which, after some collecting, become

$$x_2 = (ea + fc)x_0 + (eb + fd)y_0$$
$$y_2 = (ga + hc)x_0 + (gb + hd)y_0$$

$$\begin{pmatrix} x_2 \\ y_2 \end{pmatrix} = \begin{pmatrix} ea + fc & eb + fd \\ ga + hc & gb + hd \end{pmatrix}\begin{pmatrix} x_0 \\ y_0 \end{pmatrix}$$

Consequently, we are now able to write that we require:

$$\begin{pmatrix} e & f \\ g & h \end{pmatrix}\begin{pmatrix} a & b \\ c & d \end{pmatrix} = \begin{pmatrix} ea + fc & eb + fd \\ ga + hc & gb + hd \end{pmatrix}$$

We write this as $M_1 M_2 = M_3$.

Now what is the rule for multiplication of matrices?

All the matrices we are discussing at the moment are called 2×2 matrices, that is, they have two rows and two columns.

The final result here is

$$\begin{array}{cc} \downarrow \text{first column} & \downarrow \text{second column} \end{array}$$

$$\begin{array}{l} \text{first row} \longrightarrow \\ \text{second row} \longrightarrow \end{array} \begin{pmatrix} ea + fc & eb + fd \\ ga + hc & gb + hd \end{pmatrix}$$

Now try to see where $ea + fc$ has come from; e and f are in the first row of M_1, and a and c are in the first column of M_2. So it looks as if the element in the first row and first column position of M_3 is gained by multiplying the first column of M_2 by the first row of M_1. But what do we mean by this multiplication?

How does one multiply $\begin{pmatrix} e & f \\ . & . \end{pmatrix}$ and $\begin{pmatrix} a & . \\ c & . \end{pmatrix}$?

To produce the results we need, this multiplication will obviously have to be *defined*. The result is a product, but not the usual type, so we give it a special name of inner product or scalar product. Furthermore, we are dealing with sets of elements in a definite order, e.g. we would expect a different result from

$$\begin{pmatrix} f & e \\ . & . \end{pmatrix}\begin{pmatrix} a & . \\ c & . \end{pmatrix} \text{ or } \begin{pmatrix} e & f \\ . & . \end{pmatrix}\begin{pmatrix} c & . \\ a & . \end{pmatrix}$$

Though possibly not from

$$\begin{pmatrix} f & e \\ . & . \end{pmatrix}\begin{pmatrix} c & . \\ a & . \end{pmatrix}$$

Definition 11.1

The **scalar** product of two ordered sets is defined as the sum of the products of corresponding elements, and is only defined if the sets are equivalent. For example,

$$\text{set } A = \{a_1, a_2, a_3, a_4\}; \quad \text{set } B = \{b_1, b_2, b_3, b_4\},$$
$$\text{set } C = \{c_1, c_2\}$$

The scalar product of A and B written A.B is given by

$$A.B = a_1b_1 + a_2b_2 + a_3b_3 + a_4b_4$$

But note, the product AC is *not* defined, since the sets A and C are not equivalent. Returning to our product above, we see the following:

Scalar product of first row M_1 and first column M_2.

Scalar product of first row M_1 and second column M_2.

$$\begin{pmatrix} e & f \\ g & h \end{pmatrix}\begin{pmatrix} a & b \\ c & d \end{pmatrix} = \begin{pmatrix} ea + fc & eb + fd \\ ga + hc & gb + hd \end{pmatrix}$$

Scalar product of second row M_1 and first column M_2.

Scalar product of second row M_1 and second column M_2.

Look at the following matrices:

$$A = \begin{pmatrix} 2 & 3 \\ 1 & 4 \end{pmatrix}; \quad B = \begin{pmatrix} -1 & 3 \\ 6 & 2 \end{pmatrix}; \quad C = \begin{pmatrix} -3 & 4 \\ 2 & -1 \end{pmatrix}$$

Example 1

To find AB

$$\begin{pmatrix} 2 & 3 \\ 1 & 4 \end{pmatrix}\begin{pmatrix} -1 & 3 \\ 6 & 2 \end{pmatrix} = \begin{pmatrix} -2 + 18 & 6 + 6 \\ -1 + 24 & 3 + 8 \end{pmatrix}$$
$$= \begin{pmatrix} 16 & 12 \\ 23 & 11 \end{pmatrix}$$

Example 2

To find BA

$$\begin{pmatrix} -1 & 3 \\ 6 & 2 \end{pmatrix}\begin{pmatrix} 2 & 3 \\ 1 & 4 \end{pmatrix} = \begin{pmatrix} -2 + 3 & -3 + 12 \\ 12 + 2 & 18 + 8 \end{pmatrix}$$
$$= \begin{pmatrix} 1 & 9 \\ 14 & 26 \end{pmatrix}$$

The first thing to note is that BA $\neq$ AB. So the multiplication of matrices is not commutative (see page 15).

Example 3

To find AC

$$\begin{pmatrix} 2 & 3 \\ 1 & 4 \end{pmatrix}\begin{pmatrix} -3 & 4 \\ 2 & -1 \end{pmatrix} = \begin{pmatrix} -6 + 6 & 8 - 3 \\ -3 + 8 & 4 - 4 \end{pmatrix}$$
$$= \begin{pmatrix} 0 & 5 \\ 5 & 0 \end{pmatrix}$$

With these three examples we have demonstrated the multiplication of the matrices, but the reader should now notice how we have ceased to speak of the cartesian plane. That is, it is no longer necessary to refer to our diagrams to see what is going on, since we can quite easily interpret the final result if we want to. Knowing how to multiply the matrices enables us to carry out any transformation we like without once suffering the restriction of having to locate the intermediate changes. This process of abstraction is, of course, one of the powers of mathematics.

Exercise 11.3

1. Using the matrices of the above examples determine the following products to show that
$$AC \neq CA$$
$$BC \neq CB$$

2. Using the matrices of the operations in page 237, obtain the following products and identify the operation:

(a) $M_{y-0} M_{x-0}$ (b) $M_{x-0} M_{y-0}$
(c) $M_{x-0} R(0, +90°)$ (d) $R(0, +90°) R(0, -90°)$

3. Determine the following products:

(a) $\begin{pmatrix} 3 & 5 \\ 1 & 2 \end{pmatrix}\begin{pmatrix} 2 & -5 \\ -1 & 3 \end{pmatrix}$ (b) $\begin{pmatrix} 2 & 13 \\ 1 & 7 \end{pmatrix}\begin{pmatrix} 7 & -13 \\ -1 & 2 \end{pmatrix}$ (c) $\begin{pmatrix} 4 & 3 \\ 9 & 7 \end{pmatrix}\begin{pmatrix} 7 & -3 \\ -9 & 4 \end{pmatrix}$

4. Examine the results for Question 3 (a), (b), (c), then try to find a matrix $\begin{pmatrix} a & b \\ c & d \end{pmatrix}$ such that $\begin{pmatrix} 3 & 2 \\ 1 & 1 \end{pmatrix}\begin{pmatrix} a & b \\ c & d \end{pmatrix} = \begin{pmatrix} 1 & 0 \\ 0 & 1 \end{pmatrix}$.

5. Given the matrices $H = \begin{pmatrix} 1 & 0 \\ 0 & -1 \end{pmatrix}$; $V = \begin{pmatrix} -1 & 0 \\ 0 & 1 \end{pmatrix}$; $R = \begin{pmatrix} -1 & 0 \\ 0 & -1 \end{pmatrix}$.

$I = \begin{pmatrix} 1 & 0 \\ 0 & 1 \end{pmatrix}$ and the operation X, meaning matrix multiplication; complete the multiplication table of Fig. 11.13.

2nd matrix

1st matrix	X	I	H	V	R
	I				
	H				
	V		R		
	R				

Fig. 11.13

The table is completed in the following manner:

The product $HV = \begin{pmatrix} 1 & 0 \\ 0 & -1 \end{pmatrix}\begin{pmatrix} -1 & 0 \\ 0 & 1 \end{pmatrix} = \begin{pmatrix} -1 & 0 \\ 0 & -1 \end{pmatrix} = R$

V is the first matrix and H is the second matrix. (Recall the manner in which the transformations were carried out on the ordered pair $\begin{pmatrix} x_0 \\ y_0 \end{pmatrix}$.) The result R has been entered in the table. (See Fig. 6.12 for hints.)

6. We shall assume that the associative law holds for matrix multiplication. Using the matrices A, B, and C on page 247 show that $A(BC) = (AB)C$.

MATRIX ALGEBRA
MULTIPLICATION BY A NUMBER

You will recall that we obtained the general 2×2 matrix from the equations

$$x_1 = ax_0 + by_0$$
$$y_1 = cx_0 + dy_0$$

If we multiplied both equations by any number, n say, we would obtain

$$nx_1 = nax_0 + nby_0$$
$$ny_1 = ncx_0 + ndy_0$$

which becomes in matrix form

$$\begin{pmatrix} nx_1 \\ ny_1 \end{pmatrix} = \begin{pmatrix} na & nb \\ nc & nd \end{pmatrix} \begin{pmatrix} x_0 \\ y_0 \end{pmatrix}$$

The form of this last equation suggests two things:

(i) that the number n is a factor in the multiplication;
(ii) that a desirable simplification in the last step would be effected by suggesting that $\begin{pmatrix} nx_1 \\ ny_1 \end{pmatrix} = n \begin{pmatrix} x_1 \\ y_1 \end{pmatrix}$.

The consequence of this last suggestion would be

$$n \begin{pmatrix} x_1 \\ y_1 \end{pmatrix} = n \begin{pmatrix} a & b \\ c & d \end{pmatrix} \begin{pmatrix} x_0 \\ y_0 \end{pmatrix} = \begin{pmatrix} na & nb \\ nc & nd \end{pmatrix} \begin{pmatrix} x_0 \\ y_0 \end{pmatrix}$$

that is

$$n \begin{pmatrix} a & b \\ c & d \end{pmatrix} = \begin{pmatrix} na & nb \\ nc & nd \end{pmatrix}$$

We therefore come to the definition of the multiplication of a matrix by a real number.

Definition 11.2

In the multiplication of a matrix by the real number, n, each separate element is multiplied by n.

Example 1

$$4 \begin{pmatrix} 1 & 3 \\ 7 & 2 \end{pmatrix} = \begin{pmatrix} 4 & 12 \\ 28 & 8 \end{pmatrix}$$

Example 2

$$-7 \begin{pmatrix} 1 & -2 \\ -6 & 4 \end{pmatrix} = \begin{pmatrix} -7 & 14 \\ 42 & -28 \end{pmatrix}$$

ADDITION

The above work should give us a clue as to how to define the addition of matrices. For consider

$$x_1 = ax_0 + by_0 \quad \text{added to} \quad 2x_1 = 2ax_0 + 2by_0 \text{ producing}$$
$$y_1 = cx_0 + dy_0 \qquad\qquad\quad 2y_1 = 2cx_0 + 2dy_0$$

$$3\begin{pmatrix} x_1 \\ y_1 \end{pmatrix} = \begin{pmatrix} 3a & 3b \\ 3c & 3d \end{pmatrix}\begin{pmatrix} x_0 \\ y_0 \end{pmatrix}$$

In matrix form this is

$$\begin{pmatrix} x_1 \\ y_1 \end{pmatrix} + \begin{pmatrix} 2x_1 \\ 2y_1 \end{pmatrix} = \begin{pmatrix} a & b \\ c & d \end{pmatrix}\begin{pmatrix} x_0 \\ y_0 \end{pmatrix} + \begin{pmatrix} 2a & 2b \\ 2c & 2d \end{pmatrix}\begin{pmatrix} x_0 \\ y_0 \end{pmatrix}$$

A comparison of these last results suggests that

$$\begin{pmatrix} a & b \\ c & d \end{pmatrix} + \begin{pmatrix} 2a & 2b \\ 2c & 2d \end{pmatrix} = \begin{pmatrix} 3a & 3b \\ 3c & 3d \end{pmatrix}$$

Which leads naturally to an expected definition of matrix addition.

Definition 11.3

The sum of two matrices is equal to the matrix of the sum of corresponding elements. (We are of course discussing 2×2 matrices only.)

Example 1

$$\begin{pmatrix} 3 & 9 \\ 4 & 7 \end{pmatrix} + \begin{pmatrix} 4 & 7 \\ 6 & 5 \end{pmatrix} = \begin{pmatrix} 7 & 16 \\ 10 & 12 \end{pmatrix}$$

Example 2

$$\begin{pmatrix} 0 & -1 \\ 4 & 3 \end{pmatrix} + \begin{pmatrix} 7 & 1 \\ -4 & 17 \end{pmatrix} = \begin{pmatrix} 7 & 0 \\ 0 & 20 \end{pmatrix}$$

Example 3

$$\begin{pmatrix} a & b \\ c & d \end{pmatrix} + \begin{pmatrix} p & q \\ r & s \end{pmatrix} = \begin{pmatrix} a+p & b+q \\ c+r & d+s \end{pmatrix}$$

Definition 11.4

The matrix whose elements are all **zero** is called the zero or null matrix. If N is the null matrix rather than write $N = \begin{pmatrix} 0 & 0 \\ 0 & 0 \end{pmatrix}$ we simply write $N = \mathbf{0}$.

Consider the addition of the null matrix to any other matrix. Clearly we have

$$\begin{pmatrix} a & b \\ c & d \end{pmatrix} + N = \begin{pmatrix} a+0 & b+0 \\ c+0 & d+0 \end{pmatrix} = \begin{pmatrix} a & b \\ c & d \end{pmatrix}$$

In other words, $A + N = A$ for any matrix A.

This means that N is the identity element for the operation of matrix addition in the set of matrices (see page 29). It should be clear now that we are leading up to an examination of the possibility of matrices forming a group under matrix addition (see page 123).

The general matrix $\begin{pmatrix} a & b \\ c & d \end{pmatrix}$ added to $\begin{pmatrix} -a & -b \\ -c & -d \end{pmatrix}$ yields $\begin{pmatrix} 0 & 0 \\ 0 & 0 \end{pmatrix}$ the identity element. This means that $\begin{pmatrix} -a & -b \\ -c & -d \end{pmatrix}$ is the inverse of $\begin{pmatrix} a & b \\ c & d \end{pmatrix}$ with respect to matrix addition.

Now we need to consider associativity with three general matrices B, F, J respectively.

$$\begin{pmatrix} a & b \\ c & d \end{pmatrix} + \begin{pmatrix} e & f \\ g & h \end{pmatrix} + \begin{pmatrix} i & j \\ k & l \end{pmatrix} = \begin{pmatrix} a & b \\ c & d \end{pmatrix} + \begin{pmatrix} e+i & f+j \\ g+k & h+l \end{pmatrix} =$$

$$\begin{pmatrix} a+e+i & b+f+j \\ c+g+k & d+h+l \end{pmatrix} = \begin{pmatrix} a+e & b+f \\ c+g & d+h \end{pmatrix} + \begin{pmatrix} i & j \\ k & l \end{pmatrix} =$$

$$\begin{pmatrix} a+e+i & b+f+j \\ c+g+k & d+h+l \end{pmatrix}$$

Thus we have proved that B + (F + J) = (B + F) + J.

A summary of this (in keeping with page 123, Definition 6.4) is:

1. The set of 2 × 2 matrices M is closed under the operation +.

2. The associative law holds in M, so that if A, B, C are any three elements of M, then A + (B + C) = (A + B) + C.

3. There exists an identity element N (the null matrix) which belongs to M such that for any element A of M we have N + A = A = A + N.

4. Every element of M has an inverse with respect to + such that if A is an element of M, then its inverse A* is such that A + A* = N = A* + A.

Consequently, the set of matrices forms a group under the operation of matrix addition.

We can in fact go a little further, because since A + B = B + A for any two matrices, then the group is commutative.

SUBTRACTION

Definition 11.5

We define matrix subtraction as the inverse operation to matrix addition, i.e. subtracting a matrix is the same as adding its additive inverse.

Thus $\begin{pmatrix} a & b \\ c & d \end{pmatrix} - \begin{pmatrix} p & q \\ r & s \end{pmatrix} = \begin{pmatrix} a & b \\ c & d \end{pmatrix} + \begin{pmatrix} -p & -q \\ -r & -s \end{pmatrix} = \begin{pmatrix} a-p & b-q \\ c-r & d-s \end{pmatrix}$

Example 1

$$\begin{pmatrix} 4 & 7 \\ 9 & 3 \end{pmatrix} - \begin{pmatrix} 6 & 5 \\ 4 & 10 \end{pmatrix} = \begin{pmatrix} 4 & 7 \\ 9 & 3 \end{pmatrix} + \begin{pmatrix} -6 & -5 \\ -4 & -10 \end{pmatrix} = \begin{pmatrix} -2 & 2 \\ 5 & -7 \end{pmatrix}$$

Of course with practice (or without it) we no longer bother to write the intermediate stage.

Example 2

$$\begin{pmatrix} 1 & 2 \\ 7 & 4 \end{pmatrix} - \begin{pmatrix} 3 & 2 \\ 19 & 1 \end{pmatrix} = \begin{pmatrix} -2 & 0 \\ -12 & 3 \end{pmatrix}$$

We now see also a clear indication of what we would mean by equal matrices: if A − B = 0, then we say that A is equal to B or simply A = B.

That is, two matrices are equal if, *and only if*, their corresponding elements are equal.

e.g. if
$$\begin{pmatrix} a & b \\ c & d \end{pmatrix} = \begin{pmatrix} e & f \\ g & h \end{pmatrix}, \text{ then } \begin{aligned} a &= e \\ b &= f \\ c &= g \\ d &= h. \end{aligned}$$

MATRIX EQUATIONS

For the moment we shall deal with those involving only addition and subtraction.

Example 1

If X is a 2 × 2 matrix, solve the following equation:

$$X + \begin{pmatrix} 4 & 9 \\ 2 & 7 \end{pmatrix} = \begin{pmatrix} 8 & 5 \\ 6 & 1 \end{pmatrix}$$

We merely add the additive inverse of $\begin{pmatrix} 4 & 9 \\ 2 & 7 \end{pmatrix}$ to both sides of the equation

$$X + N = \begin{pmatrix} 8 & 5 \\ 6 & 1 \end{pmatrix} + \begin{pmatrix} -4 & -9 \\ -2 & -7 \end{pmatrix} = \begin{pmatrix} 4 & -4 \\ 4 & -6 \end{pmatrix}$$

Example 2

If A, B, and X are 2 × 2 matrices. Solve the following equation

$$X + A = B$$

Adding the additive inverse of A to both sides we obtain

$$\begin{aligned} X + A + A^* &= B + A^* \\ X + N &= B + A^* \\ X &= B + A^* = B - A. \end{aligned}$$

Example 3

Solve X = 3A if
$$A = \begin{pmatrix} 0 & 9 \\ 7 & -2 \end{pmatrix}$$

Straight away we may write
$$X = \begin{pmatrix} 0 & 27 \\ 21 & -6 \end{pmatrix}$$

Example 4

Solve 3X = A with A of Example 3.

Here we write
$$\tfrac{1}{3}(3X) = X = \tfrac{1}{3}A = \begin{pmatrix} 0 & 3 \\ \tfrac{7}{3} & -\tfrac{2}{3} \end{pmatrix}$$

Exercise 11.4

1. Solve the equation $4X = \begin{pmatrix} 8 & -16 \\ 72 & 0 \end{pmatrix}$, where X is a 2 × 2 matrix.

2. Simplify $3\begin{pmatrix} 8 & 4 \\ 3 & 2 \end{pmatrix} + 4\begin{pmatrix} 8 & 4 \\ 3 & 2 \end{pmatrix}$.

3. Simplify $13\begin{pmatrix} 6 & 9 \\ 0 & 4 \end{pmatrix} - 7\begin{pmatrix} 6 & 9 \\ 0 & 4 \end{pmatrix}$.

4. Find the additive inverse of: (a) $\begin{pmatrix} 3 & 6 \\ 1 & 2 \end{pmatrix}$; (b) $\begin{pmatrix} 1 & -3 \\ -9 & 4 \end{pmatrix}$.

5. If $A = \begin{pmatrix} 1 & 0 \\ 1 & 3 \end{pmatrix}$, $B = \begin{pmatrix} 6 & -7 \\ 4 & 1 \end{pmatrix}$, $C = \begin{pmatrix} 4 & -7 \\ 6 & -2 \end{pmatrix}$,

obtain the values of:

(a) $A + B + C$ (b) $A - B + C$ (c) $-A - B - C$ (d) $A + B - C$

A GROUP OF MATRICES WITH RESPECT TO MULTIPLICATION

Unfortunately the set of matrices, M, does not form a group with respect to matrix multiplication because not all matrices have multiplicative inverses. But we can derive some useful practice from studying a small set of matrices given by

$$H = \begin{pmatrix} 1 & 0 \\ 0 & -1 \end{pmatrix}; \quad V = \begin{pmatrix} -1 & 0 \\ 0 & 1 \end{pmatrix}; \quad R = \begin{pmatrix} -1 & 0 \\ 0 & -1 \end{pmatrix}; \quad I = \begin{pmatrix} 1 & 0 \\ 0 & 1 \end{pmatrix}$$

This was Question 5 of Exercise 11.3 and yielded a multiplication table given by Fig. 6.12 which we repeat here. The reader is left to fill in the table as an exercise.

2nd matrix

	X	I	H	V	R
1st matrix	**I**				
	H				
	V	R			
	R				

Fig. 11.13

From this table we notice that:

1. The set {IHVR} is closed with respect to matrix multiplication X.

2. The associative law holds.

3. There exists an identity element which belongs to {IHVR}; it is in fact I.

4. Each element has a multiplicative inverse thus:

$$H \times H = I$$
$$V \times V = I \quad \text{each element is its own inverse}$$
$$R \times R = I$$
$$I \times I = I$$

5. Furthermore, the group is commutative with respect to X.

There is clearly an isomorphism between the tables of 6.12 and 11.13 which would have enabled us to deduce the commutative group property for matrix multiplication in this set, had we bothered to notice it before writing out the five observations above.

It shows incidentally that the rotations discussed for Fig. 6.7 could have been related to the rotations and reflections on page 237.

THE INVERSE MATRIX (2 × 2)

The inverse matrices of the above group were extremely easy to find. In actual fact we hardly found them at all, but merely identified them from the Table 11.13.

The question to be asked now is how to go about finding the inverse of any matrix (assuming it to be possible).

For a start we need to know the identity matrix for multiplication. This we have encountered once or twice already, especially in the example above. We shall use I to denote this identity matrix. We have only checked it so far in the above set {IHVR}, but a general test is really no obstacle, because consider $\begin{pmatrix} 1 & 0 \\ 0 & 1 \end{pmatrix}\begin{pmatrix} a & b \\ c & d \end{pmatrix}$. This is clearly $\begin{pmatrix} a & b \\ c & d \end{pmatrix}$. (See page 246.)

Also $\begin{pmatrix} a & b \\ c & d \end{pmatrix}\begin{pmatrix} 1 & 0 \\ 0 & 1 \end{pmatrix} = \begin{pmatrix} a & b \\ c & d \end{pmatrix}$, so that for any matrix A whatsoever we may write IA = A = AI. It follows from this that I is the identity matrix for matrix multiplication.

We now have the following definition:

Definition 11.6

The multiplicative inverse of any matrix A (if it exists) written A^{-1} is such that

$$A^{-1}A = I = AA^{-1}$$

where I is the identity matrix. A matrix which does not have a multiplicative inverse is called a **singular** matrix. It is clear that a matrix which does have an inverse will be a non-singular matrix.

To get a little more help in our search for the inverse, let us turn back to Exercise 11.3, Question 3.

One of the examples was $\begin{pmatrix} 3 & 5 \\ 1 & 2 \end{pmatrix}\begin{pmatrix} 2 & -5 \\ -1 & 3 \end{pmatrix} = \begin{pmatrix} 1 & 0 \\ 0 & 1 \end{pmatrix}$. Say this is AB = I.

Now try BA. This is $\begin{pmatrix} 2 & -5 \\ -1 & 3 \end{pmatrix}\begin{pmatrix} 3 & 5 \\ 1 & 2 \end{pmatrix} = \begin{pmatrix} 1 & 0 \\ 0 & 1 \end{pmatrix}$.

We have now shown that BA = I = AB.

Examination of Definition 11.6 shows that we have obtained $A^{-1} = B$.

That is, the inverse of $\begin{pmatrix} 3 & 5 \\ 1 & 2 \end{pmatrix}$ is $\begin{pmatrix} 2 & -5 \\ -1 & 3 \end{pmatrix}$. This looks promising, because an inspection reveals one or two possible clues for constructing an inverse matrix. The elements in the leading diagonal are apparently interchanged, while the other elements are multiplied by −1.

Another example was $\begin{pmatrix} 4 & 3 \\ 9 & 7 \end{pmatrix}\begin{pmatrix} 7 & -3 \\ -9 & 4 \end{pmatrix} = \begin{pmatrix} 1 & 0 \\ 0 & 1 \end{pmatrix}$. Say this is CD = I.

Now try DC. This is $\begin{pmatrix} 7 & -3 \\ -9 & 4 \end{pmatrix}\begin{pmatrix} 4 & 3 \\ 9 & 7 \end{pmatrix} = \begin{pmatrix} 1 & 0 \\ 0 & 1 \end{pmatrix}$.

We have now shown that DC = I = CD.

Therefore $\qquad\qquad\qquad\qquad$ **D = C⁻¹**

That is, the inverse of $\begin{pmatrix} 4 & 3 \\ 9 & 7 \end{pmatrix}$ is $\begin{pmatrix} 7 & -3 \\ -9 & 4 \end{pmatrix}$ and the possible rule mentioned already, seems to hold once again.

Now let us try to find the inverse of $\begin{pmatrix} 3 & 5 \\ 1 & 4 \end{pmatrix}$. This looks very much like the first one we tried. Applying the discovered rule, we shall multiply the given matrix by $\begin{pmatrix} 4 & -5 \\ -1 & 3 \end{pmatrix}$ to see what happens.

Thus
$$\begin{pmatrix} 3 & 5 \\ 1 & 4 \end{pmatrix}\begin{pmatrix} 4 & -5 \\ -1 & 3 \end{pmatrix} = \begin{pmatrix} 7 & 0 \\ 0 & 7 \end{pmatrix} = 7I$$
$$\begin{pmatrix} 4 & -5 \\ -1 & 3 \end{pmatrix}\begin{pmatrix} 3 & 5 \\ 1 & 4 \end{pmatrix} = \begin{pmatrix} 7 & 0 \\ 0 & 7 \end{pmatrix} = 7I$$

We get the same result each time, and the form of the result shows we are on the right lines. An idea is to multiply by $\frac{1}{7}$.

$$\tfrac{1}{7}\begin{pmatrix} 3 & 5 \\ 1 & 4 \end{pmatrix}\begin{pmatrix} 4 & -5 \\ -1 & 3 \end{pmatrix} = \begin{pmatrix} 3 & 5 \\ 1 & 4 \end{pmatrix}\tfrac{1}{7}\begin{pmatrix} 4 & -5 \\ -1 & 3 \end{pmatrix} = \begin{pmatrix} 3 & 5 \\ 1 & 4 \end{pmatrix}\begin{pmatrix} \frac{4}{7} & -\frac{5}{7} \\ -\frac{1}{7} & \frac{3}{7} \end{pmatrix} = I$$

We deduce therefore that the inverse of

$$\begin{pmatrix} 3 & 5 \\ 1 & 4 \end{pmatrix} \text{ is } \begin{pmatrix} \frac{4}{7} & -\frac{5}{7} \\ -\frac{1}{7} & \frac{3}{7} \end{pmatrix}$$

The rule almost worked—but where has the 7 come from?

Perhaps an inspection of a general case might help.

To find the inverse of $\begin{pmatrix} a & b \\ c & d \end{pmatrix}$, apply the rule and evaluate the product.

$$\begin{pmatrix} a & b \\ c & d \end{pmatrix}\begin{pmatrix} d & -b \\ -c & a \end{pmatrix} = \begin{pmatrix} ad - bc & 0 \\ 0 & ad - bc \end{pmatrix} = (ad - bc)\begin{pmatrix} 1 & 0 \\ 0 & 1 \end{pmatrix}$$

We now see that the inverse of $\begin{pmatrix} a & b \\ c & d \end{pmatrix}$ is $\begin{pmatrix} \dfrac{d}{ad - bc} & \dfrac{-b}{ad - bc} \\ \dfrac{-c}{ad - bc} & \dfrac{a}{ad - bc} \end{pmatrix}$; $(ad \neq bc)$.

And also we see that if $ad - bc = 0$, the inverse will not exist since we would not be able to divide by $ad - bc$.

Example 1

Obtain the multiplicative inverse of $\begin{pmatrix} 4 & 3 \\ 9 & 8 \end{pmatrix}$.

The inverses are now merely written down. In this case the equivalent to $ad - bc$ is $(4 \times 8) - (3 \times 9) = 5$.

Therefore the required inverse is $\begin{pmatrix} \frac{8}{5} & -\frac{3}{5} \\ -\frac{9}{5} & \frac{4}{5} \end{pmatrix}$

Check				$\begin{pmatrix} 4 & 3 \\ 9 & 8 \end{pmatrix}\begin{pmatrix} \frac{8}{5} & -\frac{3}{5} \\ -\frac{9}{5} & \frac{4}{5} \end{pmatrix} = I$

Similarly,				$\begin{pmatrix} \frac{8}{5} & -\frac{3}{5} \\ -\frac{9}{5} & \frac{4}{5} \end{pmatrix}\begin{pmatrix} 4 & 3 \\ 9 & 8 \end{pmatrix} = I$

Example 2

Obtain the multiplicative inverse of $\begin{pmatrix} 6 & 9 \\ 2 & 3 \end{pmatrix}$

Here the equivalent to $ad - bc$ is $(6 \times 3) - (9 \times 2) = 0$, so that this is a singular matrix, i.e. it has no inverse.

Example 3

Obtain the multiplicative inverse of $\begin{pmatrix} -7 & 1 \\ 3 & 4 \end{pmatrix}$.

In this case $ad - bc = -28 - 3 = -31$.

The inverse matrix is $-\dfrac{1}{31}\begin{pmatrix} 4 & -1 \\ -3 & -7 \end{pmatrix}$.

Notice that in this last example we have put the fractional part outside for the sake of tidiness.

We could also tidy up a little further and conclude with

$$\frac{1}{31}\begin{pmatrix} -4 & 1 \\ 3 & 7 \end{pmatrix}$$

Exercise 11.5

1. Obtain the multiplicative inverses of the following matrices (if they exist):

(a) $\begin{pmatrix} 1 & 1 \\ 0 & 1 \end{pmatrix}$ 		(b) $\begin{pmatrix} 1 & 0 \\ 1 & 1 \end{pmatrix}$ 		(c) $\begin{pmatrix} 1 & 1 \\ 1 & 1 \end{pmatrix}$ 		(d) $\begin{pmatrix} -1 & 0 \\ 0 & 1 \end{pmatrix}$

(e) $\begin{pmatrix} 3 & 1 \\ 6 & 4 \end{pmatrix}$ 		(f) $\begin{pmatrix} 2 & 1 \\ 9 & 3 \end{pmatrix}$ 		(g) $\begin{pmatrix} 17 & 51 \\ 3 & 9 \end{pmatrix}$ 		(h) $\begin{pmatrix} 4 & 6 \\ -6 & 9 \end{pmatrix}$

2. Given that $I = \begin{pmatrix} 1 & 0 \\ 0 & 1 \end{pmatrix}$, $A = \begin{pmatrix} 0 & 1 \\ -1 & 0 \end{pmatrix}$, $B = \begin{pmatrix} -1 & 0 \\ 0 & -1 \end{pmatrix}$, $C = \begin{pmatrix} 0 & -1 \\ 1 & 0 \end{pmatrix}$

complete the multiplication table of Fig. 11.14.

X	I	A	B	C
I				
A			C	
B				
C			A	B

Fig. 11.14

3. State the multiplicative inverse for each of the matrices in Question 2.

4. State whether the set {I, A, B, C} of Question 2 forms a group under matrix multiplication.

5. Name the proper sub-group in the set {I, A, B, C} under matrix multiplication.

6. Examine the Tables 11.14 and 6.16. Is there an isomorphism?

SIMULTANEOUS EQUATIONS

The inverse matrix provides us with a useful means of solving two simultaneous equations, though to be fair the matrix method is sometimes more trouble than it is worth. The usual method is much more straightforward and few sensible people would bother with matrices for two equations. The matrix method does, however, have the power of being perfectly general and the type of method which is readily stored in a computer. We consider numerical cases first.

Example 1

Solve the two simultaneous equations

$$3x + 4y = 32$$
$$2x + 3y = 23$$

Solution

Writing this in matrix form we have

$$\begin{pmatrix} 3 & 4 \\ 2 & 3 \end{pmatrix} \begin{pmatrix} x \\ y \end{pmatrix} = \begin{pmatrix} 32 \\ 23 \end{pmatrix}$$

Now the inverse of $\begin{pmatrix} 3 & 4 \\ 2 & 3 \end{pmatrix}$ is $\begin{pmatrix} 3 & -4 \\ -2 & 3 \end{pmatrix}$, and we multiply the above equation by this inverse.

Thus

$$\begin{pmatrix} 3 & -4 \\ -2 & 3 \end{pmatrix} \begin{pmatrix} 3 & 4 \\ 2 & 3 \end{pmatrix} \begin{pmatrix} x \\ y \end{pmatrix} = \begin{pmatrix} 3 & -4 \\ -2 & 3 \end{pmatrix} \begin{pmatrix} 32 \\ 23 \end{pmatrix}$$

This now becomes

$$\begin{pmatrix} 1 & 0 \\ 0 & 1 \end{pmatrix} \begin{pmatrix} x \\ y \end{pmatrix} = \begin{pmatrix} x \\ y \end{pmatrix} = \begin{pmatrix} 96 & -92 \\ -64 & +69 \end{pmatrix} = \begin{pmatrix} 4 \\ 5 \end{pmatrix}$$

$\therefore \quad x = 4, y = 5$ is the solution.

Example 2

Solve the two simultaneous equations.

$$x + 4y = 1$$
$$-x + 3y = 13$$

Solution

Written in matrix form we have

$$\begin{pmatrix} 1 & 4 \\ -1 & 3 \end{pmatrix} \begin{pmatrix} x \\ y \end{pmatrix} = \begin{pmatrix} 1 \\ 13 \end{pmatrix}$$

The multiplicative inverse of $\begin{pmatrix} 1 & 4 \\ -1 & 3 \end{pmatrix}$ is $\frac{1}{7}\begin{pmatrix} 3 & -4 \\ 1 & 1 \end{pmatrix}$

$$\frac{1}{7}\begin{pmatrix} 3 & -4 \\ 1 & 1 \end{pmatrix} \begin{pmatrix} 1 & 4 \\ -1 & 3 \end{pmatrix} \begin{pmatrix} x \\ y \end{pmatrix} = \begin{pmatrix} x \\ y \end{pmatrix} = \frac{1}{7}\begin{pmatrix} 3 & -4 \\ 1 & 1 \end{pmatrix} \begin{pmatrix} 1 \\ 13 \end{pmatrix}$$

$$\therefore \quad \begin{pmatrix} x \\ y \end{pmatrix} = \frac{1}{7}\begin{pmatrix} -49 \\ 14 \end{pmatrix} = \begin{pmatrix} -7 \\ 2 \end{pmatrix}$$

$x = -7, y = 2$ is the solution.

Suppose we now look at the general pair of equations.

$$\begin{array}{ll} ax + by = e & \text{or} \quad \begin{pmatrix} a & b \\ c & d \end{pmatrix} \begin{pmatrix} x \\ y \end{pmatrix} = \begin{pmatrix} e \\ f \end{pmatrix} \\ cx + dy = f \end{array}$$

It follows that $\begin{pmatrix} x \\ y \end{pmatrix} = \begin{pmatrix} a & b \\ c & d \end{pmatrix}^{-1} \begin{pmatrix} e \\ f \end{pmatrix}$

We now have a method for solving all pairs of simultaneous equations. But the reader should beware of using such a heavy method when lighter manipulation will suffice. As an illustration of this, look back at Example 2 above. We could quite easily have added both equations to obtain

$$7y = 14$$
$$\therefore \quad y = 2$$

Substitute $y = 2$ in the first equation to obtain

$$x + 8 = 1$$
$$\therefore \quad x = -7$$

The result being $x = -7$, $y = 2$ as before.

There are occasions when the method will not work. Clearly when the inverse doesn't exist.

For example $2x + 3y = 18$ (i)
                        $4x + 6y = 36$ (ii)

In matrix form this is $\begin{pmatrix} 2 & 3 \\ 4 & 6 \end{pmatrix} \begin{pmatrix} x \\ y \end{pmatrix} = \begin{pmatrix} 18 \\ 36 \end{pmatrix}$

But, $\begin{pmatrix} 2 & 3 \\ 4 & 6 \end{pmatrix}$ has no inverse because $(2 \times 6) - (3 \times 4) = 0$.

There are usually one of two reasons for this:

1. The two equations are really one and the same equation. Notice here that (ii) is double (i), so that there are not two equations at all.

2. The equations may be inconsistent. An example of inconsistent equations are

$$x + y = 1 \qquad \text{(i)}$$
$$x + y = 0 \qquad \text{(ii)}$$

For case 1. we may offer infinitely many solutions, e.g.

x	0	9	1	2	3	etc.
y	6	0	$5\frac{1}{3}$	$4\frac{2}{3}$	4	

For case 2. there is no solution.

Exercise 11.6

Solve the pairs of simultaneous equations in Questions 1, 2, and 3 by two methods.

1. $4x + y = 9$ (i)
                        $-x + y = -1$ (ii)

2. $7x + 3y = 10$ (i)
                        $4x + 2y = 6$ (ii)

3. $13x + 7y = 3$ (i)
                        $9x + 4y = 2$ (ii)

4. Solve the simultaneous equation

$$3x + 4y = 8 \qquad \text{(i)}$$
$$27x + 36y = 50 \qquad \text{(ii)}$$

5. If T is the transformation given by

$$x_1 = 3x_0 + 4y_0$$
$$y_1 = 7x_0 + 10y_0$$

obtain the inverse transformation T^{-1} and calculate which point (x_0, y_0) gave rise to the point $x_1 = 1$, $y_1 = 2$. (If you are unable to do this have a look at the solution then try Question 6.)

6. If T is the transformation given by
$$x_1 = x_0 + y_0$$
$$y_1 = x_0 + 2y_0$$
obtain the inverse transformation T^{-1} and calculate the point (x_0, y_0) which gave rise to the point $x_1 = 8, y_1 = 9$.

In this chapter we have concentrated on 2×2 matrices but, in general, matrices are any shape and size. Some larger matrices can be found on page 188. Figs. 9.3 and 9.4 show 3×3 matrices and Fig. 9.7 shows a 6×6 matrix, but not all matrices are square.

The general matrix having r rows and c columns is referred to as an $r \times c$ matrix, e.g. a 2×3 matrix has 2 rows and 3 columns, but a 3×2 matrix has 3 rows and 2 columns. However, we are only interested in $1 \times 2, 2 \times 2$, and 2×1 matrices here.

If we look back to pages 246 and 247 we see that we gave a meaning to the multiplication of a 2×2 matrix by a 2×1 matrix in the expression

$$\begin{pmatrix} a & b \\ c & d \end{pmatrix}\begin{pmatrix} x_0 \\ y_0 \end{pmatrix} = \begin{pmatrix} ax_0 + by_0 \\ cx_0 + dy_0 \end{pmatrix} \tag{i}$$

We can illustrate this with a simple numerical example as follows:

$$\begin{pmatrix} 4 & 9 \\ 7 & 3 \end{pmatrix}\begin{pmatrix} 5 \\ 2 \end{pmatrix} = \begin{pmatrix} 4 \times 5 + 9 \times 2 \\ 7 \times 5 + 3 \times 2 \end{pmatrix} = \begin{pmatrix} 38 \\ 41 \end{pmatrix}$$

Again from (i) it is easy to see that $(a \ b)\begin{pmatrix} x_0 \\ y_0 \end{pmatrix} = (ax_0 + by_0)$ is the meaning to be given for the multiplication of a 1×2 matrix by a 2×1 matrix, the answer being a 1×1 matrix, e.g. $(8 \ 6)\begin{pmatrix} 4 \\ 10 \end{pmatrix} = (92)$, the single element being the result of the scalar product of a single row and a single column.

Now consider BA $= \begin{pmatrix} 4 \\ 10 \end{pmatrix}(8 \ 6)$. The first row of B consists of the single element 4 and the first column of A consists of the single element 8, and their scalar product is 32. It follows that the complete multiplication gives BA $= \begin{pmatrix} 32 & 24 \\ 80 & 60 \end{pmatrix}$. The reader might get the 'feel' of this last result by considering the product $\begin{pmatrix} 4 & 0 \\ 10 & 0 \end{pmatrix}\begin{pmatrix} 8 & 6 \\ 0 & 0 \end{pmatrix}$.

Not all matrix products are possible. For example in the following exercise DC and FE do not exist at all. In DC it is not possible to form the scalar product of the single element 6 and the two elements of column $\begin{pmatrix} 1 \\ 7 \end{pmatrix}$.

Exercise 11.7
Given that C $= \begin{pmatrix} 1 & 2 \\ 7 & 3 \end{pmatrix}$, D $= \begin{pmatrix} 6 \\ 8 \end{pmatrix}$, E $= (13 \ 10)$, F $= \begin{pmatrix} 2 & 6 \\ 4 & 7 \end{pmatrix}$ prove the following results:
1. EC $= (83 \ 56)$; EF $= (66 \ 148)$; ED $= (158)$.
2. CD $= \begin{pmatrix} 22 \\ 66 \end{pmatrix}$; FD $= \begin{pmatrix} 60 \\ 80 \end{pmatrix}$; DE $= \begin{pmatrix} 78 & 60 \\ 104 & 80 \end{pmatrix}$.

CHAPTER TWELVE

TRANSFORMATION GEOMETRY

In the previous chapter we discussed certain transformations of the plane as a way not only of introducing the matrix but also of giving it some particular meaning. Naturally enough, in typical mathematical fashion, we were able to abstract the matrix from its geometrical beginnings and carry out a short study of matrix algebra. Some of the later results produced were given a geometrical interpretation just to confirm one or two intuitive notions.

In this chapter we shall be concerned with pure geometry. Some people think geometry is only pure if there are no algebraic undertones; geometry, however, means different things to different people. The majority of us have our roots well and truly grounded in what is commonly called 'Euclidean geometry'. This is a logical development of the application of the congruence of plane figures. Its critics suggest that its almost total reliance on the congruence of triangles as the method of proof confines the subject in a manner which is not compatible with the outlook of modern mathematics. The implication is that Euclidean geometry is a static subject as opposed to a transformation geometry which, as the name suggests, is essentially a geometry of changing shape, size, and position wherein we study the relations between figures in the transformation. In other words, it is a dynamic subject in keeping with our modern outlook. It must be admitted, however, that in this book we cannot possibly give such an interesting topic more than a cursory glance.

REFLECTIONS

As we saw in Chapter Eleven, page 231, the reflection in a line maps a plane on to itself. (Mapping a plane means mapping all the points in the plane.) Furthermore, we have also seen that the map of a straight line after reflection is also a straight line—referred to as the *image line*. During the reflection each point on the mirror line remains exactly where it was. Any point or line which is untouched by a transformation is called an *invariant*. We also give this name to any property whatsoever which remains unaffected by the transformation.

Consider, therefore, reflection in the line 1. We shall use a standard notation of M_l to represent a reflection in the mirror line 1. In the following diagrams we shall want to show the effect of any transformation of the plane, and one of the simplest, yet most effective, is to show the transformation of the letter F. This has in fact become a standard demonstration figure.

As you know, we simply choose points such as A_0 or B_0 on the figure F and from them draw perpendiculars to 1, and extend an equal distance to the other side to arrive at the image point A_1 or B_1.

By studying Fig. 12.1 we see that in any reflection, distances remain unchanged, e.g. $\overline{A_0B_0} \cong \overline{A_1B_1}$ and also the measures of angles remain unchanged, e.g. the angle of 30° which has been shown in the figure. We therefore call distances and angles invariants for the operation of reflection (note that a transformation is an operation). This also means that a figure and its image are congruent. Looking at Fig. 12.1 we see this tells us that

$$m(\angle A_0OL) = m(\angle A_1OL)$$

where L is the intersection of the line A_0A_1 with 1.

Since distances are invariant, it should be clear that ratios of distances are invariant as well. For example, if B_0 is the midpoint of A_0C_0, then B_1 will be the midpoint of A_1C_1.

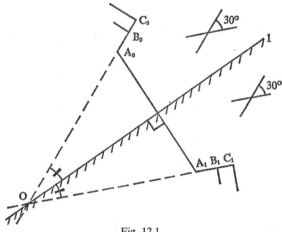

Fig. 12.1

Definition 12.1

Any transformation which leaves both shape and size unchanged is called an **isometry**.

There is an interesting type of problem we can solve by reflection, as follows:

Example 1

The line 1 in Fig. 12.2 represents a river. A man leaving point B must take water from the river and deliver it to point A. What is his shortest path?

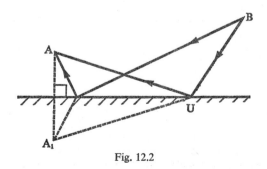

Fig. 12.2

In Fig. 12.2 we have made two suggestions, namely paths BEA and BUA. Actual measurements on the diagram will show that these paths are not equal in length and also that there is nothing about them which suggests that one of them is likely to be the shortest path. The reader is left to insert the point E.

A_1 is the reflection image in 1 of the point A, and this enables us to say that

$$\overline{EA} \cong \overline{EA_1} \quad \text{and} \quad \overline{UA} \cong \overline{UA_1}$$

Using these results we have

$$BEA = \text{length of } BEA_1$$
$$BUA = \text{length of } BUA_1$$

and both of these lengths to A_1 are obviously longer than the straight line segment BA_1, so this must be the length of the shortest path. As you can see in Fig. 12.3, we obtain length BOA as the required path.

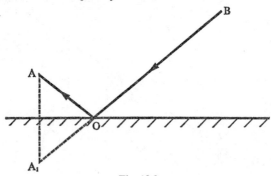

Fig. 12.3

How, then, would you construct the shortest path from A to B? This shows that if we are given one point and its image in the line we can then determine the image of any other point without the aid of drawing perpendiculars.

For example, the answer to the question just asked could be to repeat the construction of Fig. 12.3 and use the fact that the path is the same both ways. But we also see that continuation of AO until $\overline{OB} \cong \overline{OB_1}$ yields the image of B—an image constructed simply by knowing the position of A and A_1. It is natural to inquire next into the possibility of keeping track of two reflections.

In Fig. 12.4 we have two mirror lines l and k. We mean by two reflections that F will be reflected in k then the image obtained is to be reflected in l. The interesting point to emerge from this construction is that if we change the order of reflecting by taking l first followed by k second a different final result will be obtained.

This is symbolized by $M_l M_k \neq M_k M_l$ (a result which is not unexpected from our knowledge of matrices). Another way of stating this is to remark that the combination or product of two reflections is not commutative.

A second look at Fig. 12.4 shows that the acute angle α between the mirror lines is equal to $\theta_1 + \theta_2$ (the exterior angle of a triangle is equal to the sum of the two interior and opposite angles).

So that in two reflections the figure is turned through an angle equal to twice the acute angle between the mirror lines. We notice also that in one diagram $M_l M_k$ takes the figure anti-clockwise, while $M_k M_l$ takes the figure clockwise. It would be interesting to find out under what arrangement would $M_l M_k = M_k M_l$.

From the observation we have just made (about the amount of turning that the figure does) we can see that if we could make $M_l M_k$ give a half-turn (through 180°)

and M_kM_l also give a half-turn to the figure, then the final position would be the same in both cases. This is solved by placing the lines l and k at right-angles.

Since we are discussing the combination of two reflections, we should inquire into the effect of repeating the same reflection, i.e. M_kM_k or M_lM_l. We write these as M_k^2 or M_l^2 and see that their effect is to bring the figure back to the starting position. For example, in Fig. 12.3 M_l maps A on to A_1 and M_l maps A_1 on to A. Therefore M_l^2 maps A on to A.

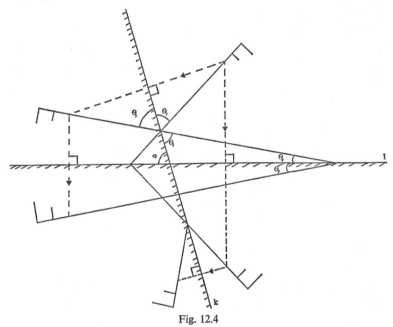

Fig. 12.4

Definition 12.2

Any transformation which maps each point of the plane on to itself is called the identity transformation and written I.

We have now that $M_lM_l = I$, and our knowledge of operations from Chapter Six, tells us that M_1 must be its own inverse, as expected. This will be written M_1^{-1} and referred to as the **inverse transformation.**

Definition 12.3

Any transformation which is its own inverse is called an **involution.**

As already remarked, a reflection does not change distances or shapes, so that the object figure is always congruent to its image, consequently it is sometimes possible to choose a mirror line such that a reflection maps the figure on to itself. Clearly there is something exceptional about this line, so we give it a special name in the following definition:

Definition 12.4

When a reflection maps a figure on to itself the mirror line is called an **axis of symmetry** for the figure. The simplest way of obtaining an axis of symmetry is to fold the figure along a line. If by folding we find that one half coincides exactly with the other half, then the line concerned is an axis of symmetry. (In all these folding situations we understand that the figure has been drawn on a piece of transparent paper, otherwise it is difficult to test for coincidence.)

Here are some examples:

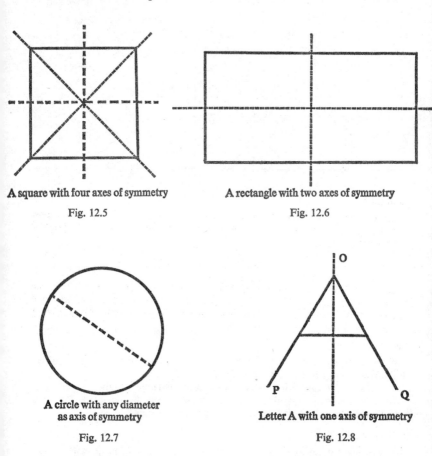

A square with four axes of symmetry

Fig. 12.5

A rectangle with two axes of symmetry

Fig. 12.6

A circle with any diameter as axis of symmetry

Fig. 12.7

Letter A with one axis of symmetry

Fig. 12.8

These simple results can often be put to good use. In finding the centre of a circle we merely have to fold the circle so that both halves coincide with one another, then repeat this with another fold. Where the two folds intersect one another will be the centre of the circle. Similarly for finding the centre of a square or rectangle. Fig. 12.8 is particularly interesting because it enables us to bisect an angle without the use of the usual instruments. All we have to do is fold the figure so that $\overline{OP}$ coincides with $\overline{OQ}$ and the fold will be the required bisector of the angle.

Example 2

Fig. 12.9 shows a circle, centre O, with an arc PQ of the circle such that $m(\angle POQ) = 60$. Show that any arc of the circle having the same measure as $\widehat{PQ}$ will yield an angle at the centre whose measure is 60 in the same manner as $m(\angle POQ)$.

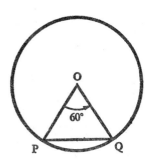

Fig. 12.9

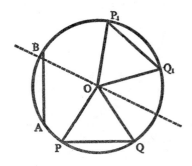

Fig. 12.10

Solution

All we have to do is fold the circle about any diameter (i.e. make the fold pass through O). Since size and shape are invariant, we obtain $\triangle POQ \cong \triangle P_1OQ_1$, and consequently

$$\widehat{P_1Q_1} \cong \widehat{PQ} \quad \text{and} \quad m(\angle P_1OQ_1) = 60.$$

If we were actually given that $\widehat{AB} \cong \widehat{PQ}$ already on the circle then we fold the figure so that P is mapped on to A and Q is mapped on to B the result will then follow as before. (This is a theorem that is usually stated as *equal arcs of a circle subtend equal angles at the centre*.)

Exercise 12.1

ABCDEFGHIJKLMNOPQRSTUVWXYZ

1. Using the printed alphabet above, state which letters have only:

    (i) one axis of symmetry;
    (ii) two axes of symmetry;
    (iii) three axes of symmetry;
    (iv) four axes of symmetry, or more.

2. How many axes of symmetry has an equilateral triangle?
3. Draw the axes of symmetry for Fig. 12.9. What does this axis of symmetry do to the chord $\overline{PQ}$?
4. Given any line segment drawn on a piece of tracing paper, by folding only, how would you:

    (i) find the middle point of the line?
    (ii) obtain a perpendicular to the line through a point in the line which is not the middle point?

5. Given a blank piece of paper, how would you obtain an angle whose measure was 45° by folding the paper only?

6. Sketch the reflections of the following figures in the given mirror lines.

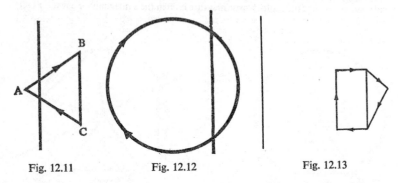

Fig. 12.11   Fig. 12.12   Fig. 12.13

7. In each of the figures for Question 6 the arrow indicated a clockwise direction. In what direction do the arrows point in the images?

8. Fig. 12.14 shows a circle, centre O, lying on a line through A. A tangent is drawn from A to touch the circle at T. By using a reflection show that the other tangent from A to the circle will be the same length as $\overline{AT}$ and equally inclined to $\overline{AO}$.

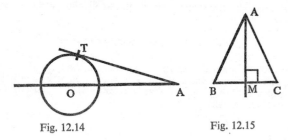

Fig. 12.14   Fig. 12.15

9. $\triangle ABC$ is an isosceles triangle with $\overline{AB} \cong \overline{AC}$. The line $\overline{AM}$ bisects the angle $\angle BAC$. By using a reflection prove that $m(\angle ABC) = m(\angle ACB)$. (This is almost the same as Question 8 without the circle.)

10. Using Fig. 12.3, if A and B are 2 km and 3 km respectively from the river and also 5 km apart, what is the length of the shortest path BOA? (*Hint:* Use Pythagoras's theorem twice.)

TRANSLATIONS

Definition 12.5

A **translation** is a one–one mapping in which the distance between each corresponding pair of points is not only the same but *also* in the same parallel direction. A train running along a pair of parallel straight lines is an example of a translation. In 5 seconds each point of the train will be moved the same distance in the same direction. In view of the definition, we only need to know about any one pair of corresponding points then we will know the complete translation.

Suppose in Fig. 12.16 we know that the F is to be translated such that A_0 and A_1 are a pair of corresponding points. Since a translation is isometric, we need only draw F at A_1 as indicated so that pairs like (B_0, B_1) are such that $\overline{B_0B_1} \cong \overline{A_0A_1}$ and $\overrightarrow{B_0B_1} \parallel \overrightarrow{A_0A_1}$.

Fig. 12.16

In Fig. 12.4 we discovered the fact that after successive reflection in two mirror lines the original figure is turned through an angle which is twice the angle between the two mirror lines.

A similar result emerges when we carry out the reflection in parallel mirror lines as shown in Fig. 12.17.

In the lower part of Fig. 12.17 we carry out the composite reflection M_lM_k (i.e. reflection in mirror line k is taken first).

From the suggested measurements on the diagram we see that the final image is $\daleth_2$, which is a distance $a + a + c - a + c - a = 2c$ away from $\daleth_0$. In the upper part of the same figure we carry out the composite reflection M_kM_l. That is, we reflect in the mirror line l first to produce an image $\sqsubset_3$

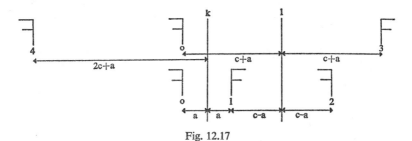

Fig. 12.17

and then reflect in mirror k line to produce the final image $\daleth_4$, which is a distance of $2c + a - a = 2c$ away from $\daleth_0$.

So we see that in each case the distance between the object and the final image is twice the distance between the mirror lines. Notice also that since the position $\daleth_4$ is different from position $\daleth_2$, we show once again that in general $M_lM_k \neq M_kM_l$ (page 262). But, another fact which emerges is that we could have mapped $\daleth_0$ on to $\daleth_2$ and $\daleth_0$ on to $\daleth_4$ by a straightforward translation in the first place. We have therefore the interesting result that:

The product of reflections in each of two parallel mirror lines, distance c apart, is equivalent to a translation of the same object through a distance $2c$.

However, this does raise the very important question of translation in which direction? Here we only mentioned distance? To identify the complete translation we must specify the direction concerned. Quantities which specify not only size or distance but also direction are called vectors, and it is to an examination of these that we now turn, because a knowledge of how to combine vectors will tell us all we need to know about combining translations.

VECTORS

There are several physical quantities which need both magnitude and direction to identify them completely. Displacement, velocity, acceleration, momentum, force (e.g. weight) are the most common. Displacement means distance and direction as in Fig. 12.17, i.e. we can speak of a distance of 2 metres without specifying in which direction, but we agree that when discussing a 'displacement' of 2 metres we must add a direction such as North East and so on. Just as we have a special name for quantities such as velocity, so we have a special name for quantities which are completely represented by a real number. Such quantities are called **scalars**. Distance, speed, mass, energy, time are five very common scalars. The reader must note that it is by definition that we agree to call distance and speed scalars, whereas displacement and velocity are called vectors. The comparison of speed and velocity is most easily brought about by considering motion in a circle. We can speak of travelling round a circle with a constant speed of 30 kmh^{-1} but not a constant, velocity of 30 kmh^{-1}. The reason is that because velocity is a vector, to say that it is constant means that not only is its magnitude constant, i.e. 30 kmh^{-1} but also its direction is constant, and this certainly cannot be the case if it travels in a circular path.

Definition 12.6

Any quantity which is completely represented by a real number is called a **scalar**.

Definition 12.7A

Any quantity which needs both magnitude and direction to represent it completely is called a **vector**.

We may clearly represent a vector by drawing a line segment and using an arrow-head to indicate the direction intended. If we insert the arrow-head at the end of the line segment we shall have drawn an arrow with the specified length and direction. From this it follows that we may completely identify a vector by an *arrow*. To distinguish vectors from scalars we use the line-segment notation.

For example, $\overline{PQ}$ means a vector whose magnitude is represented by the distance between P and Q and whose direction is in the order of writing, namely P to Q. The magnitude of the vector PQ is written $| \overline{PQ} |$, e.g. suppose $\overline{PQ}$ is a vector representing 30 m.p.h. in the North direction, then $| \overline{PQ} | = 30$.

We see here how we may offer a second definition for a vector.

Definition 12.7B

A vector is represented by a directed line segment. The notation of vectors is illustrated in Fig. 12.18.

Here we show two vectors $\overline{OA}$ and $\overline{OB}$ as adjacent sides of a parallelogram (a parallelogram has equal and parallel opposite sides). Now $\overline{OA}$ and $\overline{BC}$ are equivalent vectors, similarly $\overline{OB}$ and $\overline{AC}$ are equivalent vectors, but we could not say that $\overline{OA}$ and $\overline{CB}$ are equivalent vectors because their directions are opposite.

Sometimes a single letter is used for a vector and printed in bold type or with a wave underneath the letter. The latter notation has been used in Fig. 12.19.

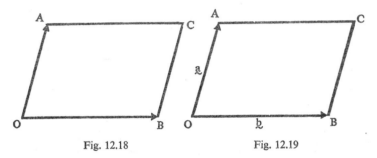

Fig. 12.18   Fig. 12.19

Clearly, from these diagrams there must be something special about the point C and its relation to the two vectors $\overline{OB}$ and $\overline{OA}$. We could arrive at C by obeying the instructions of the two vectors $\overline{OB}$ and $\overline{OA}$. What this means is if we add the results of the two translations represented by $\overline{OA}$ and $\overline{OB}$ we shall have carried out the equivalent translation $\overline{OC}$. Thus $\overline{OC}$ is said to be the **resultant** of $\overline{OA}$ and $\overline{OB}$. In other words, the parallelogram is used to define vector addition.

VECTOR ADDITION

We write the last statement as $\overline{OC} = \overline{OA} + \overline{OB}$ or $c = a + b$. Notice that we use $+$ for vector addition, but it is no longer the same as the $+$ for real numbers such as $3 + 4 = 7$. Whenever we see something like $a + b$ we know that in order to get the resultant we must construct a parallelogram with adjacent sides a and b, then construct the diagonal of the parallelogram through their point of intersection (in the case of Fig. 12.19 this means O). This diagonal represents the resultant vector in magnitude and direction. Of course, there is no need to draw the complete parallelogram each time. The $\triangle OBC$ would be adequate to obtain the resultant $\overline{OC}$ in Fig. 12.19.

COMMUTATIVITY

Using the diagram of Fig. 12.19 we see that $\overline{AC} = \text{b}$ and $\overline{BC} = \text{a}$. It follows therefore that it makes no difference to the final result of arriving at C whether we go via A first or via B first. This entitles us to write

$$\text{a} + \text{b} = \text{c} = \text{b} + \text{a}$$

which confirms that vectors obey the commutative law with respect to the operation of vector addition.

What meaning can we now give to $\text{a} + \text{a}$? Fig. 12.19 suggests that this is $\overline{OA} + \overline{BC}$. Our definition suggests that it could be $\overline{OA}$ plus any equivalent vector; in which case we can choose an $\text{a} = \overline{AE}$ and place it end on to $\overline{OA}$. The result we may clearly write as

$$\text{a} + \text{a} = 2\text{a}.$$

Similarly $n\text{a} + m\text{a} = (n + m)\text{a}$. Since 2, n, and m are scalars, we call this procedure the multiplication of a vector by a scalar. To summarize: $k\text{a}$ is a vector in the same direction as a but having k times its magnitude.

A particular value for k is zero. When k takes this value we call $k\text{a}$ the **null** vector and write it as O. There is no direction associated with the null vector. The null vector is clearly such that if a is any vector, then

$$O + \text{a} = O = \text{a} + O$$

and is therefore the identity element for vector addition.

Note: We have to define: (i) equivalent vectors (congruent directed line segments); (ii) vector addition (by the parallelogram law); (iii) product of a vector and a scalar, in order to calculate with vectors. Some authors regard (ii) as an essential part of the definition of a vector. We shall rest content with definition 12.7B, bearing in mind that for the purposes of calculation a vector is subject to (i), (ii), and (iii).

ADDITIVE INVERSE

If $\text{a} + \text{b} = O$, then b is said to be the inverse of a (see page 58). We agree to write the additive inverse of a as $-\text{a}$.

Thus,

$$\text{a} + (-\text{a}) = O$$

Once again refer to Fig. 12.19 and see that $\text{a} + (-\text{a})$ means a translation to A and back again to achieve the same result as the null vector. We therefore arrive at the following definition:

Definition 12.8

If b is any vector, then $-\text{b}$ is a vector having the same magnitude but the opposite direction.

For example,

$$\overline{OA} = -\overline{AO}$$
$$\overline{OB} = -\overline{BO}$$
$$-k\overline{OA} = k\overline{AO}$$
$$-k\overline{OB} = k\overline{BO}$$

SUBTRACTION

As we have done before, we define subtraction as the inverse operation to addition. This means that any vector subtraction such as

$$m\underset{\sim}{a} - n\underset{\sim}{b} \text{ is equal to } m\underset{\sim}{a} + (-n\underset{\sim}{b})$$

In other words, subtracting a vector is the same as adding its inverse. Diagrammatically we appeal to Fig. 12.20.

In this case $\overline{OD} = -\underset{\sim}{b}$, so that $\overline{OD} + \underset{\sim}{a} = \overline{OE}$. But from the diagram we can see that the figures OAED and OACB are congruent, and consequently $\overline{OE} = \overline{BA}$, so that in any parallelogram one diagonal gives a vector sum of its sides and the other diagonal gives a vector difference. Thus $\overline{OC} = \underset{\sim}{a} + \underset{\sim}{b}$, $\overline{AB} = \underset{\sim}{b} - \underset{\sim}{a}$.

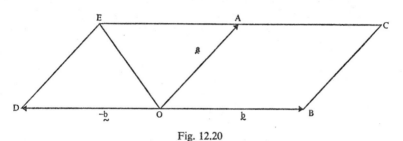

Fig. 12.20

ASSOCIATIVITY

What we wish to show here is that vector addition is associative.

$$(\underset{\sim}{a} + \underset{\sim}{b}) + \underset{\sim}{c} = \underset{\sim}{a} + (\underset{\sim}{b} + \underset{\sim}{c})$$

Similar inquiries have been carried out on page 15.
We demonstrate the truth of this law in Fig. 12.21.

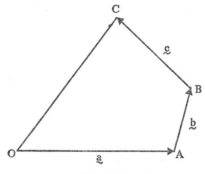

Fig. 12.21

272 New Mathematics Made Simple

As already remarked on page 269, there is no need to draw a complete parallelogram to obtain the resultant each time. Here we see at once that

$$a + b = \overline{OB} \quad \text{and} \quad b + c = \overline{AC}$$

Therefore,
$$(a + b) + c = \overline{OB} + c$$
$$= \overline{OC}$$

also,
$$a + (b + c) = a + \overline{AC}$$
$$= \overline{OC}$$

showing that we obtain the same result regardless of which pair of vectors are associated as a first step.

$$\therefore \quad (a + b) + c = a + (b + c)$$

which means that we may write $a + b + c$ without any possible ambiguity, i.e. everyone will obtain the same result of $\overline{OC}$ no matter how they start.

Summarizing the results so far, if a, b, and c are any elements in the set of vectors, then

(1) the set is closed under the operation of vector addition, i.e. $a + b$ is always another vector;

(2) the associative law holds for vector addition, i e.

$$(a + b) + c = a + (b + c);$$

(3) the null vector 0 is the identity element for vector addition and belongs to the set;

(4) each element a has an inverse $-a$ for vector addition.

Reference to page 123, Definition 6.4, confirms that the set of vectors forms a group for the operation of vector addition.

Furthermore, since

(5) $$a + b = b + a$$

the group is commutative.

Since any translation is completely represented by a vector, it follows that the set of all translations form a group under the operation $+$.

Example 1

Find the result of carrying out a translation of 2 km East followed by a translation 3 km South. Show how the letter F is translated.

Solution

Taking a starting-point, or origin O, let OA represent 2 km East and AB represent 3 km South.

The result of these two translations is to arrive at B as shown. We could just as well have gone from O to B direct, a fact which is summarized in the statement.

$$\overline{OB} = \overline{OA} + \overline{AB}$$

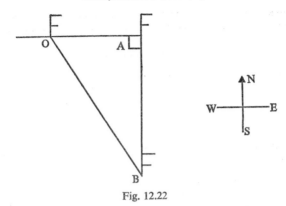

Fig. 12.22

But in order to specify $\overline{OB}$ we need to state its magnitude and direction. Measurement from a scale diagram shows that the magnitude of $\overline{OB}$ represents 3·6 km. The direction of $\overline{OB}$ will be given by stating $m(\angle AOB)$, which is 56 to the nearest degree (or 56° 19' using tangent tables). We have thus shown that the equivalent translation is 3·6 km East 56° 19' South.

Example 2

A boat set North at a speed of 10 kmh⁻¹ is swept off course by a 3-kmh⁻¹ NE tide. Draw a velocity diagram and give a true velocity for the boat.

Solution

Taking O as a starting-point, let $\overline{OB}$ represent the 10-kmh⁻¹ effort of the boat and $\overline{OA}$ the effort of the 3-kmh⁻¹ tide.

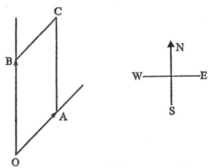

Fig. 12.23

Completing the parallelogram we find by measurement that the actual direction of travel for the boat is given by

$$|\overline{OC}| = 12·3$$
and $$m(\angle BOC) = 10 \text{ by measurement}$$

Our final conclusion is that the resultant velocity of the boat is 12·3 kmh⁻¹ N 10° E.

Example 3

A passenger in a train travelling at 60 kmh⁻¹ walks perpendicular to the train's motion by crossing a compartment at 4 kmh⁻¹ to shut a window. What is the passenger's resultant velocity?

Solution

Taking O as origin, let $\overline{OA}$ represent the train's velocity and $\overline{OB}$ the velocity of the passenger across the compartment.

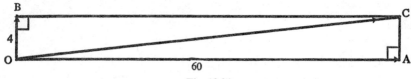

Fig. 12.24

Completing the parallelogram (here a rectangle), we see the resultant velocity of the passenger is given by $\overline{OC}$. By measurement $|\overline{OC}| = 60 \cdot 1$ kmh⁻¹ and $m(\angle COA) = 4$ approx. We describe our result as $60 \cdot 1$ kmh⁻¹ at $4°$ to the direction of the train's motion.

Example 4

In the accompanying diagram $\overline{OA} = a$, $\overline{OB} = b$, $\overline{OC} = 3a$, $\overline{OD} = 3b$. Determine $\overline{AB}$ and $\overline{CD}$. What do the results entitle you to conclude about $\overline{AB}$ and $\overline{CD}$?

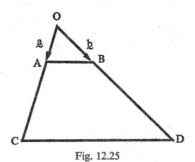

Fig. 12.25

Solution

From the diagram we see that
$$\overline{AB} = \overline{AO} + \overline{OB}$$
$$= -a + b$$
$$\overline{CD} = \overline{CO} + \overline{OD}$$
$$= -3a + 3b$$
$$\therefore \quad \overline{CD} = 3\overline{AB}$$

The line segments $\overline{CD}$ and $\overline{AB}$ are parallel and also

$$\frac{|\overline{CD}|}{|\overline{AB}|} = \frac{|\overline{CO}|}{|\overline{AO}|} = \frac{|\overline{DO}|}{|\overline{BO}|} = \frac{3}{1}$$

Example 5

Find the resultant of $\overline{AB} + \overline{BC} + \overline{CD} + \overline{DE}$

Solution

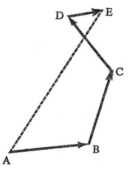

Fig. 12.26

From the diagram $\overline{AB} + \overline{BC} + \overline{CD} + \overline{DE}$

$$= (\overline{AB} + \overline{BC}) + \overline{CD} + \overline{DE}$$
$$= \overline{AC} + \overline{CD} + \overline{DE}$$
$$= (\overline{AC} + \overline{CD}) + \overline{DE}$$
$$= \overline{AD} + \overline{DE}$$
$$= \overline{AE}$$

In fact, there is no need to draw a diagram in cases like this. The solution is obvious after some inspection. For example:

(i) $\overline{XY} + \overline{YZ} + \overline{ZT} + \overline{TV} = \overline{XV}$
(ii) $\overline{LN} + \overline{NO} + \overline{OS} = \overline{LS}$
(iii) $\overline{AB} + \overline{BC} + \overline{CD} + \overline{DE} + \overline{EF} + \overline{FG} + \ldots + \overline{YZ} = \overline{AZ}$

Exercise 12.2

1. Fig. 12.27 is a rhombus (opposite sides parallel and all equal) with $\overline{OA} = \underset{\sim}{a}$ and $\overline{OB} = \underset{\sim}{b}$. Express the following in terms of $\underset{\sim}{a}$ and $\underset{\sim}{b}$:

(i) $\overline{BC}$ (ii) $\overline{AC}$ (iii) $\overline{OC}$ (iv) $\overline{AB}$ (v) $\overline{BA}$ (vi) $\overline{CA}$

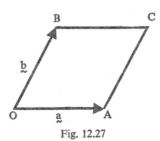

Fig. 12.27

2. The sides of the Fig. 12.27 are all of equal length. Why isn't $\underset{\sim}{a} = \underset{\sim}{b}$?

3. In Fig. 12.28 X and Y are the midpoints of the sides OA and OB. By putting $\overline{OX} = a$ and $\overline{OY} = b$ and finding $\overline{AB}$ and $\overline{XY}$, prove that $\overline{XY}$ is parallel to $\overline{AB}$ and $\overline{XY} = \frac{1}{2}\overline{AB}$.

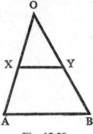

Fig. 12.28

4. Write down the resultant of the following vector sums:

(i) $\overline{AP} + \overline{PQ} + \overline{QR}$

(ii) $\overline{AS} + \overline{ST} + \overline{TN} + \overline{NA}$

(iii) $\overline{LM} + \overline{MN} + \overline{NO} + \overline{OT}$

5. Fig. 12.29 is a set of four congruent rectangles having $\overline{OB} = b$ and $\overline{OA} = a$. Express the following in terms of a and b:

(i) $\overline{OT}$ (ii) $\overline{OU}$ (iii) $\overline{OY}$ (iv) $\overline{OW}$ (v) $\overline{OX}$ (vi) $\overline{OV}$

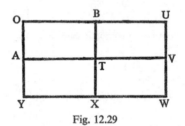

Fig. 12.29

6. Using the diagram of Question 5. Determine the following resultants in terms of a and b:

(i) $\overline{OY} + \overline{YX}$ (ii) $\overline{OY} + \overline{YW}$ (iii) $\overline{OY} + \overline{BU}$

(iv) $\overline{OT} + \overline{TV} + \overline{VU}$ (v) $\overline{BV} + \overline{XA}$

7. If the letter F starts at the origin O in Fig. 12.29 and is given the following translations, state the point of the diagram at which it finally arrives.

(i) a (ii) $a + 2b$ (iii) $b + a$ (iv) $a + b$ (v) $2a + 2b$

COMPONENTS

Having discussed the composition of vectors, we shall now discuss their decomposition. That is, instead of single vectors in many directions we shall decompose them into many vectors in one or two directions. For a measure of the total result we need a **unit vector**.

Definition 12.9

A **unit vector** is a vector with any direction, but whose magnitude is 1, i.e. $\underline{i}$ is a unit vector if, and only if, $|\underline{i}| = 1$.

Taking the cartesian plane of Fig. 12.30, we have $\underline{i}$ as the unit vector along the *x*-axis and $\underline{j}$ as the unit vector along the *y*-axis. The reader should mark in the origin O and the point A (4, 0).

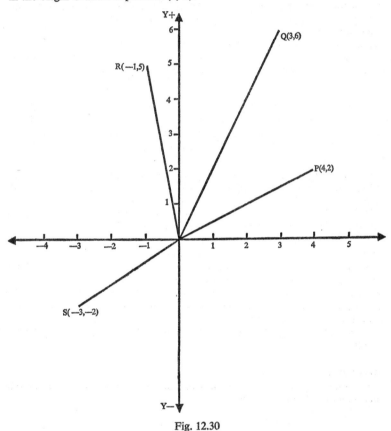

Fig. 12.30

Let P be the point (4, 2), then OP is a vector which may be written in the form

$$\overline{OP} = 4\underline{i} + 2\underline{j}$$

$\overline{OA} = 4\underline{i}$, is called the component of $\overline{OP}$ in the direction OX⁺ and $\overline{AP} = 2\underline{j}$ is called the component of $\overline{OP}$ in the direction OY⁺. The whole process is called 'expressing the vector $\overline{OP}$ as a sum of its components in the direction OX⁺ and OY⁺'. If we wish to reverse the process, then $\overline{OP} = 4\underline{i} + 2\underline{j}$ enables us to draw the diagram as for Fig. 12.30 and so obtain the single vector $\overline{OP}$.

Alternatively, since $\triangle OAP$ is a right-angled triangle, we may apply Pythagoras' theorem to obtain

$$| \overline{OP} |^2 = 4^2 + 2^2$$
$$= 20$$
$$\therefore \quad | \overline{OP} | = \sqrt{20} = 4.472$$

and for those who have trigonometry at their fingertips we obtain the direction of $\overline{OP}$ by using

$$m \, (\angle POA) = \tan^{-1} \tfrac{2}{4}$$
$$= 26° \, 34'$$

But as the reader will have already noticed, we intend to continue working from scale diagrams.

It will be appreciated that we are now able to express a vector as an ordered pair. The vector OP is fully represented by stating the ordered pair (4, 2), meaning a vector whose x-component is 4 and whose y-component is 2. The method is very powerful in finding the resultant of several vectors.

Example 1

Find the sum of the vectors $\overline{OP}, \overline{OQ}, \overline{OR}, \overline{OS}$ which have been given in Fig. 12.30.

Solution

We express each vector as a sum of its separate components.
This process gives us the following results:

$$\overline{OP} = 4i + 2j$$
$$\overline{OQ} = 3i + 6j$$
$$\overline{OR} = -i + 5j$$
$$\overline{OS} = -3i - 2j.$$

Let the resultant be $\overline{OK}$. Since all the vectors pass through O, then the resultant certainly will.

Therefore $\quad \overline{OP} + \overline{OQ} + \overline{OR} + \overline{OS} = (4 + 3 - 1 - 3)i + (2 + 6 + 5 - 2)j$

$$\overline{OK} = 3i + 11j.$$

This tells us that K is the point (3, 11). By measurement it will be found that $|\overline{OK}| = 11.4$, and the direction is 75° to the x-axis.

Example 2

Find the vector sum of

$$(2, -3); (6, -7); (10, -2); (-13, 9); (-6, 8).$$

Solution

The equivalent sum is

$$2i - 3j + 6i - 7j + 10i - 2j - 13i + 9j - 6i + 8j = -i + 5j$$

This is the same vector as $\overline{OR}$ in Fig. 12.30.

In the next example we show how to construct the resultant. But first of all we note that if we are given the components of a vector we are able to draw the vector anywhere in the plane. In Fig. 12.30 any line parallel to and having the same length as $\overline{OP}$ will have components of 4i and 2j.

Example 3

Construct the sum of vectors (1, 2), (4, −3), (−2, 4) using an 'end-on' diagram such as Fig. 12.26.

Solution

Insert the origin O for Fig. 12.31.

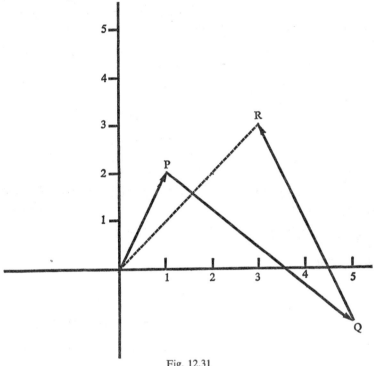

Fig. 12.31

Starting at O, we draw $\overline{OP} = 1\mathbf{i} + 2\mathbf{j}$.

Starting at P, we draw $\overline{PQ} = 4\mathbf{i} - 3\mathbf{j}$.

Starting at Q, we draw $\overline{QR} = -2\mathbf{i} + 4\mathbf{j}$.

We have so far seen that $\overline{OP} + \overline{PQ} + \overline{QR} = \overline{OR}$, which is the resultant or equivalent vector. From the diagram we see immediately the direction of the equivalent vector. Check that as before $\overline{OR} = 3\mathbf{i} + 3\mathbf{j}$ and is inclined at 45° to the *x*-axis.

Exercise 12.3

1. Given that $\overline{OP} = (2, 4)$, $\overline{OR} = (6, 7)$, $\overline{OS} = (-2, -3)$, $\overline{OT} = (-5, 6)$, obtain the resultant ordered pair of:

   (i) $\overline{OP} + \overline{OR} + \overline{OS} + \overline{OT}$

   (ii) $2\overline{OP} - \overline{OR}$

   (iii) $3\overline{OP} + \overline{OS} + \overline{OT}$

   (iv) $\tfrac{1}{2}\overline{OP} + \overline{OR} + \overline{OS} + \overline{OT}$

2. Construct the sum by using an 'end-on' diagram for the vectors.

$$(1, 3), (2, -4), (-3, 5)$$

3. A letter F stands at the origin of co-ordinates O and is then given two translations whose vectors are given by

$$a = 2i + 3j \quad \text{and} \quad b = 6i + 7j$$

What is the single equivalent translation and where in the plane does the letter F finally arrive?

4. If in Question 3 the letter F had started at the point (4, 3) instead of the origin, where would its final position be after receiving the same translation a and b in succession?

ROTATIONS

To perform a rotation in the plane we must have: (i) a point about which to perform the rotation called the **centre of rotation**, and (ii) the direction of the rotation either clockwise or anti-clockwise. Having selected a centre of rotation, the rotation of a figure is carried out by keeping each point of the figure a constant distance from the centre. In Fig. 12.32 F_0 is rotated 90° clockwise about 0 into the position F_1. Likewise the odd-shaped figure has been rotated through 90° about 0. Notice that all we need is a point and a radial line in the body. The spine of F and its base was all we needed to fix its final position.

Fig. 12.32

If we look back at Fig. 12.4 we remember that the effect of the two successive reflections in the mirror lines was to turn the figure F through twice the angle between the mirror lines. But we now pay extra attention to the fact that the orientation of the figure was unchanged, i.e. F was still the right way round in the sense that an arrow made out of its spine and top $\ulcorner\!\!\rightarrow$ would remain pointing in a clockwise position after the two reflections, and also that the base point of the letter F is always the same distance from the point of intersection of the mirrors at 0. It follows therefore that we could have obtained the final position in each case by one simple rotation of the figure about 0 as the centre of rotation and once again a rotation through an angle equal to twice the angle between the mirror lines.

We could, of course, have proved the example of Fig. 12.10 by rotating the circle about the centre 0. Notice once again the image and object figures are congruent, that is, size and shape are invariant, while 0 is the only fixed point in the transformation of rotation.

ROTATION OF A LINE

To obtain the image of a straight line when the transformation is a rotation is not as straightforward as often suggested. To obtain the image of line 1_0 in Fig. 12.33 with the operation $R(0, -\theta°)$, select any two points A_0 and B_0 on 1_0. We then rotate the triangle OA_0B_0 through $\theta°$ about 0 to image points A_1 and B_1. The join of A_1B_1 gives the image of 1_0 in the transformation $R(0, -\theta°)$.

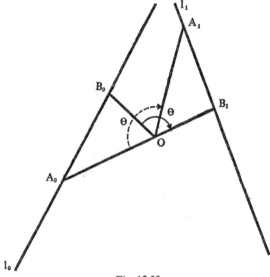

Fig. 12.33

Looking back on the examples given so far, the reader will notice that nearly all of them concern themselves with ideas and results from Euclidean geometry. It would be a pity if the reader were to get the idea that transformation geometry was just another way of proving the already familiar results of Euclid. On the other hand, by discussing the possibly more familiar results, we do manage to compare the style of the two geometries, and some practice in using the basic ideas is always time well spent.

For problems more closely related to transformation geometry consider the following. Fig. 12.34 shows a square KLMN with one vertex L lying on the straight line *l*.

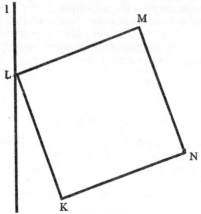

Fig. 12.34

If *l* and K are kept fixed and L moves along *l*, on what path does N move? The size of the square changes with the position of L as shown in Fig. 12.35.

Solution

One way of attempting to solve the problem is to try a succession of new positions hoping that a pattern of results will emerge which give us a clue to the expected path of N. We have made three attempts in Fig. 12.35, and an examination of the results shows that N, N_1, N_2 all lie on a straight line. This is sufficient to give us an idea, but not really to prove that the pattern of N is a straight line.

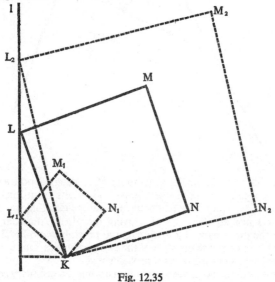

Fig. 12.35

Consider now a transformation R(K, −90°) of *l* onto *l₁* and let the image be given in Fig. 12.36. Under this transformation KN is the image of KL, the angle θ between LK and *l* remains invariant so that *l₁* is the image of *l* and, furthermore, *l₁* is perpendicular to *l* (because we rotated the figure through 90°).

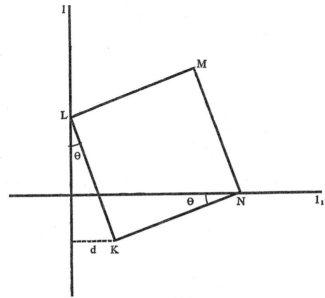

Fig. 12.36

Now since N is the image of L, it follows that as L moves along *l*, then N moves along *l₁*. Our constructions of Fig. 12.35 confirm what we have proved in Fig. 12.36, namely that the path of N is a straight line perpendicular to *l* the same distance from K as *l*. We can put this result to very good use in the following typical problem of transformation geometry:

Example 2

l and *n* are two fixed non-perpendicular straight lines and K is a fixed point. Construct a square, KLMN, such that L lies on *l* and N lies on *n*, using the fixed point K as one of the vertices (as suggested by the lettering).

Solution

Fig. 12.37 shows the arrangement we require, but the trouble is that we do not know how to construct the diagram yet!

Working on the basis of our experience with Example 1, we shall carry out the transformation R(K, −90°). Since the rotation is through −90°, we know that N, being the image of L, will move on the image of the line *l*, namely *l₁*.

We obtain the image line *l₁* as in Example 1. and note that it cuts *n* at N.

Since N is the image of L, then $\overline{KL} \cong \overline{KN}$ and $m(\angle LKN) = 90$, and we can complete the square. (*Note:* The intersection $l_1 \cap n$ gives N first. We do not know where L is until we have obtained N.)

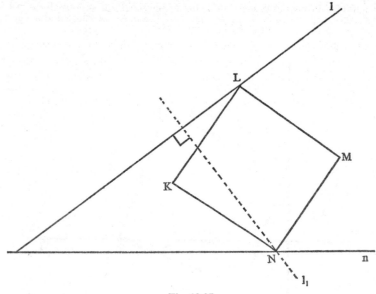

Fig. 12.37

Exercise 12.4

1. Show that if l and n in Fig. 12.37 had been two perpendicular lines, the construction of the square was not possible unless K was the same distance from both lines in which case an infinite number of squares KLMN are possible.

2. With K, l, and n, as for Fig. 12.37, construct an equilateral triangle KLN such that $L \in l$ and $N \in n$. (*Hint:* Carry out the transformation R(K, $-60°$).)

3. Under what conditions would Question 2 be impossible?

4. Following the method of Question 2, construct an isosceles triangle having $\overline{KL} \cong \overline{KN}$ and $m(\angle LKN) = 40°$ with $L \in l$ and $N \in n$ as before. (*Hint:* Try R(K, $-40°$).)

5. R(0, $\theta°$). R(0, $\alpha°$) means the application of two rotations in succession, R(0, $\alpha°$) first then R(0, $\theta°$) second. What is the equivalent rotation?

6. What does the result of Question 5 tell us about the set of rotations about the single point 0?

7. What is the inverse transformation for R(0, $\theta°$)?

8. What is the identity rotation?

9. Is the set of rotations about the point 0 as centre of rotation a group for the operation of multiplication or successive application?

10. If your answer to Question 9 is 'yes', then is the group a commutative one?

RADIAL EXPANSION OR ENLARGEMENT

The matrix for this type of transformation has already been discussed (page 236). We state the definition once again.

Definition 12.10

In any radial expansion or enlargement, a point P and its image P_1 lie in a straight line through a fixed point O and on the same side of O such that

$$\overline{OP_1} = k\overline{OP}, \text{ where } k \text{ is constant.}$$

This transformation is often referred to as a **dilatation** and k the **constant of enlargement**. We represent the enlargement by E(0, k). We see from $\overline{OP_1} = k\overline{OP}$ that if $k = 1$ we get the identity transformation in which everything remains unchanged.

Unlike the other transformations we have discussed, the object is no longer congruent to its image. Lengths are changed, but angles and ratios of lengths remain the same, hence parallel lines remain parallel, thus retaining similarity of the object and image figures. This does not seem to be much, but in fact we obtain some very interesting constructions from this particular transformation. The diagram of Fig. 12.38 shows the results of E(0, 3).

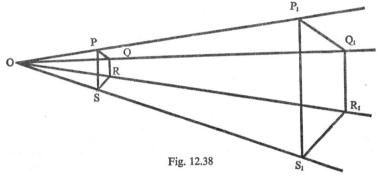

Fig. 12.38

We may, of course, choose $0 < k < 1$, e.g. $k = \frac{1}{2}$, in which case we obtain a smaller figure $P_2Q_2R_2S_2$. All three figures are similar. (The reader is left to insert the third figure.)

Example 1

Construct a circle to touch two given fixed lines *l* and *n* and pass through a fixed point K.

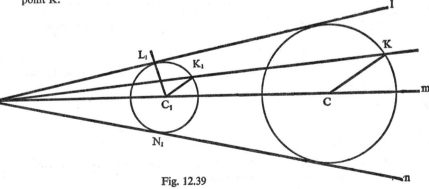

Fig. 12.39

Solution

(*a*) Let the line *m* be the bisector of the angle between *l* and *n* which meet at O. Draw any circle with centre C_1 on *m*, which touches *l* and *n* as shown.

(*b*) Join OK and let this line cut the circle at K_1 (see diagram).

(*c*) Join K_1C_1.

(*d*) Draw a line through K parallel to K_1C_1 to cut *m* at C.

C is the centre of the required circle. All we have done here is to carry out a radial enlargement of the first random circle which was drawn. If the first circle touched *l* at L_1 and *n* at N_1, then the second circle will touch *l* at L and *n* at N, where L and N are the images of L_1 and N_1 in the transformation.

Example 2

Given a segment of a circle, construct a square standing on the chord of the segment and having two vertices on the arc.

Solution

(*a*) Find the centre of the chord at O.

(*b*) Construct any small square as shown with the perpendicular *l* as axis of symmetry. (*l* passes through O the mid point of the chord.)

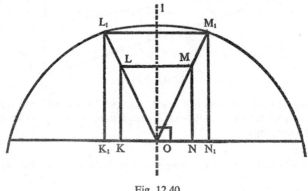

Fig. 12.40

(*c*) Use O as the fixed point in a radial enlargement.

(*d*) $K_1L_1M_1N_1$ is the required figure.

We could have drawn KLMN much larger and then carried out a radial shrinkage to get $K_1L_1M_1N_1$.

Example 3

Given any fixed $\triangle ABC$, construct a $\triangle LMN$ such that L lies on $\overline{BC}$, M lies on $\overline{AC}$, and N lies on $\overline{AB}$ with

$$m(\angle LMN) = 40$$
$$m(\angle MLN) = 60$$

Solution
Illustrated in Fig. 12.41.

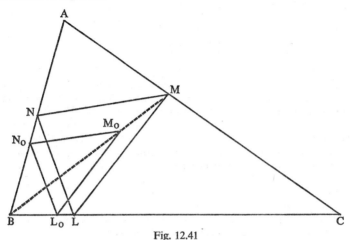

Fig. 12.41

(*a*) We start by choosing N_0 and L_0 anywhere on AB and BC respectively.
(*b*) Construct $\triangle L_0 M_0 N_0$ such that $m(\angle L_0 N_0 M_0) = 80$ and $m(\angle N_0 L_0 M_0) = 60$.
(*c*) Join BM_0 and produce to cut AC at M.
(*d*) Draw $\overline{MN} \| \overline{M_0 N_0}$ and $\overline{ML} \| \overline{M_0 L_0}$.

$\triangle LMN$ is the enlargement of $\triangle L_0 M_0 N_0$ with B as fixed point, i.e. N is the image of N_0 etc.

Exercise 12.5
1. In Example 1 above construct a second circle through K to touch *l* and *n*. (*Hint:* OK cuts the first circle in two points.)
2. Following Example 2, construct a rectangle whose length is twice its width and has two vertices on the chord and two vertices on the arc of a semicircle.
3. In the manner of Example 3 above make $\triangle NML$ an equilateral triangle, but with one side parallel to a side of $\triangle ABC$.
4. In the manner of Example 3 above, construct a square having two vertices on BC and one each on $\overline{AB}$ and $\overline{AC}$.

CHAPTER THIRTEEN

ELEMENTARY TOPOLOGY

Following the work of Chapter Twelve we now come to the most intriguing transformation of them all—the topological transformation. Once again we study those properties of figures which remain invariant. We shall no longer be concerned with any measurements of line segments or angles; congruence or similarity will not be considered, neither will notions of parallel lines. The removal of these constraints provides us with an extremely flexible type of transformation. So flexible in fact that some people refer to the subject as

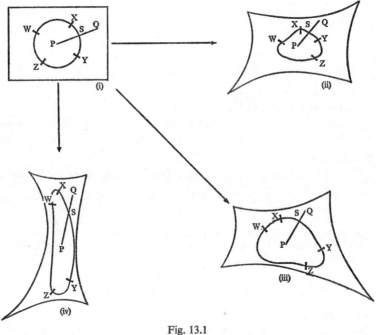

Fig. 13.1

'rubber-sheet geometry', and we feel that the elementary aspects of topology are best introduced in the manner of a rubber-sheet geometry.

Fig. 13.1 gives a typical example of what we call a topological transformation of a circle drawn on a very elastic rubber sheet.

The reader might be tempted to think that by continuing to stretch any curve we shall eventually open up a few gaps or holes. We consider a curve to be just like a piece of elastic; as long as we do not stretch the elastic beyond its breaking point it will remain in one piece without any gaps.

A curve without any gaps is called a **continuous curve**, hence any topological transformation must preserve **continuity**. Or to put it another way, anything which is connected must not be disconnected. (This is for curves in a surface such as the surface of a sphere or the top of a table or the sheet of paper on which these words are printed.)

A curve like the circle is called a **simple closed curve**, because when it is drawn on a plane or a sphere it separates the surface into two regions which, for want of better definition at this stage, we may call inner and outer regions.

The point about this last remark is that if we consider the equator of a spherical earth as our curve, then each of the two regions of the sphere has equal claims to being called the inner region.

Returning to Fig. 13.1 we observe that in the transformations:

(*a*) continuity is preserved;
(*b*) inner regions remain inner regions (the point P confirms this);
(*c*) outer regions remain outer regions (the point Q confirms this);
(*d*) intersection points are preserved (the point S confirms this);
(*e*) the order of points on a curve is preserved (W, X, Y, and Z in that order).

Notice also that S remains between X and Y as well as P and Q. All these are **invariants**, and a transformation which has these invariant properties is called a **topological transformation**. If we break continuity (disconnect a curve), we create a new figure, i.e. a new topological figure. Likewise if we were to make a hole in the surface, or indeed fill one in, we would be creating a new figure. Also if we were to fold the figure over we would change the order of the points on a curve. We therefore come to the following definition:

Definition 13.1

A topological transformation is any transformation of a figure which maintains continuity and is made without the use of cutting, folding, or perforating. In fact, cutting is allowed, providing the surface is stuck together again in the same position as before. We shall not deal with any transformations which involve cutting.

Definition 13.2

Figures which can be derived from one another by a topological transformation are said to be **topologically equivalent.**

Definition 13.3

Any figure which is cut, folded, or perforated forms a new topological figure.

Bearing in mind that we consider our diagrams to be drawn on very elastic sheets of rubber, let us examine Fig. 13.2.

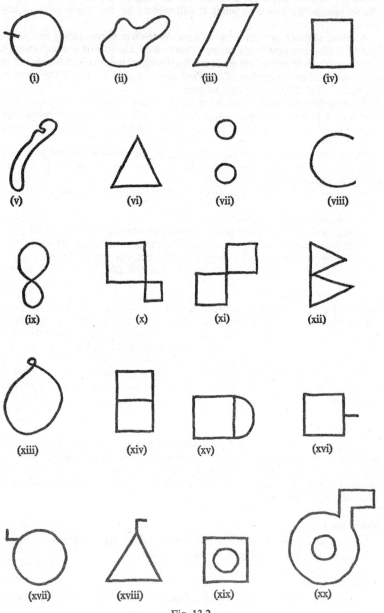

Fig. 13.2

Before reading on, try to divide the twenty figures into sets of topologically equivalent figures. Your results should be:

A = {(ii) (iii) (iv) (v) (vi)} Each figure is a simple closed curve.
B = {(ix) (x) (xi) (xii) (xiii)} Each figure could be stretched on to the other. The curves are not simple closed curves because they separate the plane into three regions.
C = {(xiv) (xv)}
D = {(xvi) (xvii) (xviii)}
E = {(xix) (xx)}

The figure (vii) is a different topological figure from all the others: for example, we would have to cut the curves of A in half and then join the relevant loose ends to obtain (vii), and since cutting is not allowed in topological transformations, then clearly (vii) ∉ A. Furthermore, no matter how we stretch the members of B, what is connected cannot be disconnected so that (vii) ∉ B.

The figure (viii) is obtained from (ii) by a cut, and so cannot be topologically equivalent. It is worth noting that (xiv) and (ix) are not topologically equivalent. The curve of (ix) cuts itself at one point of intersection only, whereas the curve in (xiv) intersects itself over the whole of the 'central' line.

Any true sketch of a figure is usually a topologically equivalent figure. A true map of an area taken by an aerial photograph might be like that given in Fig. 13.3.

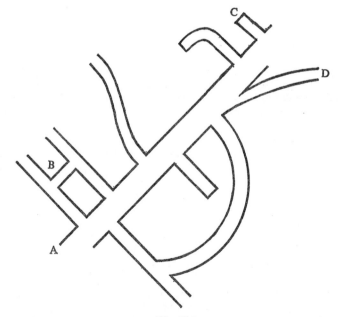

Fig. 13.3

But if you were at A giving directions to someone who wished to get to B, C, or D you would probably say no more than

> For B—First left then first right
> C—Fifth left
> D—Fourth right

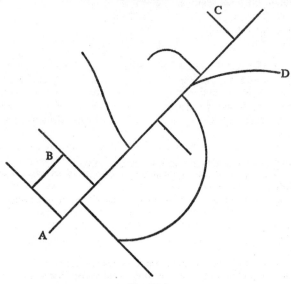

Fig. 13.4

And if you were feeling particularly helpful on that day you might draw a sketch like Fig. 13.4. The angles between the lines would be irrelevant, and so would the distances. What is important is the *order* of the turnings or, more precisely, the certainty that Figs. 13.4 and 13.3 are topologically equivalent.

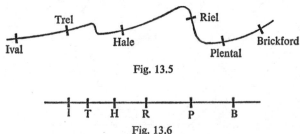

Fig. 13.5

Fig. 13.6

A railway map is another good example of topological equivalence. The true journey is probably like Fig. 13.5, but a descriptive sketch which gives the important information regarding the order of the stations would probably be like Fig. 13.6. The wiring diagram for a car is another illustration of a topological equivalence.

Exercise 13.1

1. Fit the following figures into the sets obtained from Fig. 13.2:

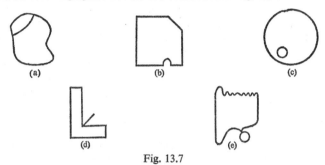

Fig. 13.7

2. Which of the figures in Question 1 are simple closed curves?

3. Using the alphabet printed on page 265, divide the capital letters into sets of topologically equivalent figures.

4. Using the figures of Fig. 13.2, how could we make (viii) topologically equivalent to:

    1. (iv)?
    2. (xvii)?
    3. (xiii)?
    4. How could (xi) be made topologically equivalent to (vii)?
    5. Are any of these transformations topological transformations?

5. A simple closed curve separates a plane into two regions. The Fig. 13.8 is a simple closed curve. The point A is clearly inside the curve. A test for this is to draw a straight line through A and count the number of intersections of the line with the curve.

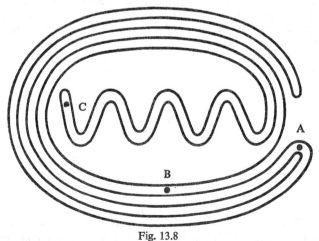

Fig. 13.8

The number of intersections being one, what does this tell you about the position of A. Is it inside the curve or outside?

Suppose the number of intersections had been two? What would this have meant? Determine whether B and C are inside or outside the curve.

NETWORKS

Any system of arcs which connect a set of points on a surface is to be called a network. Fig. 13.9 illustrates three networks and the descriptive details follow below:

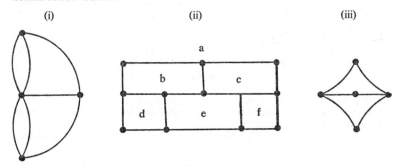

Fig. 13.9

The points of the networks indicated by ● will be called **vertices** (another name is **node**). Any join of two vertices will be called an **arc**. (Even if the join is a straight line, we shall call it an arc because they are topologically equivalent anyway.) A node always lies at the end of an arc. Insert the missing node on the central diagram of Fig. 13.9.

A vertex is called odd or even according to whether an odd or even number of arcs are joined to the vertex.

Examining the diagrams of Fig. 13.9 we tabulate the network information as shown.

Network	V Vertices	R Regions	A Arcs	V + R − A	Odd V	Even V	Unicursal
(i)	4	5	7	2	4	0	
(ii)	12	6	16	2	8	4	
(iii)	5	3	6	2	2	3	

The fifth column $V + R - A = 2$ is Euler's network formula and is one of the properties of networks which remain invariant in any topological transformation. Obviously it will be a little more convincing if we examine the diagrams of Fig. 13.10 to see if it works for more networks.

UNICURSAL NETWORKS

When drawing these networks for the first time one always tries eventually to see if the complete network could be drawn without taking the pencil off the paper and without travelling along any arc more than once. Having managed this in some cases, the next trick to try is to see whether one can start and finish at the same vertex. After this has been done we would probably look for some clue in the network which would save the tedium of trial-and-error attempts. All the information that a network can offer is entered in the table above; it follows, therefore, that the clue will be there somewhere. The reader will be left to discover the rule for himself in Exercise 13.2. In the meantime here is a definition of what we are looking for in a network:

Definition 13.4

A network is said to be **traversable** if the complete network can be drawn by travelling without any disconnection along each arc once and once only. If we can arrive back at a starting-point we say that the network is **unicursal.**

Exercise 13.2

1. Examine the networks of Fig. 13.10 and enter the network information in the table on page 294.

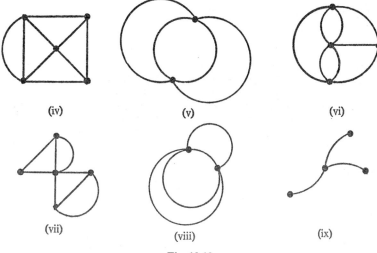

(iv)   (v)   (vi)

(vii)   (viii)   (ix)

Fig. 13.10

2. After trial and error state whether the networks are unicursal or not.
3. You should notice that (i) is not traversable but (vi) is traversable.
4. What addition could be made to (vii) to convert it into a unicursal network?
5. Which of the above figures are topologically equivalent?
6. Explain why a network with two vertices could not have one, and only one, odd vertex?
7. Explain why a network with three vertices could not have one, and only one, odd vertex.

Questions 6 and 7 lead up to Euler's result that in any network there is either an even number of odd vertices or there are none at all. Another of Euler's results is that a curve is only traversable if there are either two odd vertices or none. You can check this out in the above table. He also showed that only in the case of a unicursal network with no odd vertices is it possible to complete the tracing of the network by returning to the starting-point. In the above figures only (v) and (viii) can be completed in this manner. Try it.

NETWORKS FOR MAPS

Any map divides its space into regions and similar to our interest in networks, we ask ourselves if it is possible to traverse the map by passing through the boundaries of the regions once, and once only.

Some years ago a great deal of interest was generated in the problem illustrated by (ii) Fig. 13.9, and a fairly large cash prize was offered for the solution to the following problem:

Example 1

The diagram shows the plan of a house which has 16 sections of wall each containing either one window or one door. Taking a long enough ball of string, is it possible to pass the string through every window and door once, and once only?

Typical attempts are given below where *x* marks the missed opening.

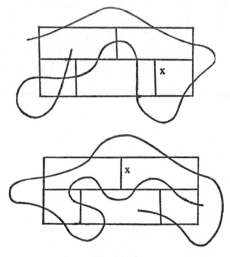

Fig. 13.11

The reader should make one or two attempts at solving the problem before reading any further. A moment's thought about the problem will reveal that all we are trying to do is traverse a network of connexions between the six regions of the figure 13.9 which have been labelled *a*, *b*, *c*, *d*, *e*, *f*. Notice that *a* is the complete outside region.

We accordingly set about drawing this network.

Solution

Starting with *a* this is connected by two arcs to *b*, *c*, *d*, *f* and one arc to *e*.

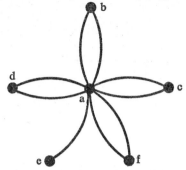

Consider *e*: connected to *a*, *b*, *f* and two others, which we leave to the reader.

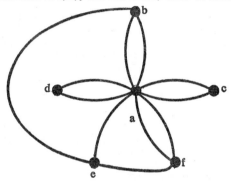

Thirdly, we consider *b*: connected twice with *a* and once with each of *c*, *d*, *e*.

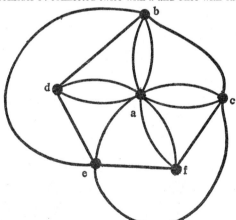

Fourthly, we consider *c*: connected twice with *a* and once with each of *b*, *e*, *f*.

Fig. 13.12

Finally, d and f: their connexions have already been inserted by considering the other connexions. In order to see if a solution is possible we examine the final network to see if it is unicursal. We observe that a, b, c, e are odd vertices. Since there are 4 odd vertices, Euler's rule tells us that the network is neither traversable nor unicursal, and consequently the problem has no solution.

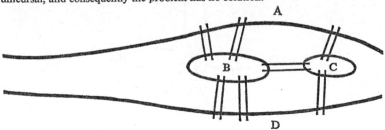

Fig. 13.13

Example 2

This is the Koenigsberg Bridge problem which was given to Euler. Apparently there used to be seven bridges which connected the two islands B and C with themselves and the banks. The problem posed by the locals was whether it was possible to go for a walk which traversed each bridge once, and once only. Set about 1736, Euler showed that the walk was impossible, and it is believed that his study of this particular problem gave rise to the subject of topology as discussed in this chapter.

Solution

You will probably have already recognized the nature of the problem. We have to find out if the network of the map of Fig. 13.13 is unicursal or not.

When making the network of a map we have seen that each region becomes a vertex, so there will be four vertices a, b, c, and d to correspond to the regions A, B, C, and D.

We observe the following connexions:

a to b twice, a to c once;
d to b twice, d to c once;
b to c once.

Draw the network yourself, then check with the correct network (i) of Fig. 13.9, wherein we see that there are 4 odd vertices. The network is not unicursal. The required walk is impossible.

Example 3

Show that constructing another bridge to connect A to D makes the network traversable.

Solution

We now need a new network, but instead of starting all over again we shall add the extra connexion to the network of the previous example. The extra connexion is from a to d. The result is (vi) Fig. 13.10. This converts the network into having 2 odd vertices.

The network is traversable, starting at either b or c and ending at c or b.

In each of the three examples it was necessary to identify the connexions before drawing the network. It would be tidier if we were to tabulate them in some manner. We accordingly complete a matrix for the networks in Fig. 13.14. The reader will readily recognize the procedure from the relation tables of Chapter Nine.

If we insert an extra row at the bottom of each table we can total the number of connexions to each vertex and pick out a possible unicursal network long before its construction. Indeed, we can sum these totals and by halving the grand total discover the number of connexions in the network. If we use the Euler relation $V + R - A = 2$ we shall be able to find out the number of regions without constructing the network at all.

	a	b	c	d	e	f
a	0	2	2	2	1	2
b	2	0	1	1	1	0
c	2	1	0	0	1	1
d	2	1	0	0	1	0
e	1	1	1	1	0	1
f	2	0	1	0	1	0
Total	9	5	5	4	5	4 = 32

	a	b	c	d
a	0	2	1	0
b	2	0	1	2
c	1	1	0	1
d	0	2	1	0
	3	5	3	3 = 14

	a	b	c	d
a	0	2	1	1
b	2	0	1	2
c	1	1	0	1
d	1	2	1	0
	4	5	3	4 = 16

Fig. 13.14

Exercise 13.3

1. Through Paris runs the River Seine, which contains in mid-river the two islands which were the site of the ancient city. These islands are connected by 12 bridges both to each other and to the river banks. A map of the arrangement is given in Fig. 13.15.

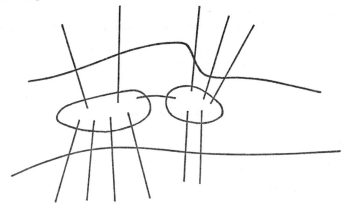

Fig. 13.15

Construct a network and its matrix for this map and determine whether a unicursal walk is possible. If the starting-point is the larger island, where is the finishing-point?

2. If one bridge is removed from the islands in Question 1 a unicursal walk is possible. Which bridge is to be removed?

3. We made the Koenigsberg bridge walk possible by adding an extra bridge connecting A with D. In fact, there were six possible bridges we could have built to make a solution possible. Using the totals given in the network matrix of Fig. 13.14, explain why there are six alternatives.

4. What do you notice about the matrices of any network for a map?

5. Given the matrices of two networks in Fig. 13.16. Determine the number of arcs and the number of regions of each network without a diagram.

```
0 1 2 2          0 1 1 1 3 1
1 0 1 1          1 0 2 1 0 1
2 1 0 1          1 2 0 3 1 0
2 1 1 0          1 1 3 0 1 0
    (i)          3 0 1 1 0 1
                 1 1 0 0 1 0
                     (ii)
```

Fig. 13.16

6. Are the networks of Question 5 unicursal?

TREES

Most of the networks we have dealt with so far have all contained more than one region. In fact, the single occasion on which we have met a network with only one region is in (ix) of Fig. 13.10. Such a network is called a **tree**. The reason is obvious if we consider the appearance of a tree in winter, each branch or twig ends in a final vertex, say, but at no time do the branches connect up to enclose a region. It is obvious that in a tree we must have one more vertex than there are arcs. Of immediate interest is whether Euler's result of $V + R - A = 2$ will still hold.

We noticed that (ix) in Fig. 13.10 satisfied this result, but will every tree satisfy the result?

Now for any tree $R = 1$. So Euler's result becomes $V - A = 1$ or $V = A + 1$, which is obviously true, so a tree satisfies Euler's result.

BETTI NUMBERS

Any network may be converted into a tree by removing the necessary arcs, which reduce the number of regions in a network to one. The network of Fig. 13.17 is reduced to a tree by the removal of the arcs which have been struck through. The number of arcs which have to be removed from a network to convert it into a tree is called the Betti number of the network. (After Enrico Betti, 1823–92.)

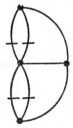

Fig. 13.17

In the case of Fig. 13.17 the Betti number $B = 4$. This seems to be a hit or miss affair, so we ought to see if we can make a more accurate determination of the number if possible. The number of vertices is fixed and the number of regions has to be reduced to one. These are the definite facts of our alteration.

From Euler's result we write

$$V + R - A = 2$$

and therefore $\quad V + R - 2 = A_N \quad$ (A_N arcs in network)

But for the tree which we hope to produce we shall have $R = 1$

$$\therefore \quad V + 1 - 2 = A_T \quad (A_T \text{ arcs in tree})$$

Since V is the same in both equations, we see that

$$A_N - A_T = R - 1$$

and this is the number of connexions we must cut out and is therefore the Betti number of the network.

We have therefore as a general formula that the Betti number of a network is given by $B = R - 1$.

Check: in Fig. 13.17 $R = 5$ $\therefore$ $B = 4$, which we have already suggested in the diagram. Incidentally, the solution is not unique, as the second diagram of Fig. 13.17 shows.

EXTENSION OF EULER'S RESULT

So far we have discussed topological figures which have all been drawn in the surface of a plane. We have discovered that Euler's result of $V + R - A = 2$ is an invariant for any topological transformation in the plane. As a preparation for extending our ideas to three dimensions and other spaces we consider Euler's result applied to some of the standard geometrical three-dimensional figures.

The joins of 4 points in space form a tetrahedron diagram (i) in Fig. 13.18. This figure has an inside or an outside, so that it divides space into two sections, but remember that Euler's result is for networks and regions formed by the joins or connexions in the networks.

In three dimensions those 'regions' become faces of the figures, so we change R to F to indicate that we are dealing with such three-dimensional networks. Also the arcs now become edges of the figure and we change A to E.

For (i) in Fig. 13.18 we therefore have

$$V = 4, \quad E = 6, \quad F = 4, \quad V + F - E = 2 \text{ as before}$$

Diagram (ii) is a cuboid, and we have

$$V = 8, \quad E = 12, \quad F = 6, \quad V + F - E = 2 \text{ as before}$$

Both of these figures are examples of what are called **polyhedra**, and like two-dimensional figures they may undergo topological transformations under the same conditions as before. Diagrams (ii) and (iii) are topologically equivalent, but of course the restriction of drawing three-dimensional networks on two-dimensional paper is very difficult to overcome visually, even

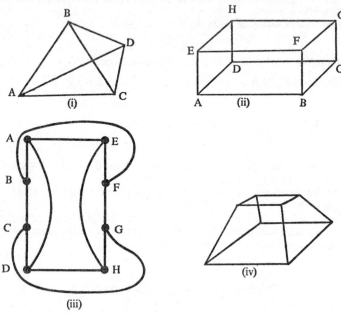

Fig. 13.18

though the two-dimensional network is topologically equivalent. Topological equivalents of these figures may also be drawn on simply connected surfaces, such as the sphere.   (a simply connected surface is one which is

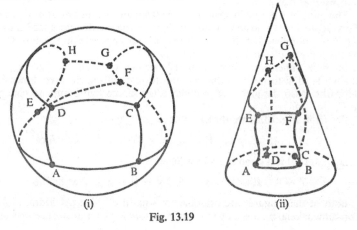

Fig. 13.19

separated into two separate regions by a simple closed curve). For example,
the cuboid network figure may be drawn topologically as in (i) Fig. 13.19 on
the sphere or on the surface of the cone. Both sphere and cone are topological equivalent surfaces (a closed cylinder, the surface of a broom-handle,
etc., are also topologically equivalent to the sphere).

Let us accept that the network of any polyhedra, like the cube or tetrahedron, may be drawn on a simply connected surface like the sphere or
cone. By so considering our network we are better able to suggest a way to
prove Euler's result. We start by assuming that on any network $V + F - E = 2$, and we discuss every possible way of adding to the network. For each
addition there will be no change in Euler's result. Complete the fourth and
fifth entries in the table below.

Addition to network	Old Picture	New Picture	V	F	E	V+F–E
Add edge			1	0	1	—
Add one edge to join two vertices			0	1	1	—
Add one edge to an edge, not at a vertex			2	0	2	—
Add one edge to two edges, not at vertices			2	1		—
Add one vertex to an existing edge			1	0	1	—

These are all the different additions which can be made to an existing network in which we have assumed $V + F - E = 2$. But all networks start
from one vertex and one face (region) and no edges, and since $V + F - E = 2$ to start with, further addition to complete the network will leave this
result unchanged as we have proved above.

Notice that we have not considered networks in which vertices are connected to themselves, i.e. a network which consists of a simple closed curve
with one vertex on its curved edge. Now we have proved Euler's result for
networks on a simply connected surface, but since we accept that the network
of polyhedra may be drawn on such a surface, it follows that the proof above
holds for a polyhedra network.

Exercise 13.4

1. Obtain the Betti number of each of the networks in Fig. 13.10.
2. Which of the following surfaces are topologically equivalent? Sphere, cube,
tetrahedron, square-based pyramid, cuboid, cylinder, a doughnut, a sphere with a
hole right through.

3. In Fig. 13.18 (iv) is the network for a truncated square-based pyramid. Give two other solids whose vertices give a topologically equivalent network.

4. If we allow networks which contain vertices connected with themselves by simple closed curves, what are the other two possibilities for adding to a network which were not considered in the proof of Euler's result?

5. Which of the networks of Fig. 13.18 are unicursal?

6. Draw the network of a double square-based pyramid and show that it is unicursal.

SURFACES

If we draw a simple closed curve on surfaces encountered so far in this chapter, then the surface is separated into two regions, one inner region and one outer region. Strictly speaking, we could consider the surface to be really divided into three regions or sets of points. One inside the curve, one on the curve, and one outside the curve, but we are content with our two-region description here.

A surface which was always separated in this manner by a simple closed curve was called a *simply-connected surface.* The two surfaces of Fig. 13.19 were simply connected, and so is the plane surface of this piece of paper. At first sight it appears that these surfaces are all topologically equivalent, until we actually try to obtain a cuboid or sphere from this piece of paper which you are reading at the moment. Some of the edges would have to be stitched

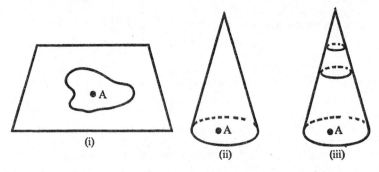

Fig. 13.20

together, and this is not allowed in a topological transformation, since no edges were stitched before the transformation.

We have already referred to the difficulty of deciding which is the inside and outside regions of a curve (page 289). The surfaces in Fig. 13.20 offer an interesting illustration of the difficulty.

A simple closed curve is drawn in the plane surface of (i) Fig. 13.20, and a point A is taken in the inner region. If we now carry out a topological transformation which shrinks the curve it will always contain A in its inner region.

Turning our attention to (ii), we have taken the base edge for the simple closed curve and then in (iii) shrunk the curve up to the apex of the cone—the point A no longer being a member of the inner region. Clearly this procedure has revealed a topological difference between the two surfaces, for in a plane surface it was impossible to shrink the curve without containing A. We may think of this difference in another way. On a sphere or its topological equivalent a simple closed curve may be shrunk to any point on the surface whatsoever. On a plane surface this is not the case.

We have thus shown that not all simply connected surfaces are topologically equivalent.

Definition 13.5

A surface is said to be simply connected if, and only if, any simple closed curve on the surface can be shrunk to an interior point without leaving the surface.

In the above definition we are only suggesting that we may shrink the simple closed curve to be as small as we please; we are not suggesting that it becomes the point—for this would not be a topological transformation.

On the basis of this definition a sphere is a simply connected surface, and so is any other surface which is topologically equivalent to the sphere, such as the cone and cuboid, and so is this piece of paper.

An example of a non-simply-connected surface is a torus—the doughnut-shaped figure of 13.21 (i). The dotted simple closed curve could not be shrunk to a point in the surface, and come to that neither could the dashed curve.

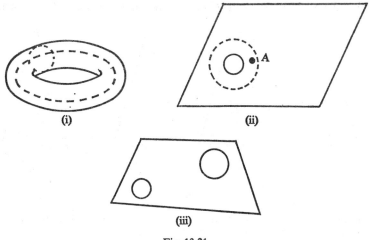

Fig. 13.21

The surface of a piece of paper with a hole cut in (ii) is not a simply-connected surface, because the dotted simple closed curve cannot be shrunk to the point A without leaving the surface and crossing the hole.

It is possible to change a non-simply-connected surface to one which is by inserting cuts. For example, (ii) above will become simply connected by making a cut from a point on the inner edge to the outer edge. The surface (iii) above clearly needs two cuts to make it simply connected.

CLASSIFICATION OF SURFACES

The use of Betti numbers provides some form of classification of networks which seems possible in the case of surfaces. That is, a classification in terms of cutting. Unfortunately such a classification is not altogether satisfactory, because it proves to be so very elusive for the more difficult type of surface. We have defined a cut as a disconnexion of a surface which starts on an edge and finishes on an edge. This means that any plane simply-connected surface will be divided into pieces by such a cut, whereas a figure such as (ii) Fig. 13.21 (i.e. an annulus) could still be left in one piece by cutting from the inner edge to the outer one. We may therefore classify a surface by a Betti number, which is the maximum number of cuts which a surface will sustain before becoming more than one piece.

Thus a plane rectangle or disc has B = 0, an annulus B = 1, a plane with two holes B = 2.

This is all very well for surfaces with clearly defined edges, but what about the sphere or torus? In this case we allow ourselves to make a puncture (not a hole, because this would remove a disc of the surface), but we do not consider the edge of the puncture as an edge of the surface.

Thus the torus may be cut along both curves shown in Fig. 13.21 (i) without becoming two pieces, so that B = 2 for the torus but clearly B = 0 for the sphere. There are, however, many surfaces which make the detection of possible cuts no easy matter.

FURTHER CLASSIFICATION

A further classification is made in the following manner:

1. Any surface such as a sphere is called a **bounded** surface.

2. A plane is not considered to be a bounded surface. When we draw rectangles and discs we call these plane surfaces, but they are finite, i.e. it is possible to draw them completely, they are in fact a sub-region of a plane.

3. A **boundary** is the curve or curves of separation of one region from another. Thus a plane has no boundary and is unbounded. On the other hand, a sphere also has no boundary but is nevertheless bounded. We say a boundary is the set of **curves of separation** in the case of an annulus, i.e. the separation curves of the annulus are its boundary with the rest of the plane.

4. A surface is said to be closed if it is bounded but has no boundary, e.g. (i) the surface of a solid cube is closed; (ii) a plane rectangle is not closed because it has a boundary. A layman test for this is that a surface is closed if we can travel in any direction in the surface without coming to a boundary.

An examination of the above classification should impress the reader with the limitations of having to rely on visualizing the surface each time. These difficulties are typical of our intuitive approach to the subject.

Sooner or later it has to be organized on a firmer mathematical basis, and since the elements under discussion are always those sets of points which are called surfaces, it follows that any mathematical formulation of topology must concern itself with the language of point set theory to become general topology. This is quite beyond the scope of the present work, so we conclude with the next exercise and the following experiments.

Exercise 13.5

1.

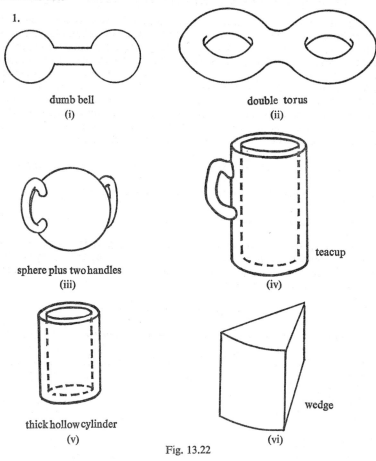

dumb bell
(i)

double torus
(ii)

sphere plus two handles
(iii)

teacup
(iv)

thick hollow cylinder
(v)

wedge
(vi)

Fig. 13.22

(a) Divide the above 6 surfaces into sets of topologically equivalent surfaces.

(b) Which surfaces above are topologically equivalent to

   (i) a sphere;

   (ii) a torus?

(c) State the Betti number of each surface above.

(d) Which of the above surfaces is closed?

(e) Which of the above surfaces are simply connected?

2. Illustrate with sketches how a torus can be transformed into a teacup.

3. Which of the following surfaces are topologically equivalent to a surface in Question 1?

(a) A jacket with no buttonholes?

(b) A cup such as (iv) in Question 1 but having two handles?

(c) A pair of trousers?

(d) A slipper?

EXPERIMENTS IN TOPOLOGY: THE MOEBIUS STRIP

For all the surfaces we have used so far we have restricted ourselves to considering just one side. For example, in the case of a sphere we have only dealt with the outer side, and when speaking of a plane sheet of paper we have only referred to one side as being the surface under consideration. At first sight it would appear that all surfaces are two-sided, but as Moebius (1790–1868) showed, there are surfaces with only one side. The simplest of such surfaces is the Moebius Strip. (For all the following experiments the reader would find it an advantage to buy a roll of gumstrip paper 5 cm wide.)

Cut a 30-cm strip and stick the ends together after giving the strip a half twist. The diagrams of Fig. 13.23 should make this clear.

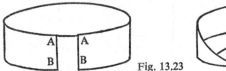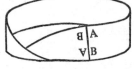

Fig. 13.23

Letter the ends as shown on both sides of the strip (this will mean 4 As and 4 Bs), then stick together so that the As and Bs come together as shown in the final figure. This is a Moebius Strip. Now make a mark on the edge and then follow the edge round the strip until arriving back at the mark. Only one edge! Now draw a line straight down the middle of the strip—only one side!

We have therefore a one-edged surface which has only one side. Now take a pair of scissors and make a perforation in the strip, then cut it down the middle. You should obtain a single loop having two edges and two whole twists. A further cut down the middle will give—try it first—two separate but interlocked loops, thus indicating that the Betti number of a Moebius strip is $B = 1$ and also indicating the type of cutting necessary to establish the Betti numbers.

There are many combinations of cuts and twists which yield unpredictable results.

Exercise 13.6

1. Starting with strips about 30 cm long, carry out the instructions of the table and complete the result column.

No. of half twists	Type of cut	Result
0	central	2 loops untwisted
1	central	1 loop—2 complete twists
1	$\frac{1}{3}$ from the edge	
2	central	
2	$\frac{1}{3}$ from the edge	
3	central	
3	$\frac{1}{3}$ from the edge	
4	central	
4	$\frac{1}{3}$ from the edge	

2.

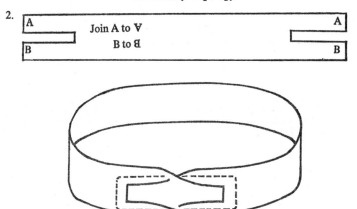

Join A to ∀
B to ઘ

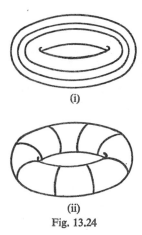

Give each of the above strips a half twist in the same direction then connect up. Cut round the hole as shown by the dotted line.

3. Repeat Question 2, but this time give A a twist in the opposite direction to B. Cut round the hole as before.

THE PUNCTURED TORUS

The best example for a punctured torus is the inner tube and valve of a car tyre. Imagine that we have stripes painted round the inner hole on both

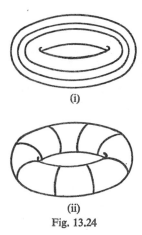

(i)

(ii)
Fig. 13.24

sides of the torus surface. Then by threading the tube through the valve (not very likely, but a good topological transformation) we shall produce the change in the direction of the stripes as shown in Fig. 13.24.

Exercise 13.7

1. Take a pair of striped pyjamas with a draw-string top and stitch the bottoms of the legs together. You now have a punctured torus with the draw-string controlling the valve opening as in Fig. 13.25.

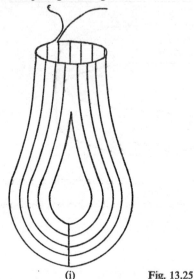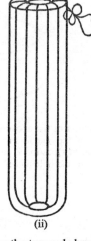

(i)   **Fig. 13.25**   (ii)

Reach through the valve down the leg and grab where the two ends have been stitched. Pull up through the valve. Close the valve by drawing the string tight. You will now have the figure (ii) of Fig. 13.25, which is the topological equivalent of Fig. 13.24 (ii).

THE FOUR-COLOUR PROBLEM

In colouring a map, two countries or regions which have a common frontier or boundary must be given two different colours. The same colour may be used for countries whose frontiers have only one point in common as in (i) below. We inquire now into the minimum number of colours necessary to colour any map which is drawn on a simply connected surface.

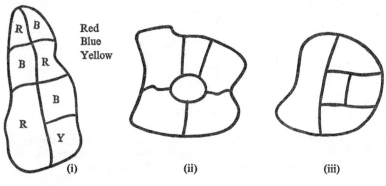

Red
Blue
Yellow

(i)   (ii)   (iii)

.The first map has been coloured to give the idea of what is meant by colouring the map. Copy the other two and try to colour them using four colours only.

It is always possible to colour these maps using only four colours, but no one has ever proved that four colours is sufficient for every map. (It has been proved for five colours.)

ANSWERS

Exercise 1.1

1. {January, June, July}
2. {September, April, June, November}
3. {of, in}
4. {e, x, c, l, n, t}
5. {June}
6. {3, 7, 8}
7. The set of the first 4 letters of the English alphabet
8. The set of vowels in the English alphabet
9. The set of the last 3 letters of the English alphabet
10. The set of ways of travelling from one place to another
11. The set of garden flowers
12. The set of birds
13. $\in$ 14. $\in$ 15. $\notin$ 16. $\notin$ 17. $\notin$ 18. $\in$

Exercise 1.2

1. $\neq$ 2. $=$ 3. $\neq$ 4. $=$ 5. $\neq$
6. $\neq$ 7. $=$ 8. $\neq$ 9. $\neq$ 10. $\neq$

Exercise 1.3

Answers may vary

1. {a, b, c, d}
   {w, x, y, z}

2. {1, 2, 3, 4, 5, 6}
   {2, 4, 6, 8, 10, 12}

Exercise 1.4

1. $\not\subset$ 2. $\not\subset$ 3. $\not\subset$ 4. $\subset$ 5. $\subset$ 6. $\subset$
7. {x, y}, {x}, {y}, ϕ
8. {a, b, c, d}, {a, b, c}, {a, b, d}, {a, c, d}, {b, c, d}, {a, b}, {a, c}, {a, d}, {b, c}, {b, d}, {c, d}, {a}, {b}, {c}, {d}, ϕ
9. If a set contains n members, then it contains 2^n subsets, 32. (Remember that $2^0 = 1$.)
10. T, T, T, F, T, T, F, T

Exercise 1.5

1. 8   2. 7   3. 7   4. 12   5. 0   6. 4
7. 1   8. 0

Exercise 1.6

1. $J \cup K = \{1, 2, 3, 5, 7, 9\}$
2. $K \cup M = \{2, 3, 4, 5, 6, 7, 8, 9\}$
3. $K \cup N = \{0, 3, 5, 7, 9\}$
4. $J \cup M = \{1, 2, 3, 4, 6, 8\}$
5. $J \cup N = \{0, 1, 2, 3, 5, 9\}$
6. $M \cup N = \{0, 2, 4, 5, 6, 8, 9\}$
7. $N \cup M = \{0, 2, 4, 5, 6, 8, 9\}$
8. $K \cup K = K$

Exercise 1.7

1. $C \cap D = \{2, 3\}$
2. $D \cap C = \{2, 3\}$
3. $C \cap E = E$
4. $C \cap F = \phi$
5. $D \cap E = \{3\}$
6. $D \cap F = \{7, 8\}$
7. $E \cap F = \phi$
8. $E \cap E = E$

(i) {f, g}   (ii) {r}   (iii) {f, g, r, s, t}   (iv) ϕ

Exercise 1.8

1. T   2. T   3. F   4. T   5. T   6. T
7. T   8. T   9. F   10. T   11. T   12. F
13. F   14. Meaningless, s is an element not a set   15. Meaningless

Exercise 1.9

1. $\{a, k\}$
2. $\{z, w\}$
3. ϕ
4. $\{a, k, w, x, y, z\}$
5. $\{b, c, d, e, m, u\}$
6. $\{x, y\}$
7. $\{m, u\}$
8. $\mathscr{E}$
9. $\mathscr{E}$
10. ϕ
11. (i) Using elements in Fig. 1.7 $(A \cup B)' = \{b, c, d, e\}$
    (ii) Using elements in Fig. 1.7 $(A \cap B)' = \{x, y, z, w, m, u, b, c, d, e\}$

Exercise 1.10

1. $\{a, b, c, d, e, f, g\}$
2. $\{a, b, c, d, e, f, g\}$
3. $\{c\}$
4. $\{e\}$
5. ϕ
6. ϕ
7. $\{c, e, f, g\}$
8. $\{c, e\}$

Exercise 1.11

1. $\{a, b, c, e\}$
2. $\{e\}$
3. $\{a, b, e, f\}$
4. $\{c, e, f, g\}$
5. $\{a, b, c, d, e\}$
6. $\{a, b, c, d, e, f\}$
7. $\{a, b, c\}$
8. $\{e, f\}$
9. 4, 3, 4, 4
10. 2, 1, 6

Exercise 1.12

1. (i) a, b; (ii) c; (iii) e; (iv) x, y; (v) z; (vi) d; (vii) p, q; (viii) r, s
2. z, x, y, p, q regions (v), (iv), and (vii)
3. d region (vi)
4. r, s region (viii)

Exercise 1.13

1. $n(A \cup B) = n(A) + n(B) - n(A \cap B)$ $\therefore$ $n(A \cap B) = 13$
2. 3; 10; 6; 37
3. 13; 37
4. (*a*) (viii) becomes 93 (v) becomes 67
   (*b*) (ii) becomes 12 (iii) becomes 44 (x) becomes 39
5. (i) Some men are lazy
   (ii) All people are either lazy or men
        Some men are the only people who are not lazy
   (iii) All people are lazy
   (iv) All men are lazy
   (v) The only lazy people are men
   (vi) Some people are neither lazy nor men
6. (i) The set of whole numbers which are divisible by 15
   (ii) The set of whole numbers which are divisible by 30
   (iii) The set of whole numbers divisible by 3 (because any number divisible by 6 is also divisible by 3, i.e. $S \subset T$)
7. (i) $D \cap T \neq \phi$
   (ii) $D \cap T' \neq \phi$
   (iii) $T \cap C = \phi$
   (iv) $C \subset T'$

Supplementary Exercise Chapter One

1. (i) T; (ii) F; (iii) F; (iv) T; (v) F (check with page 2); (vi) T; (vii) T; (viii) T; (ix) T; (x) T
2. (i) T; (ii) F; (iii) T; (iv) T; (v) T; (vi) T; (vii) T; (viii) F
3. (i) $A \subset \mathscr{E}$, T; (ii) $B \subset A$, F; (iii) $B \cap A = \phi$, T; (iv) $B \cap A = \phi$, T; (v) $B \cap A = \phi$, T
4. (i) farmers who keep cows, sheep, and goats, i.e. all three
   (ii) farmers who keep either cows or sheep or both
   (iii) farmers who keep sheep as well as goats but not cows
   (iv) farmers who keep all four
   (v) farmers who keep sheep and goats but not cows or pigs

5. $173 = 110 + 55 + 67 - (20 + x) - (11 + x) - (16 + x) + x$
$\therefore \quad x = 6$
$\therefore \quad n(\text{M} \cap \text{F}) = 26, n(\text{M} \cap \text{S}) = 17, n(\text{F} \cap \text{S}) = 22$
$n\{\text{complaints about two or more items}\} = 53$

Exercise 2.1

1. No 2. Yes 3. No 4. No
5. No 6. $3 + 7 = 7 + 3$ 7. $9 + 1 = 1 + 9$ 8. $a + b = b + a$
9. $14 + 6 = 6 + 14$ 10. $4 + 11 = 11 + 4$

Exercise 2.2

1. Yes 2. Yes 3. Yes 4. No, as any tea drinker will know
5. $5 + (7 + 6) = (5 + 7) + 6$ 6. $17 + (15 + 32) = (17 + 15) + 32$
7. $(9 + 8) + 7 = 9 + (8 + 7)$ 8. $13 + (12 + 6) = (13 + 12) + 6$
9. $(72 + 31) + 46 = 72 + (31 + 46)$

Exercise 2.3

1. A 2. C 3. C 4. C, A 5. C
6. C, A 7. A 8. C 9. C, A 10. A

Exercise 2.4

1. $4 + 6 = 10$ 2. $8 + 3 = 11$ 3. $2 + 6 = 8$

Exercise 2.5

1. Open your eyes 2. Sit down (or lie down)
3. Return from work 4. Open this book
5. Take 5 steps backward 6. Tie your shoelace
7. Wake up 8. Switch the light off

Exercise 2.6

1. 15 2. 754 3. 69 4. 26 5. 756 6. 39 7. 312
8. r 9. 6 10. 4 11. 8 12. 8 13. 9 14. 9

Exercise 2.7

1. $10 - 4 = 6$ 2. $9 - 6 = 3$ 3. $9 - 7 = 2$ 4. $13 - 7 = 6$

Exercise 2.8

1. 182 2. 1 907 3. 188

Exercise 2.9

1. 9 2. 8 3. 6 4. 24

Exercise 2.10

1. $5 \times 9 = 9 \times 5$ 2. $7 \times 8 = 8 \times 7$ 3. $31 \times 7 = 7 \times 31$
4. $12 \times 6 = 6 \times 12$ 5. $9 \times 17 = 17 \times 9$ 6. $p \times q = q \times p$
7. $\square \times \triangle = \triangle \times \square$

Exercise 2.11

1. $(3 \times 7) \times 5 = 3 \times (7 \times 5)$ 2. $4 \times (2 \times 3) = (4 \times 2) \times 3$
3. $(6 \times 3) \times 2 = 6 \times (3 \times 2)$ 4. $17 \times (8 \times 5) = (17 \times 8) \times 5$
5. $(13 \times 9) \times 3 = 13 \times (9 \times 3)$ 6. A 7. C
8. C 9. C, A 10. C 11. C, A 12. A

Exercise 2.12

1. $7 \times (2 + 5) = (7 \times 2) + (7 \times 5)$ 2. $4 \times (3 + 6) = (4 \times 3) + (4 \times 6)$
3. $(4 + 5) \times 3 = (4 \times 3) + (5 \times 3)$ 4. $(8 + 9) \times 6 = (8 \times 6) + (9 \times 6)$
5. $(5 \times 2) + (7 \times 2) = (5 + 7) \times 2$ 6. $(6 \times 3) + (8 \times 3) = (6 + 8) \times 3$

Exercise 2.13

1. $1200 < n < 2100$, 1755   2. $5600 < n < 7200$, 6150
3. $1500 < n < 2400$, 2262   4. $3600 < n < 5000$, 4042
5. $12\,000 < n < 20\,000$, 17 249   6. $28\,000 < n < 40\,000$, 37 884
7. $40\,000 < n < 54\,000$, 42 952   8. $30\,000 < n < 42\,000$, 38 324

Exercise 2.14

1. 6   2. 7   3. 6   4. 9   5. 9   6. 7
7. 5   8. 4   9. 5   10. 7   11. 9   12. 4

Exercise 2.15

1. 8   2. 13   3. 13   4. 11   5. 12   6. 6
7. 13   8. 12   9. 19

Exercise 2.16

1. 3 $r2$ 2. 3 $r3$ 3. 2 $r1$ 4. 6 $r1$ 5. 4 $r4$
6. 5 $r3$ 7. 13 $r3$ 8. 9 $r1$ 9. 11 $r1$

Exercise 3.1

1. (-7) 2. (-15) 3. 0   4. (-12) 5. 0   6. (-306)

Exercise 3.2

1. $(+4)$ 2. $(+8)$ 3. $(+5)$ 4. $(+14)$ 5. $(+22)$
6. $(+35)$ 7. $(+3)$ 8. $(+12)$ 9. $(+21)$

Exercise 3.3

1. (-5) 2. (-12) 3. (-12) 4. (-13) 5. (-19)
6. (-35) 7. (-57) 8. (-59) 9. (-128) 10. (-122)

Exercise 3.4

1. (-3) 2. (-9) 3. (-6) 4. (-3) 5. $(+21)$
6. $(+12)$ 7. (-2) 8. $(+2)$ 9. (-12) 10. $(+104)$
11. $(+157)$ 12. (-78) 13. (-17) 14. (-71)

Exercise 3.5

1. $(+11)$ 2. (-20) 3. $(+4)$ 4. (-9) 5. (-21)
6. (-5) 7. $(+5)$ 8. $(+21)$ 9. (-31) 10. (-113)
11. $(+257)$ 12. (-12)

Exercise 3.6

1. (-12) 2. (-16) 3. (-24) 4. (-35) 5. (-120)
6. (-120) 7. (-104) 8. (-273) 9. (-2247) 10. (-1025)
11. (-750) 12. (-2898)

Exercise 3.7

1. $(+15)$ 2. $(+28)$ 3. (-28) 4. (-28) 5. $(+28)$
6. $(+120)$ 7. (-120) 8. (-120) 9. $(+120)$ 10. $(+105)$
11. (-366) 12. $(+366)$ 13. (-1608) 14. $(+630)$

Exercise 3.8

1. 0 | 2. (+54) | 3. (−1066) | 4. 0 | 5. (−315)
6. 0 | 7. (−624) | 8. (−108) | 9. (+144) | 10. (−900)
11. (+90) | 12. (+1)

Exercise 3.9

1. (−5) | 2. (−2) | 3. (−8) | 4. (−9) | 5. (−9)
6. (−9) | 7. (+5) | 8. (+25) | 9. (−25) | 10. (−25)
11. (−32) | 12. (−12)

Exercise 3.10

1. (+17) | 2. (+1) | 3. (−1) | 4. (+7) | 5. (+7)
6. (−12) | 7. (+12) | 8. (−12) | 9. (+12) | 10. (−13)
11. (+13) | 12. (+9) | 13. (−9) | 14. (+9)

Exercise 3.11

1. −4 | 2. −9 | 3. 13 | 4. 13 | 5. −1 | 6. −1
7. 1 | 8. 1 | 9. −21 | 10. −21 | 11. −4 | 12. −$\frac{1}{4}$

Exercise 4.1

1. F | 2. T | 3. T | 4. T | 5. F | 6. F
7. T | 8. T | 9. T | 10. F

Exercise 4.2

1. 28 | 2. 36 | 3. 17 | 4. 32 | 5. 4 | 6. 19
7. 5 | 8. 54 | 9. 1 | 10. 0
11. (30 − 12) ÷ (3 × 2)
12. [30 − (12 ÷ 3)] × 2
13. [(30 − 12) ÷ 3] × 2
14. 30 − [12 ÷ (3 × 2)]
15. 30 − [(12 ÷ 3) × 2]

Exercise 4.3

1. T | 2. T | 3. F | 4. T | 5. F
6. T | 7. F | 8. T

Exercise 4.4

1. O | 2. T | 3. O | 4. F | 5. O | 6. F

Exercise 4.5

1. {+2} | 2. {(−3)} | 3. {(−3)(−4)(−5)} | 4. φ
5. {(+4)} | 6. {(−5), (−4), (−3), (−2), (−1), 0, (+1), (+2), (+3), (+4)}
7. {0} | 8. {(−1)} | 9. {(−2)} | 10. {(−5)}

Exercise 4.6

1. 7 | 2. 7 | 3. 9 | 4. −6 | 5. 4
6. −5 | 7. 7 | 8. −4 | 9. −7 | 10. −3

Exercise 4.7

1. 14 | 2. 7 | 3. 55 | 4. −17 | 5. 77
6. −24 | 7. 11 | 8. 8 | 9. −22 | 10. 0

Exercise 4.8

1. 40 | 2. −27 | 3. 210 | 4. 0 | 5. 28
6. −72 | 7. −25 | 8. −9 | 9. 56 | 10. 160
11. −90 | 12. −1

Exercise 4.9

1. 9	2. -7	3. -9	4. 15	5. -47
6. 9	7. 52	8. 13	9. 6	10. 0

Exercise 4.10

1. 5	2. 40	3. 9	4. -5	5. 8	6. -4
7. -3	8. 12	9. 10	10. 73	11. 9	12. 2
13. 12	14. 13	15. 2	16. 46		

Exercise 4.11

1. 2	2. 2	3. 2	4. 13	5. 7	6. 2
7. 6	8. -12	9. 13	10. 1		

Exercise 4.12

1. $n + 7$ 2. $2n - 3$ 3. $n + 2n = 3n$
4. $n + 8$ 5. $2n + 3(7 - n) = 21 - n$
6. $2n + 9$ 7. $3n + 4n = 7n$ 8. $2n + 5$
9. $4n$ 10. $n - 3$

Exercise 4.13

1. $t - 3 = 4$ 2. $12x = 32$ 3. $3k - 8 = 7$
4. $h + 7 = 51$ 5. $u \div 5 = (-9)$

Exercise 4.14

Typical solutions
1. $2n + 6 = 22$; 8   2. $n + (n + 3) = 67$; 32 and 35
3. $r + (r - 9) = 103$; Roger 56 kg and Ted 47 kg
4. $p + (p + 8) = 26$; 9 metres   5. $2(w + 12) = 44$; 10 metres
6. $n + 4n = 75$; 15
7. $5n - n = 32$ or $n - 5n = 32$; 8 or -8

Exercise 5.1

1. $\frac{17}{4}$	2. $-\frac{11}{3}$	3. $\frac{9}{8}$	4. 4	5. $\frac{20}{3}$
6. $\frac{5}{2}$	7. 6	8. -21	9. -9	10. -8

Exercise 5.2

1. $1\frac{3}{5}$	2. $4\frac{1}{10}$	3. $7\frac{2}{7}$	4. $8\frac{6}{7}$	5. $4\frac{2}{9}$
6. $9\frac{4}{9}$	7. $12\frac{4}{11}$	8. $9\frac{11}{25}$	9. $8\frac{21}{50}$	

Exercise 5.3

1. $\frac{1}{12}$	2. $\frac{1}{12}$	3. $\frac{1}{21}$	4. $\frac{1}{40}$	5. $\frac{1}{18}$
6. $\frac{1}{30}$	7. $\frac{1}{60}$	8. $\frac{1}{39}$	9. $\frac{1}{68}$	

Exercise 5.4

1. $\frac{12}{35}$	2. $\frac{10}{21}$	3. $\frac{40}{63}$	4. 0	5. $\frac{20}{39}$	6. $\frac{8}{45}$
7. $\frac{16}{39}$	8. $\frac{9}{16}$	9. $\frac{8}{3}$	10. $\frac{3}{2}$	11. $\frac{1}{8}$	12. $\frac{1}{15}$

Exercise 5.5

1. $\frac{1}{4}$	2. $\frac{2}{5}$	3. $\frac{25}{28}$	4. 12	5. $\frac{5}{12}$
6. $\frac{3}{10}$	7. $\frac{5}{9}$	8. $\frac{1}{8}$		

Answers

Exercise 5.6

1. $3\frac{3}{5}$ 2. $8\frac{1}{2}$ 3. $\frac{2}{5}$ 4. 52 5. 12
6. 6 7. 112 8. $2\frac{1}{4}$

Exercise 5.7

1. $\frac{8}{5}$ 2. $\frac{1}{7}$ 3. $-\frac{9}{4}$ 4. $\frac{12}{7}$ 5. $\frac{31}{9}$
6. $-\frac{37}{15}$ 7. $\frac{2}{7}$ 8. $-\frac{1}{9}$ 9. $-\frac{7}{47}$

Exercise 5.8

1. 2 2. $1\frac{1}{2}$ 3. $\frac{5}{24}$ 4. $1\frac{5}{11}$ 5. $1\frac{7}{9}$
6. $2\frac{3}{11}$ 7. $4\frac{1}{2}$ 8. $10\frac{1}{2}$ 9. $1\frac{2}{3}$ 10. $6\frac{2}{5}$
11. $6\frac{3}{10}$ 12. $\frac{1}{20}$ 13. 5 14. $\frac{2}{3}$ 15. $1\frac{1}{3}$

Exercise 5.9

1. 1 2. 1 3. 1 4. $\frac{7}{13}$ 5. $\frac{2}{3}$
6. $\frac{3}{4}$ 7. $\frac{2}{3}$ 8. $\frac{3}{5}$ 9. 3 10. $7\frac{2}{3}$
11. $1\frac{5}{7}$ 12. $7\frac{3}{5}$

Exercise 5.10

1. $\frac{17}{21}$ 2. $\frac{5}{9}$ 3. $\frac{9}{10}$ 4. $1\frac{5}{12}$ 5. $\frac{41}{63}$
6. $\frac{1}{3}$ 7. $\frac{35}{36}$ 8. $\frac{43}{77}$ 9. $\frac{53}{72}$ 10. $\frac{59}{60}$
11. $\frac{79}{120}$ 12. $1\frac{1}{12}$

Exercise 5.11

1. $11\frac{1}{2}$ 2. $26\frac{1}{4}$ 3. $21\frac{5}{12}$ 4. $72\frac{7}{10}$ 5. $34\frac{9}{40}$
6. $75\frac{5}{8}$ 7. $297\frac{11}{12}$ 8. $806\frac{13}{14}$ 9. $801\frac{49}{60}$

Exercise 5.12

1. $\frac{4}{9}$ 2. $\frac{1}{3}$ 3. $4\frac{2}{5}$ 4. $3\frac{1}{10}$ 5. $\frac{2}{33}$
6. $\frac{11}{35}$ 7. $\frac{1}{9}$ 8. $\frac{2}{25}$ 9. $\frac{13}{48}$ 10. $\frac{3}{40}$
11. $2\frac{5}{12}$ 12. $2\frac{1}{40}$ 13. $7\frac{13}{18}$ 14. $30\frac{32}{36}$ 15. $10\frac{38}{45}$

Exercise 5.13

1. 0·47 2. 0·8 3. 0·68 4. 0·75 5. 0·112
6. 0·65 7. $12\frac{7}{10}$ 8. $\frac{37}{10000}$ 9. $\frac{1}{8}$ 10. $\frac{3}{5}$
11. $\frac{17}{500}$ 12. $\frac{19}{400}$

Exercise 5.14

1. 55·21 2. 11·375 3. 284·3 4. 1·5222 5. 12·99069
6. 1·222291 7. 43·6 8. 169·9 9. 53·89 10. 0·039
11. 3·8957 12. 0·15057

Exercise 5.15

1. 2·72 2. 3·8 3. 202·32 4. 1·8944
5. 25·9386 6. 67·854

Exercise 5.16

1. 18·5 2. 0·28 3. 54·3 4. 14·7

Exercise 5.17

1. 0·35 2. 0·875 3. 0·275 4. 0·38
5. 2·75 6. 58·36

Exercise 5.18

1. $0.83\bar{3}$
2. $0.1\bar{1}$
3. $0.6\bar{6}$
4. $0.15\bar{15}$
5. $0.7\overline{857142}$
6. $0.5\overline{71428}$
7. (i) 35%; (ii) 87.5%; (iii) 27.5%; (iv) 38%

Exercise 5.19

1. $\frac{35}{99}$
2. $5\frac{2}{3}$
3. $\frac{44}{111}$
4. $\frac{2105}{3333}$
5. $\frac{22}{30}$
6. $2\frac{431}{1650}$
7. $0.\bar{4}$
8. $0.\overline{57}$
9. $\frac{48}{99} + \frac{52}{99} = \frac{100}{99} = 1.\overline{01}$
10. 4
11. Change to $\frac{1}{3} \times \frac{1}{3} = \frac{1}{9} = 0.\bar{1}$
12. Change to $\frac{1}{9} \times \frac{43}{90} = \frac{43}{810}$

Exercise 6.1

1. 4
2. 2
3. 4
4. 4
5. 0
6. 1
7. 1
8. 0
9. 0
10. 4
11. 4
12. The distributive law of multiplication over addition which holds for clock arithmetic.

Exercise 6.2

1. 1
2. 4
3. 4
4. 4
5. 2
6. 1
7. 4
8. 4
9. 1
10. 1
11. 0
12. 2

Exercise 6.3

1. 1
2. 3
3. 2
4. 2
5. 1
6. 2
7. 3
8. 1
9. 1
10. 4
11. 0
12. meaningless

Exercise 6.4

1. Yes, by noting that each entry of the table is an element of the set {0, 1, 2, 3, 4, 5}
2. Yes, already discussed in the text
3. Additive identity 0, Multiplicative identity 1
4. All of them $0^* = 0$, $1^* = 5$, $2^* = 4$, $3^* = 3$, $4^* = 2$, $5^* = 1$
5. $5^{-1} = 5$; $1^{-1} = 1$, only
6. (*a*) 0; (*b*) 0; (*c*) 2
7. Yes, because $3 \times (4 + 2) = (3 \times 4) + (3 \times 2)$, etc., but we need to investigate all possible forms to prove this and not just these three cases
8. (*a*) $3 - 5 = 4$; $5^* = 1$; $3 + 5^* = 4$
   (*b*) $2 - 4 = 4$; $4^* = 2$; $2 + 4^* = 4$
   (*c*) $5 - 1 = 4$; $1^* = 5$; $5 + 5 = 4$
9. (*a*) Not possible 2^{-1} doesn't exist
   (*b*) Not possible because 3^{-1} doesn't exist
   (*c*) 3
   (*d*) Not possible; 4^{-1} doesn't exist
   (*e*) 2
10. Same result as 9 (*c*)
    Missing figures are $+0$, 3, 3; $\times$, 2, 3, 2

Exercise 6.5

1. $2 + 4 = 1$ (mod 5)   meaningless mod 6
2. $4 + 2 = 1$ (mod 5)   meaningless mod 6
3. meaningless (mod 5) $5 + 2 = 1$ (mod 6)
4. $2 - 4 = 3$ (mod 5)   meaningless mod 6
5. $3 - 4 = 4$ (mod 5)   meaningless mod 6

6.

	0	1	2	3
0	0	1	2	3
1	1	2	3	0
2	2	3	0	1
3	3	0	1	2

	0	1	2	3
0	0	0	0	0
1	0	1	2	3
2	0	2	0	2
3	0	3	2	1

(a) Closed under both addition and multiplication (the table contains only elements of {0, 1, 2, 3})

(b) 0 is the additive identity; $0^* = 0$, $1^* = 3$, $2^* = 2$, $3^* = 1$

(c) Yes, the element 1

(d) $1^{-1} = 1$, $3^{-1} = 3$

(e) Yes, because where the division is possible, i.e. where a result exists the result is unique

7. (a) $2^* = 2$; (b) $3^{-1} = 3$; (c) 2^{-1} meaningless; (d) meaningless; (e) 3 exactly as for part (b)

8. (a) {1, 3}; (b) {3}; (c) {1, 3}; (d) {2}; (e) {1}

Exercise 6.6

1. (a) V; (b) H; (c) I; (d) I; (e) H; (f) V

2. Because $V \otimes V = I$ then V is its own inverse
   Because $R \otimes R = I$ then R is its own inverse

3. Yes, because the results are unique

4. (a) Yes, because all the entries in the table belong to M
   (b) Both ∩ and ∪ are operations
   (c) Yes
   (d) D is the identity element for ∪ because

$$D \cup A = A = A \cup D$$
$$D \cup B = B = B \cup D$$
$$D \cup C = C = C \cup D$$
$$D \cup D = D = D \cup D$$

   (e) A, because $A \cap A = A = A \cap A$

$$A \cap B = B = B \cap A$$
$$A \cap C = C = C \cap A$$
$$A \cap D = D = D \cap A$$

   (f) Only D, which is its own inverse
   (g) Only A which is its own inverse

5. Exactly the same as the table of Fig. 6.6

6. Monday, i.e. August 14th is zero day, October 9th is day 56
   $56 \equiv 0 \pmod 7$ October 9th is a Monday

7. $365 \equiv 1 \pmod 7$ if Monday is zero day then: (ii) Tuesday; (i) Wednesday

8. August 28th is zero day. Friend A has days off whose number is x such that

$$x \equiv 3 \pmod 6$$
$$x \equiv 4 \pmod 6$$
$$x \equiv 5 \pmod 6$$

Friend B has days off such that

$$x \equiv 2 \pmod 4$$
$$x \equiv 3 \pmod 4$$

Now November 4th is day number $68 \equiv 2 \pmod 6$ neither of them can go
$$68 \equiv 0 \pmod 4$$

November 11th is day number $75 \equiv 3 \pmod 6$ both of them can go
$$75 \equiv 3 \pmod 4$$

Exercise 6.7

1. $\{0, 3\}$; $\{0, 2, 4\}$

+	0	2	4
0	0	2	4
2	2	4	0
4	4	0	2

2.

⊗	I	H
I	I	H
H	H	I

⊗	I	V
I	I	V
V	V	I

⊗	I	R
I	I	R
R	R	I

3. Yes 4. Yes 5. No, because 0 has no inverse

6. Yes. One

×	1	4
1	1	4
4	4	1

7. (*a*) Yes; (*b*) Yes; (*c*) No

(*d*) (i) A; (ii) P; (iii) No identity element, not associative.

Note that $1 \ominus 4 = 1$ But, $4 \ominus 1 \neq 1$
$2 \ominus 4 = 2$ $4 \ominus 2 = 2$
$3 \ominus 4 = 3$ $4 \ominus 3 \neq 3$
$4 \ominus 4 = 4$ $4 \ominus 4 = 4$

That is, 4 is an identity on the right, but not on the left, so that it is not an identity element

(*e*) The groups in (i) and (ii) are isomorphic

(*f*) (1) inverse A is A; (2) inverse B is C; (3) inverse C is B; (4) inverse D is D

Since (i), (ii) are isomorphic the answers for (ii) are P, R, Q, S

(*g*)

	A	D
A	A	D
D	D	A

Exercise 7.1

1. 5 2. 10 3. 7 4. 81 5. 23
6. 30 7. $14\frac{2}{3}$ 8. $54\frac{3}{4}$ 9. $11 + \frac{6}{8} + \frac{2}{64} = 11\frac{50}{64} = 11\frac{25}{32}$
10. $7 + \frac{1}{2} + \frac{1}{4} = 7\frac{3}{4}$ 11. $108 + \frac{2}{5} + \frac{2}{25} = 108\frac{12}{25}$
12. $75 + \frac{1}{9} + \frac{7}{81} = 75\frac{16}{81}$

Exercise 7.2

1. 177_8 2. $12\,022_3$ 3. 241_6 4. $10\,001_2$ 5. $0{\cdot}270_8$
6. $0{\cdot}102_3$ 7. $177{\cdot}270_8$ 8. $12\,022{\cdot}102_3$ 9. $10\,001{\cdot}11_2$ 10. $101{\cdot}3_4$

Exercise 7.3

1. $2{\cdot}13_4$ (2 places) $2{\cdot}2_4$ (1 place)
2. $1{\cdot}111_2$ (3 places) $1{\cdot}11_2$ (2 places)
3. $4{\cdot}66_8$ (2 places) $4{\cdot}7_8$ (1 place)
4. $4{\cdot}23_5$ (2 places) $4{\cdot}3_5$ (1 place)
5. $1{\cdot}22_3$ (2 places) $2{\cdot}0_3$ (1 place)
6. $3{\cdot}0_7$ (1 place)
7. $4{\cdot}25_9$ (2 places) $4{\cdot}2_9$ (1 place)
8. $1{\cdot}13_5$ (2 places) $1{\cdot}1_5$ (1 place)
9. $5{\cdot}24_7$ (2 places) $5{\cdot}2_7$ (1 place)
10. $5{\cdot}24_6$ (2 places) $5{\cdot}3_6$ (1 place)

Exercise 7.4

1. 9, 8, 7, 6, 5   2. 6, 5, 4   3. 54_8 4. 348_9 5. 215_6
6. 1221_3 7. 2140_6
8. $103_6 = 39_{10}$, $47_8 = 39_{10}$ both the same size
9. $231_4 = 45_{10}$, $142_5 = 47_{10}$ result 2_{10}
10. $34_8 = 28_{10}$; $31_9 = 28_{10}$ result 0

Exercise 7.5

1. 15 200   2. $20\,140_6$ 3. 121   4. 144   5. $100\,011_2$
6. $15\,411_7$ 7. $14 \cdot 3_7$ 8. $15 \cdot 2_6$ 9. $20 \cdot 14_6$ 10. $1 \cdot 603_7$

Exercise 7.6

1. 122_7 2. 112_9 3. 10_5 4. 20_6 5. 11_4
6. 13_7 remainder 1   7. 41_6 remainder 31_6 8. 51_6 remainder 13_6
9. 236_7 remainder 12_7 10. 162_7 remainder 1   11. (a) $\frac{1}{3}$; (b) 11_5

Exercise 7.7

1. Five, six   2. $16 + 8 + 4$ 3. 11 g, 9 g
4. 9 g v 1 g; 27 v $(9 + 3 + 1)$; $27 + 3$ v 1; $27 + 9$ v 1
5. $45_{10} = 1200_3$;
$\quad\quad = 2000_3 - 100_3$
$\quad\quad = 10\,000_3 - 1100_3$
$\quad\quad 81$ v $(27 + 9)$
$\quad 62_{10} = 2022_3$
$\quad\quad = 2100_3 - 1$
$\quad\quad = 10\,100_3 - 1001_3$
$\quad\quad \therefore 62 = (81 + 9)$ v $(27 + 1)$
6. No. 2 g, 6 g, 7 g, etc., cannot be weighed
7. Yes, there is one odd column
   The play required is remove 1 from 7 to leave 6, 4, 2
8. Remove 3 from 4 and leave 0, 1, 1
9. (a) and (b) Remove one complete set and leave 0, 2, 2 or 0, n, n
10. $\frac{1}{7}$ 11. $\frac{1}{5}$ 12. $\frac{4}{7}$ 13. $\frac{13}{15}$

Exercise 8.1

1.—4. Construction   5. No   6. Yes   7. No
8. No   9. Yes   10. No

Exercise 8.2

1. Q   2. $\overline{RS}$ 3. $\overline{TB}$ 4. ϕ 5. B
6. $\overline{QS}$ 7. S   8. T

Exercise 8.3

1. f 2. g 3. c 4. e 5. b
6. d 7. h 8. j 9. a 10. i

Exercise 8.4

1. 6   2. No   3. Triangle   4. 2   5. 5

Exercise 8.5

1. Common side not between them
2. V
3. S or R
4. *d*
5. *a* or *c*
6. No, point P on the angle
7. $\overrightarrow{PT}$ and $\overrightarrow{PR}$

Exercise 8.6

1. Straight; obtuse; acute
2. 117°; 63°; 180°
3. $\overleftrightarrow{MN}$

Exercise 8.7

1. Indefinite number
2. Only one
3. Yes
4. Yes
5. Not necessarily

Exercise 8.8

1. 6 cm, $5\frac{1}{2}$ cm
2. 4 cm, $4\frac{1}{2}$ cm
3. 6 cm, $6\frac{1}{2}$ cm
4. 3 cm, $3\frac{9}{2}$ cm

Exercise 8.9

1. $\frac{1}{2}$m., $\frac{1}{4}$ m., $14\frac{1}{2} \pm \frac{1}{4}$ m.
2. $\frac{1}{4}$ m., $\frac{1}{8}$ m., $3\frac{1}{4} \pm \frac{1}{8}$m.
3. $\frac{1}{8}$ cm., $\frac{1}{16}$ cm., $5\frac{7}{8} \pm \frac{1}{16}$ cm.
4. 1 m., $\frac{1}{2}$ m., $9 \pm \frac{1}{2}$ m.
5. $\frac{1}{6}$ km., $\frac{1}{12}$ km., $57\frac{0}{6} \pm \frac{1}{12}$ km.
6. $\frac{1}{16}$ mm., $\frac{1}{32}$ mm., $15\frac{9}{16}$ mm. $\pm \frac{1}{32}$ mm.

Exercise 8.10

1. Yes
2. $5\frac{1}{2}$
3. Yes
4. Yes

Exercise 8.13

1. $\triangle ABC \cong \triangle EDF$
2. $\triangle PQR \cong \triangle LKJ$
3. $\triangle GEF \cong \triangle SRT$
4. $\triangle TUV \cong \triangle ZYX$

Exercise 8.14

1. S.A.S.
2. S.S.S.
3. S.A.S

Exercise 8.15

1. $\angle CAB \cong \angle DAB$
$\angle CBA \cong \angle DBA$
$\overline{AB} \cong \overline{AB}$
$\triangle ABC \cong \triangle ABD$
A.S.A

2. $\overline{ED} \cong \overline{GD}$
$\overline{FE} \cong \overline{FG}$
$\overline{DF} \cong \overline{DF}$
$\triangle FDE \cong \triangle FDG$
S.S.S.

3. $\overline{SR} \cong \overline{QP}$
$\overline{PR} \cong \overline{PR}$
$\angle SRP \cong \angle QPR$
$\triangle PRS \cong \triangle RPQ$
S.A.S.

4. $\overline{MN} \cong \overline{KN}$
$\overline{LN} \cong \overline{JN}$
$\angle MNL \cong \angle KNJ$
$\triangle MNL \cong \triangle KNJ$
S.A.S.

Exercise 8.16

1. *g*
2. *d*
3. *c*
4. 60
5. 60
6. 120
7. 120
8. 120
9. 60

Exercise 8.17

1. $\overrightarrow{CD}$ bisects ∠ACB   Given
     ∠ACD ≅ ∠BCD   Def. of bisect
     $\overline{AC}$ ≅ $\overline{BC}$ Given
     $\overline{CD}$ ≅ $\overline{CD}$ Identity congruence
     △ADC ≅ △BDC   S.A.S.

2. $\overleftrightarrow{ML} \parallel \overleftrightarrow{JK}$ Given
     ∠MLJ ≅ ∠KJL   Alt. int. angles
     $\overleftrightarrow{MJ} \parallel \overleftrightarrow{LK}$ Given
     ∠MJL ≅ ∠KLJ   Alt. int. angles
     $\overline{JL}$ ≅ $\overline{JL}$ Identity congruence
     △JKL ≅ △LMJ   A.S.A.

3. $\overline{SR}$ bisects $\overline{PQ}$ Given
     $\overline{PT}$ ≅ $\overline{QT}$ Def. of bisect
$\overline{PQ}$ bisects $\overline{SR}$ Given
     $\overline{ST}$ ≅ $\overline{RT}$ Def. of bisect
     ∠PTS ≅ ∠QTR   Vert. angles
     △PTS ≅ △QTR   S.A.S.

Exercise 8.18

1. 40   2. 70   3. 4   4. 6

5. $\overline{DE} \parallel \overline{AB}$ Given
     ∠A ≅ ∠CDE   Corr. angles
     ∠B ≅ ∠CED   Corr. angles
     ∠C ≅ ∠C   Identity congruence
     △ABC ~ △DEC   Angles of one respectively congruent to the angles of the other

Exercise 8.19

1. 90   2. 50   3. 60

Exercise 8.20

1. $\frac{x}{8} = \frac{42}{12}$; $x = 28$; 28 m   2. $\frac{10}{12} = \frac{30}{x}$; $x = 36$; 36 units

3. $\frac{x}{24} = \frac{6}{8}$; $x = 18$; 18 units

Exercise 9.1

1.

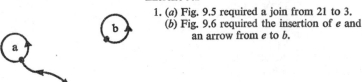

1. (a) Fig. 9.5 required a join from 21 to 3.
    (b) Fig. 9.6 required the insertion of *e* and an arrow from *e* to *b*.

2. 'lives in the same street as', etc.

3.

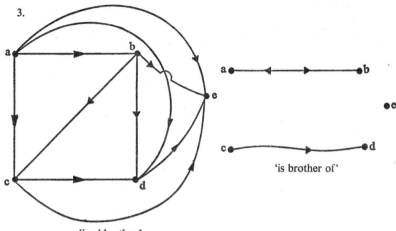

'is older than'   'is brother of'

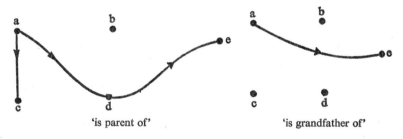

'is parent of'   'is grandfather of'

4. (*a*) $\begin{pmatrix} 0 & 1 & 0 & 0 \\ 0 & 0 & 1 & 0 \\ 1 & 0 & 0 & 1 \\ 0 & 1 & 1 & 0 \end{pmatrix}$ (*b*) 'is connected by direct steamer service to'

5.

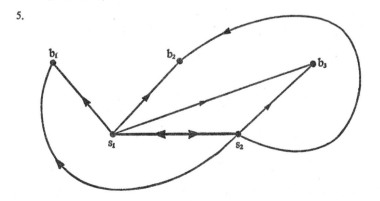

6.
$$
\begin{array}{c c c c c c c c c}
 & 2 & 3 & 5 & 7 & 21 & 35 & 42 & 17 \\
2 & 1 & 0 & 0 & 0 & 0 & 0 & 0 & 0 \\
3 & 0 & 1 & 0 & 0 & 0 & 0 & 0 & 0 \\
5 & 0 & 0 & 1 & 0 & 0 & 0 & 0 & 0 \\
7 & 0 & 0 & 0 & 1 & 0 & 0 & 0 & 0 \\
21 & 0 & 1 & 0 & 1 & 1 & 0 & 0 & 0 \\
35 & 0 & 0 & 1 & 1 & 0 & 1 & 0 & 0 \\
42 & 1 & 1 & 0 & 1 & 1 & 0 & 1 & 0 \\
17 & 0 & 0 & 0 & 0 & 0 & 0 & 0 & 1 \\
\end{array}
$$

Exercise 9.2

(i) R, AS, T   (ii) AS   (iii) R, S, T, E
(iv) AS   (v) AS   (vi) R, S, T, E
(vii) AS, T, O   (viii) —   (ix) T

Note (viii) a R $b \not\Rightarrow b$ R a *always* therefore R is neither AS nor S

(ix) a a boy, b a boy, c a girl

a R b, b R c but $c \not{R} b$ ∴ R is not symmetric S

b R a ∴ R is not antisymmetric AS

7. greater than, less than
8. Relation not complete   9. Symmetric   10. R, S, T, E

Exercise 9.3

1. {(1, 2)}, [(3, 2)], (10, 2)
   (1, 3), (3, 3), (10, 3)
   (1, 11), (3, 11), {(10, 11)}
2. The underlined pairs Question 1   3. AS, 'child of'
4. (*a*) (3, 2); (*b*) $R_1 \cup R_2 = R_1 = \{(3,2); (10, 2); (10, 3)\}$
   (*c*) $R_2 \cup R_3 = \{(1, 2); (3, 2); (10, 11)\}$; $R_2 \cap R_3 = \phi$
5. The relation is the set {(2, 1); (3, 2); (4, 3)}

Exercise 9.4

1   2

3   4

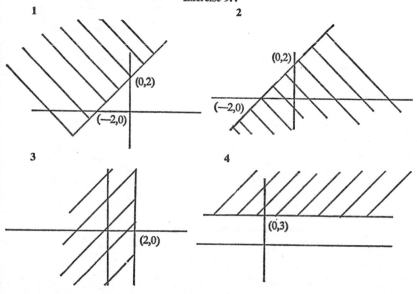

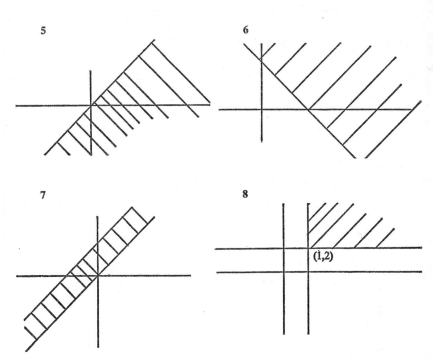

Exercise 9.5

1. (*a*)
2. (*a*) Function

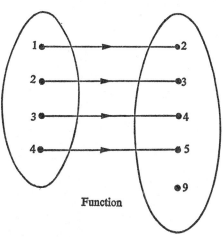

Function

(b) Not a Fn
(c)

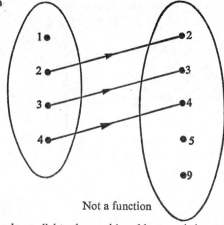

Not a function

3. 1st (no section is parallel to the *y* axis) and last   4. 1st and last   5. None

Exercise 9.6

1. No   2. (a) $x \to 7x^2$, 28; (b) $x \to 49x^2$, 196; (c) $x \to 49x$, 98; (d) $x \to x^4$, 16
3. $\{(x, y) \mid x = y^2\}$. No. Graph is   4. (a)

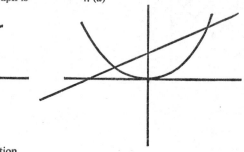

(b) Two points of intersection
5. (a) (iii)
   (b) (ii)
   (c) (i)
   (d) Shaded
region, g_3 and g_4

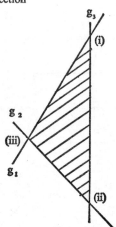

Exercise 10.1

1.

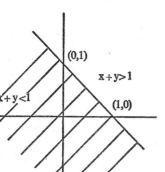

2.

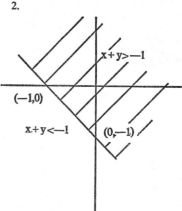

3.

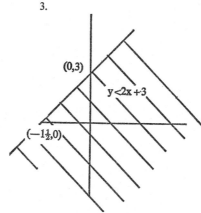

4. $\frac{x}{2} - \frac{y}{2} = 1$, or $x - y = 2$, $x - y < 2$, $x - y > 2$
5. $x = 2$, $x > 2$, $x < 2$
6. $\frac{x}{3} + \frac{y}{4} = 1$ or $4x + 3y = 12$, $4x + 3y < 12$, $4x + 3y > 12$
7. $x + y = 4$, $x + y < 4$, $x + y > 4$
8. $3x + 4y = 15$, $3x + 4y > 15$, $3x + 4y < 15$

Exercise 10.2

1. (11, 4); (12, 4); (12, 3)
2. (2, 3)
3. (2, 7); (3, 5); (3, 6)
4. (2, 7), 41
5. 28 submarines and 7 destroyers
6. (12, 10) Profit £86
7. $x + 2y \leqslant 50$; $450x + 250y \leqslant 11\,250$; profit line $50x + 50y = $ C
   Solution set. (15, 17) outlay £11 000, profit £1600
   (14, 18) outlay £10 800, profit £1600

8. $15x + 9y \leqslant 270$; $40x + 48y$ to be maximized; $4x + 5y \leqslant 100$
   Solution set. (11, 11) outlay £264; labout 99 units; profit £968
                (12, 10) outlay £270; labour 98 units; profit £960
   (10, 12) outlay £268; labour 100 units; profit £976.
   Notice that his resources limit his development to only 22 of the 30 available
   hectares and that (10, 12) is the most profitable arrangement
9. (30, 20). £7000 profit per week, (0, 50), £10 000 profit per week.

Exercise 10.3

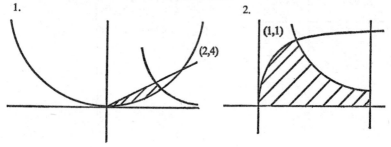

1.   2.

(2,4)

(1,1)

3. 1

Exercise 11.1

1. $a = 0, b = -1, c = 1, d = 0$; $a = k, b = 0, c = 0, d = k$
2. $(3, -2)$; $(-3, 2)$; $(-2, 3)$; $(2, -3)$; $(-3, -2)$; $(3k, 2k)$
   $(-1, -2)$; $(1, 2)$; $(-2, -1)$; $(2, 1)$; $(1, -2)$; $(-k, 2k)$
3. $\begin{pmatrix} x_1 \\ y_1 \end{pmatrix} = \begin{pmatrix} 4 & 2 \\ 1 & -3 \end{pmatrix}\begin{pmatrix} x_0 \\ y_0 \end{pmatrix}$; $\begin{pmatrix} x_1 \\ y_1 \end{pmatrix} = \begin{pmatrix} -1 & 1 \\ 1 & -1 \end{pmatrix}\begin{pmatrix} x_0 \\ y_0 \end{pmatrix}$; $\begin{pmatrix} x_1 \\ y_1 \end{pmatrix} = \begin{pmatrix} 2 & 2 \\ -3 & 1 \end{pmatrix}\begin{pmatrix} x_0 \\ y_0 \end{pmatrix}$
4. $x_1 = x_0 + 2y_0$; $x_1 = 2x_0 + 3y_0$; $x_1 = -4x_0$
   $y_1 = x_0 + y_0$ $y_1 =$ y_0 $y_1 = +4y_0$

Exercise 11.2

1. (b, d), $(2b, 2d)$, $(3b, 3d)$, $(4b, 4d)$; $\dfrac{x}{b} = \dfrac{y}{d}$; $(b, d \neq 0)$

2. (a, c), $(2a, 2c)$, $(3a, 3c)$, $(4a, 4c)$; $\dfrac{x}{a} = \dfrac{y}{c}$; $(a, c \neq 0)$

3. $x_1 = x_0 + 2y_0$, $y_1 = 3x_0 + 11y_0$; Using $y_0 = 2x_0$
   $x_1 = 5x_0$, $y_1 = 25x_0$ $\therefore$ $y_1 = 5x_1$ is the image line
4. (h, k) This transformation leaves everything as it was
5. $\begin{pmatrix} 2 & 1 \\ 3 & 0 \end{pmatrix}$

Exercise 11.3

1. $CA = \begin{pmatrix} -2 & 7 \\ 3 & 2 \end{pmatrix}$; $BC = \begin{pmatrix} 9 & -7 \\ -14 & 22 \end{pmatrix}$; $CB = \begin{pmatrix} 27 & -1 \\ -8 & 4 \end{pmatrix}$
2. (a) $\begin{pmatrix} -1 & 0 \\ 0 & -1 \end{pmatrix} = H_0$; (b) $\begin{pmatrix} -1 & 0 \\ 0 & -1 \end{pmatrix} = H_0$; (c) $\begin{pmatrix} 0 & 1 \\ 1 & 0 \end{pmatrix} = M_{y=x}$; (d) $\begin{pmatrix} 1 & 0 \\ 0 & 1 \end{pmatrix} = I$
3. (a) $\begin{pmatrix} 1 & 0 \\ 0 & 1 \end{pmatrix}$; (b) $\begin{pmatrix} 1 & 0 \\ 0 & 1 \end{pmatrix}$; (c) $\begin{pmatrix} 1 & 0 \\ 0 & 1 \end{pmatrix}$
4. $\begin{pmatrix} 1 & -2 \\ -1 & 3 \end{pmatrix}$ 5. See Fig. 6.6

Exercise 11.4

1. $X = \begin{pmatrix} 2 & -4 \\ 18 & 0 \end{pmatrix}$

2. $7\begin{pmatrix} 8 & 4 \\ 3 & 2 \end{pmatrix} = \begin{pmatrix} 56 & 28 \\ 21 & 14 \end{pmatrix}$

3. $6\begin{pmatrix} 6 & 9 \\ 0 & 4 \end{pmatrix} = \begin{pmatrix} 36 & 54 \\ 0 & 24 \end{pmatrix}$

4. $\begin{pmatrix} -3 & -6 \\ -1 & -2 \end{pmatrix}$; $\begin{pmatrix} -1 & 3 \\ 9 & -4 \end{pmatrix}$

5. (a) $\begin{pmatrix} 11 & -14 \\ 11 & 2 \end{pmatrix}$; (b) $\begin{pmatrix} -1 & 0 \\ 3 & 0 \end{pmatrix}$; (c) $\begin{pmatrix} -11 & 14 \\ -11 & -2 \end{pmatrix}$; (d) $\begin{pmatrix} 3 & 0 \\ -1 & 6 \end{pmatrix}$

Exercise 11.5

1. (a) $\begin{pmatrix} 1 & -1 \\ 0 & 1 \end{pmatrix}$; (b) $\begin{pmatrix} 1 & 0 \\ -1 & 1 \end{pmatrix}$; (c) —; (d) $\begin{pmatrix} -1 & 0 \\ 0 & 1 \end{pmatrix}$; (e) $\frac{1}{6}\begin{pmatrix} 4 & -1 \\ -6 & 3 \end{pmatrix}$;

(f) $-\frac{1}{3}\begin{pmatrix} 3 & -1 \\ -9 & 2 \end{pmatrix}$; (g) —; (h) $\frac{1}{72}\begin{pmatrix} 9 & -6 \\ 6 & 4 \end{pmatrix}$

2.

X	I	A	B	C
I	I	A	B	C
A	A	B	C	I
B	B	C	I	A
C	C	I	A	B

3. $I^{-1} = I$, $A^{-1} = C$, $B^{-1} = B$, $C^{-1} = A$

4. A commutative group   5. {I, B}   6. Yes

Exercise 11.6

1. $\begin{pmatrix} x \\ y \end{pmatrix} = \frac{1}{5}\begin{pmatrix} 1 & -1 \\ 1 & 4 \end{pmatrix}\begin{pmatrix} 9 \\ -1 \end{pmatrix} = \begin{pmatrix} 2 \\ 1 \end{pmatrix}$ Subtract (ii) from (i).
$\therefore\ 5x = 10$
$x = 2$

2. $\begin{pmatrix} x \\ y \end{pmatrix} = \frac{1}{2}\begin{pmatrix} 2 & -3 \\ -4 & 7 \end{pmatrix}\begin{pmatrix} 10 \\ 6 \end{pmatrix} = \begin{pmatrix} 1 \\ 1 \end{pmatrix}$ Subtract 3 (ii) from 2 (i).
$\therefore\ 2x = 2$
$x = 1$

3. $\begin{pmatrix} x \\ y \end{pmatrix} = -\frac{1}{11}\begin{pmatrix} 4 & -7 \\ -9 & 13 \end{pmatrix}\begin{pmatrix} 3 \\ 2 \end{pmatrix} = \begin{pmatrix} \frac{2}{11} \\ \frac{1}{11} \end{pmatrix}$ Subtract 7 (ii) from 4 (i).
$-11x = -2\quad x = \frac{2}{11}$

4. No inverse. Equations inconsistent   (ii) should have been $27x + 36y = 72$

5. The operation is $\begin{pmatrix} 3 & 4 \\ 7 & 10 \end{pmatrix}$; the inverse operation is $\frac{1}{2}\begin{pmatrix} 10 & -4 \\ -7 & 3 \end{pmatrix}$

$\therefore\ \begin{pmatrix} x_0 \\ y_0 \end{pmatrix} = \frac{1}{2}\begin{pmatrix} 10 & -4 \\ -7 & 3 \end{pmatrix}\begin{pmatrix} x_1 \\ y_1 \end{pmatrix}$

if $\begin{pmatrix} x_1 \\ y_1 \end{pmatrix} = \begin{pmatrix} 1 \\ 2 \end{pmatrix}$ then $\begin{pmatrix} x_0 \\ y_0 \end{pmatrix} = \frac{1}{2}\begin{pmatrix} 10 & -4 \\ -7 & 3 \end{pmatrix}\begin{pmatrix} 1 \\ 2 \end{pmatrix} = \begin{pmatrix} 1 \\ -\frac{1}{2} \end{pmatrix}$

6. $T^{-1} = \begin{pmatrix} 2 & -1 \\ -1 & 1 \end{pmatrix}$. $x_0 = 7$ $y_0 = 1$

Exercise 12.1

1. (i) ABCDEKMTUVWY   (ii) HIX   (iii) none   (iv) O
2. 3   3. Bisects PQ at right angles
4. (i) Fold the line so that the end points are mapped on to each other. The fold passes through the mid-point.
   (ii) Fold the line about the chosen point; the fold is the required perpendicular.
5. Fold the paper approximately down the middle then use this as a line. Fold this line on to itself to obtain a second fold at right angles. Now bisect the right angle by method of Fig. 12.8.

6.

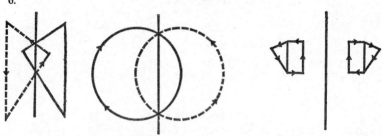

7. The directions are reversed. This is one of the features of reflection. We say that it reverses the orientation of the figure.
8. OA is the mirror line, the required results follow from using this as axis of symmetry in the final figure.
9. Use $\overline{AM}$ as mirror line $\overline{AB}$ and $\overline{AC}$ are images of each other in the line $\overline{AM}$. Since angles are invariants under reflection it follows that $m(\angle ABC) = m(\angle ACB)$.
10. $5^2 = (3 - 2)^2 + h^2$; $|\overline{A_1B}|^2 = (3 + 2)^2 + h^2 = 49$ $\therefore$ $|\overline{A_1B}| = 7$

Exercise 12.2

1. (i) $\overline{BC} = a$; (ii) $\overline{AC} = b$; (iii) $a + b$; (iv) $-a + b$; (v) $-b + a$; (vi) $-b$
2. a and b do not have the same direction
3. $\overline{XY} = -a + b$, $\overline{AB} = -2a + 2b$, $\overline{AB} = 2\overline{XY}$ and $\overline{AB} \parallel \overline{XY}$
4. (i) $\overline{AR}$; (ii) O; (iii) $\overline{LT}$
5. (i) $a + b$; (ii) $2b$; (iii) $2a$; (iv) $2a + 2b$; (v) $2a + b$; (vi) $a + 2b$
6. (i) $2a + b$; (ii) $2a + 2b$; (iii) $2a + b$; (iv) $\overline{OU} = 2b$; (v) $(a + b) + (-b - a) = 0$
7. (i) A; (ii) V; (iii) T; (iv) T; (v) W

Exercise 12.3

1. (i) $(1, 14)$; (ii) $(-2, 1)$; (iii) $(-1, 15)$; (iv) $(0, 12)$
2. $(0, 4)$

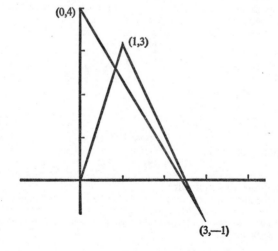

3. $a + b = 8i + 10j$ is the single translation. Final point of arrival is (8, 10).

4. (12, 13)

Exercise 12.4

1. R(K, 90°) maps l on to l_1 but l_1 is parallel to n, hence there is no intersection for N.

If K is the same distance from l and n, then l_1 coincides with n and there are an infinite number of points N in the intersection.

2. $\{N\} = l_1 \cap n$; K and N are two of the vertices of the triangle, use $\overline{KL} \cong \overline{KN}$ to obtain L.

3. It is always possible. If R(K, −60°) fails, then try R(K, +60°).

4. —

5. R(0, $\theta + \alpha$)

6. Closed with respect to successive application or 'multiplication', since we write the two rotations next to one another as for matrices for example.

7. R(0, −θ)   **8.** R(0, 0°)   **9.** Yes   **10.** Yes

Exercise 12.5

1. Construction

2. Using Fig. 12.40 put $\overline{KO} = \overline{ON} = \overline{MN}$

3. Start with $N_0L_0 \parallel AC$

4. Using Fig. 12.41; from N draw a perpendicular to $\overline{BC}$ at F. Construct a square on NF then carry out an enlargement.

Exercise 13.1

1. (a) ϵC; (b) ϵA; (c) ϵE; (d) ϵD; (e) ϵB

2. (b)

3. {A, R}; {B}; {C, G, I, J, L, M, N, S, U, V, W, Z}; {D, O}; {E, F, T, Y}; {H}; {K, X}; {P, Q}

4. 1. Connect the ends.

2. Connect one end to any other point on the curve except the other end.

3. Cross over before connecting the ends.

4. By cutting.

5. None of them are permissible topological transformations.

5. A is inside the curve. If two intersections then A was outside the curve. B and C are both outside.

Exercise 13.2

1.   (iv) 5, 6, 9, 2, 2, 3, traversable   (v) 2, 4, 4, 2, 0, 2, unicursal

  (vi) 4, 6, 8, 2, 2, 2, traversable   (vii) 5, 5, 8, 2, 4, 1, not traversable

  (viii) 2, 4, 4, 2, 0, 2, unicursal   (ix) 4, 1, 3, 2, 4, 0, not traversable

2. See table.

3. Extra arc reduces the number of odd vertices.

4. Join any two odd vertices and so convert them to even vertices.

5. (v) and (viii) Just examine the table for identical entries.

6. Both vertices must be of the same order, since two vertices lie on each join.

7. If A, B are the other two vertices, then unjoined to C they have the same order. Any subsequent join of C to A and B merely changes one vertex to odd, the other to even so there will be two odd vertices in the network or none.

Exercise 13.3

1. Using the same lettering as Example 2. There are two odd vertices, therefore a unicursal walk is impossible. If we start at b we must finish at a the other odd vertex.

2. A bridge connecting A and B. This creates a network with no odd vertices.
3. Any one connexion changes two odd vertices to even vertices, thus leaving only two odd vertices in the network. It therefore becomes traversable. There are six alternative connections.
4. They are all symmetrical about the leading diagonal.
5. (i) $V = 4$, $A = 8$, $R = 2 + A - V = 6$
   (ii) $V = 6$, $A = 17$, $R = 2 + A - V = 13$
6. (i) Two odd vertices. No. (ii) four odd vertices. No.

Exercise 13.4

1. (iv) 5; (v) 3; (vi) 5; (vii) 4; (viii) 3; (ix) 0
2. The first six; the last two 3. Cube, Cuboid
4. (i) An existing vertex to be joined to itself increasing F by 1 and E by 1.
   (ii) A new vertex to be joined to itself on an existing edge increasing F by 1 and E by 2 and V by 1.
5. None 6. All the vertices are even.

Exercise 13.5

1. (a) {(i) (vi)}; {(ii) (iii)}; {(iv) (v)}
   (b) {(i) (vi)} top eq. to sphere; {(iv) (v)} top eq. to torus
   (c) (i) 0; (ii) 5; (iii) 5; (iv) 2; (v) 2; (vi) 0
   (d) All of them (e) (i), (vi)
2. Vice versa is

3. (a) with (ii); (b) with (ii); (c) (ii); (d) with (i)

Exercise 13.6

1. In order the results are: 2 loops interlocked; 2 loops interlocked; 2 loops interlocked; 1 loop 1 knot; 2 loops interlocked 1 knot; 2 loops interlocked; 2 loops interlocked.
2. Figure forms two interlocked holes.
3. Figure separates into one strip with two holes.

Index